Annual Reports in Combinatorial Chemistry and Molecular Diversity

Volume 2

Annual Reports in Combinatorial Chemistry and Molecular Diversity

Volume 2

edited by

W.H. Moos
MitoKor
San Diego, CA 92121, U.S.A.

M.R. Pavia
Millennium Pharmaceuticals
Cambridge, MA 02139, U.S.A.

SPRINGER-SCIENCE+BUSINESS MEDIA, B.V.

A C.I.P. Catalogue record for this book is available from the Library of Congress.

ISBN 978-90-481-5227-8 ISBN 978-94-017-0735-0 (eBook)
DOI 10.1007/978-94-017-0735-0

Printed on acid-free paper

CONTENTS

Preface

The trend continues. Chemists throughout both industry and academia are utilizing combinatorial and parallel synthesis methods in many diverse fields. However, the major use of these techniques has been in the pharmaceutical industry where scientists are both generating diverse libraries of chemical structures for identifying new lead structures, as well as rapidly optimizing leads found from screening these libraries, from testing collections of historical compounds, and even for the optimization of natural products. Selected examples of these efforts are reviewed throughout this volume, with additional examples seen almost daily in the literature.

Volume 2 of 'Annual Reports in Combinatorial Chemistry and Molecular Diversity' is co-edited by Walter Moos (MitoKor) and Michael Pavia (Millennium Pharmaceuticals), and contains contributions by a large number of authors from around the globe. The purpose of this volume is to bring the reader 'up-to-date' via a timely collection of reviews and commentaries that focus on key topics in the field of combinatorial chemistry and molecular diversity.

More specifically, the present volume includes 11 chapters from thoughtleaders in the field, covering the following areas:

1. Solution phase methods – Coe and Storer
2. New solid phases – Sucholeiki
3. Advances in solid phase synthesis – Hall
4. Purification – Weller
5. Analytical – Fitch
6. Carbohydrates – Sofia
7. Patent strategies – Green
8. Patent tactics – Bozicevic
9. Advances in diversity profiling and combinatorial series design – Agrafiotis et al.
10. Comprehensive survey of chemical libraries yielding biologically active agents – Dolle
11. Compendium of solid phase chemistry publications – James.

Clearly, the repertoire of organic reactions that can be utilized in library format continues to expand. Almost every important synthetic organic reaction can be carried out in a combinatorial or parallel fashion using either solution or solid phase methods. The use of solution chemistries in parallel format or for the preparation of small mixtures continues to expand rapidly in tandem with purification methods based on solid phase reagents (see chapters by Coe & Storer and Weller). The use of solution chemistry to generate libraries has, among other plusses, the advantage of allowing chemical reactions to be performed in an environment that has been extensively studied by organic chemists for over a century. Purification using solid phase reagents offers the advantage of simple purification by filtration. It is likely that this method will become increasingly popular with the medicinal chemistry community.

However, the use of solid phase chemistry continues to be a popular method for library generation, and can be effectively used for both split/mix type or parallel syntheses. New and improved solid supports (see chapter by Sucholeiki), as well as new linker chemistries, have contributed to the broad range of chemistries that are now routinely performed on the solid phase (see chapter by Hall).

The focus in the early days of combinatorial chemistry was to generate large libraries for lead identification. The field has shifted its emphasis in the past several years to focus increasingly on preparing large numbers of

M.R. Pavia and W.H. Moos (eds.), vii–viii.

analogues of active hits. To obtain useful structure activity information, the purity and identity of the compounds must be determined at a greater level than in the past. As such, increasing efforts are taking place in the purification of combinatorial libraries prepared by all methods (see chapter by Weller), and in the analytical techniques used to study both the processes of library preparation as well as the structures of the final compounds (see chapter by Fitch).

The automation community has continued to make great strides in developing standalone instruments, with which a working chemist can rapidly prepare and purify a large number of compounds from a wide range of structural classes.

We are also seeing combinatorial and parallel methods being applied to an increasing number of structures, including the synthesis of complex carbohydrate and glycoconjugate structures (see chapter by Sofia), as well as to the synthesis of large numbers of analogues of complex natural products (see chapter by Hall).

Compounds being prepared by any of the available methods are being tested for biological activity in a wide range of assay systems that are turning up exciting new leads as enzyme inhibitors, receptor agonists and antagonists, and a range of other biologically active agents (see chapter by Dolle).

The profiling of diversity and the design of new libraries is practically a field unto itself at times, because of its rapid development and complexity. The breadth and depth of this discipline is covered encyclopedically in the chapter by Agrafiotis and colleagues.

Of course, the successes offered by these new techniques must ultimately translate to a proprietary position for the investigators involved. Since the very early days of combinatorial chemistry, a debate has been carried out as to which patent approaches apply best. The two chapters by Green and Bozicevic address aspects of both strategy and tactics in the realm of intellectual property.

As in past volumes, James, in the final chapter, provides a cumulative update of solid phase organic chemistries published in prior years, continuing the tradition of making the journal, *Molecular Diversity*, and this 'Annual' series useful references for the practicing combinatorial chemist.

The concepts pioneered in the pharmaceutical/biotechnology chemistry community are now being applied and advanced in some exciting new directions. For example, in a paper entitled 'Combinatorial Electrochemistry: A Highly Parallel, Optical Screening Method for Discovery of Better Electrocatalysts' by Reddington et al. [Science, 280 (1998) 1735–1737], the authors describe a clever method of selecting the most active electrocatalyst compositions in a large (645 member) electrode array containing the elements platinum, ruthenium, osmium, iridium, and rhodium.

Another paper, entitled 'Thermographic Selection of Effective Catalysts from an Encoded Polymer-Bound Library' by Taylor and Morken [*Science*, 280 (1998) 267–70], allows an investigator to evaluate, rapidly and simultaneously, each member of a large encoded library for its ability to catalyze a reaction in solution. Such methods are likely to have a significant impact across the chemical industry.

Where might the field of combinatorial chemistry be heading next? One intriguing direction is indicated by a recent paper by Eliseev and Nelen entitled 'Use of Molecular Recognition to Drive Chemical Evolution: Mechanisms of an Automated Genetic Algorithm Implementation' [*Chem. Eur. J.*, 4 (1998) 825–34]. The approach represents an automated implementation of a genetic algorithm mechanism in a chemical system that makes it possible to amplify a subset of compounds that possess higher affinity for a given target within an equilibrating pool of components. This method represents a thermodynamic analogue of template synthesis in which assembly of the reactants on the template leads to faster coupling of the preassembled units. The thermodynamic control described therein assures that the best binders are being formed. Such methods may be especially important for combinatorial chemistry, allowing one to combine chemistry and screening into a unified process... . Food for thought!

We are pleased to present Volume 2 of 'Annual Reports in Combinatorial Chemistry and Molecular Diversity'.

Michael R. Pavia, Cambridge, MA
Walter H. Moos, San Diego, CA

July 1998

M.R. Pavia and W.H. Moos (eds.), Annual Reports in Combinatorial Chemistry and Molecular Diversity, Vol. 2, 1–8.
KLUWER/ESCOM

Solution-phase combinatorial chemistry

Diane M. Coe[a] & Richard Storer[b]
[a] *Glaxo Wellcome Research and Development, Medicines Research Centre, Gunnels Wood Road, Stevenage SG1 2NY, U.K.*
[b] *BioChem Therapeutic Inc., 275 Boulevard Armand-Frappier, Laval, PQ, Canada H7V 4A7*

Key words: library, polymer-supported reagents, solution phase

Summary

The use of solution phase techniques has been explored as an alternative to solid-phase chemistry approaches for the preparation of arrays of compounds in the drug discovery process. Solution-phase work is free from some of the constraints of solid-phase approaches but has disadvantages with respect to purification. This article will also illustrate some of the advances made in recent years in solution phase array chemistry including using supported reagents and simple extractive protocols for the effective preparation of high quality samples.

Introduction

While this review focuses on material that has been disclosed in the primary literature, some relevant work presented at recent meetings is included to indicate the importance of emerging technologies. This chapter covers solution phase synthesis of pools of compounds and of discrete samples and the emerging field of fluorous synthesis. The use of liquid-liquid and liquid-solid extraction has been employed in both the preparation of pools and discrete samples and will be discussed at appropriate points.

Articles from two companies have reviewed their approaches to solution phase libraries. Merritt et al. [1] described the evolution of approaches at Glaxo Wellcome. The initial approach to pooled libraries was successful in identifying leads for medicinal chemistry programmes but also identified problems and prompted future efforts to be directed to discrete libraries. Garr et al. [2] reviewed the methods used at Panlabs for preparation and analysis of three classes of compounds for the Optiverse^{TM} Screening Library. A series of solution phase libraries based on 4-aminopiperidine, piperazine and 4-aminobenzylamine were synthesised by acylation, sulphonylation and N-alkylation.

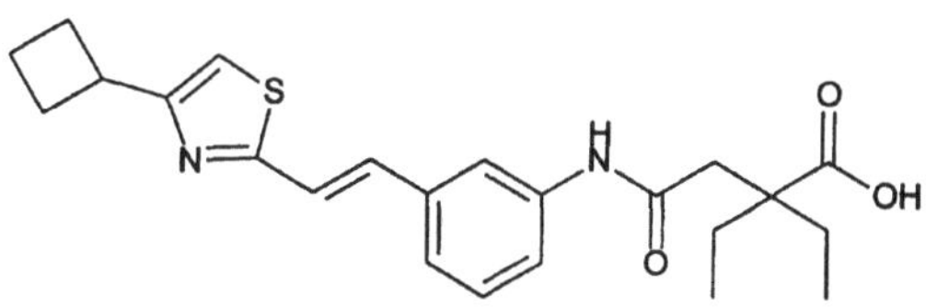

Figure 1. The structure of Ro24-5913.

Solution-phase synthesis of pools of compounds

A number of reports have described such approaches, some of which have served to give an indication of the historical development of the area. The inherent disadvantage of this approach – namely the generation of physical mixtures of compounds and the inability to employ any tagging strategy to assist in the deconvolution – has restricted the pool size to small numbers.

Maehr and Yang [3] confirmed that the Leukotriene D_4 antagonist Ro24-5913 had the optimum structural features for bioactivity using a solution based approach (Figure 1). A library of 700 compounds was prepared from 10 halomethyl ketones, 7 nitro aromatic aldehydes and 10 acyclic anhydrides. The most active compounds were identified by generation and screening of iterative sub-libraries; however, none was more potent than the original lead compound.

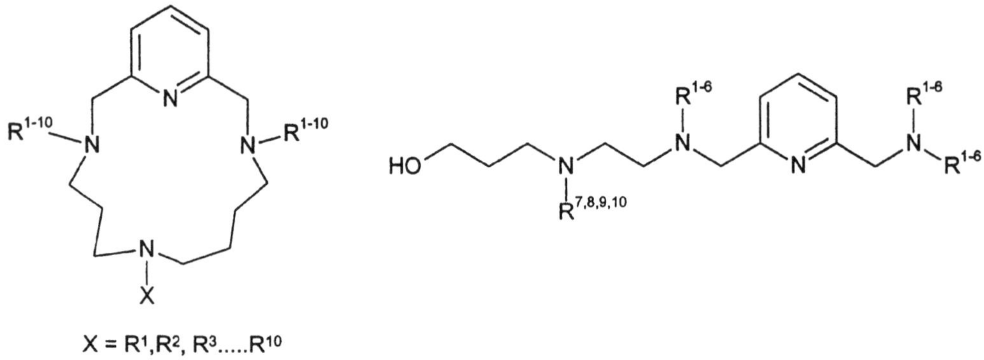

Figure 2. Polyazapyridinophane and pyridinopolyamine scaffolds.

Isis Pharmaceuticals have disclosed a series of libraries based on polyazapyridinophane and linear pyridinopolyamine scaffolds [4,5]. Pools of compounds were prepared by N-alkylation of two of the amino groups contained within the scaffold using a mixture of benzyl halides followed by deprotection and reaction of the third amino group with a single alkylating agent (Figure 2). The libraries were shown to contain compounds having antibacterial activity in the low micromolar range.

The use of rigid scaffolds was also employed by Falorni et al. [6] who prepared a number of small libraries having a diketopiperazine tetra-carboxylic acid template. The diversity elements were introduced via amide bond chemistry.

A library of β-amino alcohols was prepared by Ganesan et al. [7] by the lithium perchlorate promoted ring opening of epoxides using small pools of primary and secondary amines. The amines were divided into structural classes and representatives of each class were used in multiple pools of four. Each pool of amines was reacted with 80 epoxides to generate the library of >6000 samples.

Application of purification methods has been used to improve the quality of products in a number of examples. Boger et al. have published two papers on the preparation of 'indexed' libraries suitable for probing receptor and protein homo- and hetero-dimerisation events [8,9]. Two different linking strategies were developed; either an olefinic linker introduced via olefin metathesis or a bis-amide linker (Figure 3). The work extends their previous synthesis of libraries using liquid-liquid extraction to effect purification of intermediate stages in the library preparation.

Neuville and Zhu [10] reported the preparation of a library containing eight pools each of six aryl piperazines by nucleophilic substitution followed by deprotection and acylation under Schotten-Baumann conditions. The products were isolated in high purity by removal of excess reagents using liquid-liquid extraction.

The discovery of an antirhinoviral lead from a library of 4000 ureas was disclosed by Kaldor and co-workers [11]. Pools of 10 compounds were prepared by reaction of excess isocyanate with an equimolar mixture of 10 amines. Purification was effected by addition of aminomethylpolystyrene to remove isocyanate impurities followed by filtration and evaporation (Figure 4).

Solution-phase synthesis of discrete compounds

Parallel synthesis is now established as an integral component of lead optimisation methodology and increasing numbers of reports on the use of this approach in medicinal chemistry programmes are appearing in the literature. Jarvest et al. [12] explored the structure activity relationship at the 2-position of benzoxazinones in the search for inhibitors of herpes simplex virus-1 protease. Reaction of anthranilic acids with excess isocyanate or chloroformate afforded the 2-amino or 2-alkoxy substituted derivatives (Figure 5).

Carroll and co-workers [13] reported on the preparation of benzimidazole libraries from anilines and carboxylic acids using 2-ethoxy-1-ethoxycarbonyl-1,2-dihydroquinoline.

Application of effective parallel purification methods continues to improve the quality of products prepared using solution phase approaches. Sim and Ganesan [14] developed a one-pot three component synthesis of thiohydantoins using the reductive am-

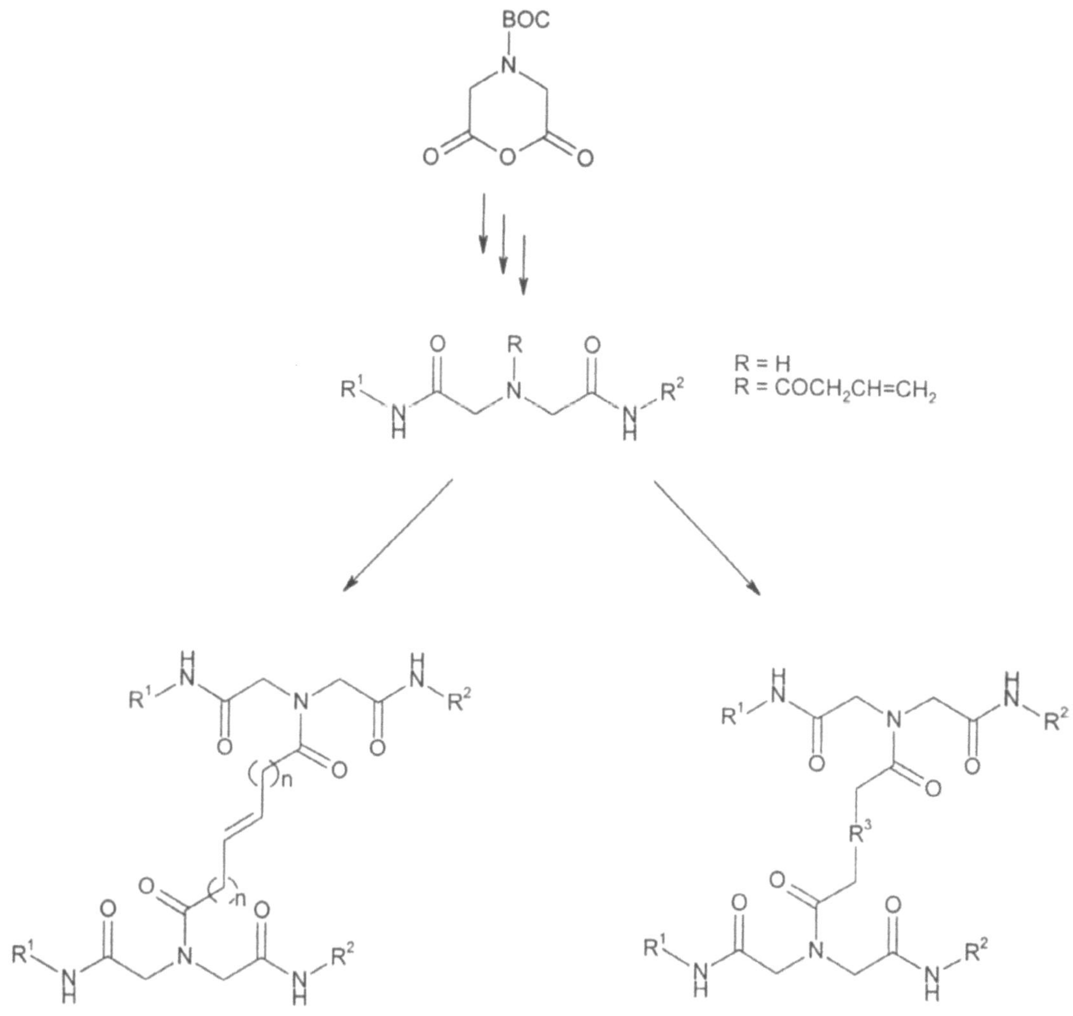

Figure 3. Preparation of libraries varying in length of linking tether.

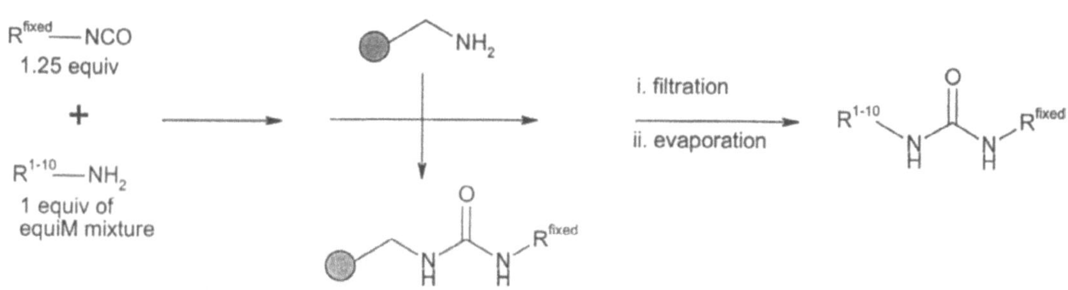

Figure 4. Use of solid supported 'covalent scavenger'.

ination of α-amino esters with aromatic aldehydes and sodium triacetoxyborohydride followed by the reaction with an isocyanate in the presence of triethylamine (Figure 6). The thiohydantions were isolated

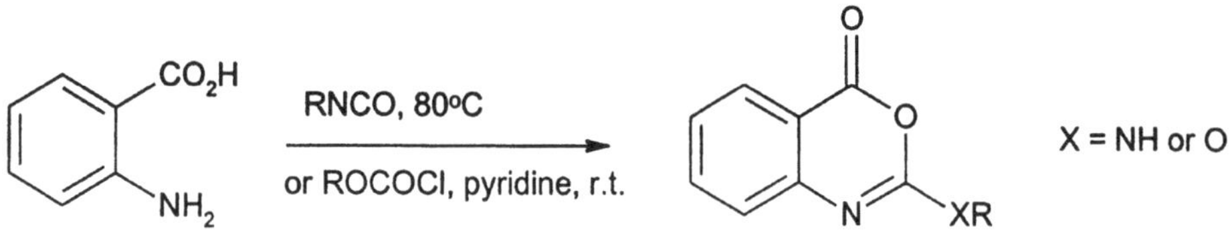

Figure 5. Synthesis of 2-hetero atom substituted benzoxazinones.

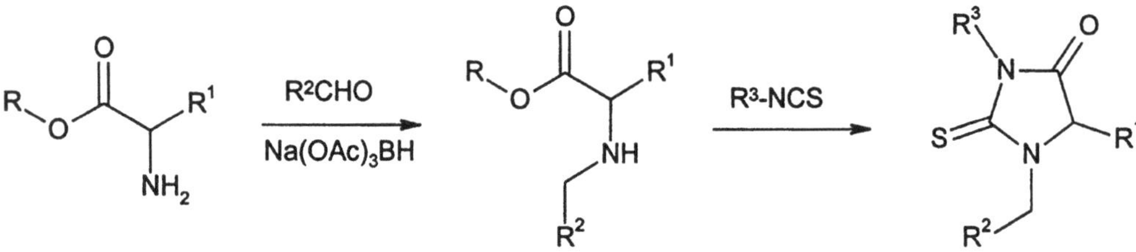

Figure 6. Synthesis of thiohydantoins.

by an aqueous work-up protocol which incorporated the addition of glycine to convert unreacted reagents into water soluble materials. The methodology was used in the preparation of an array of 600 discrete compounds.

Merritt et al. [1] and others [15] have disclosed a number of techniques for parallel phase separation which facilitates the aqueous washing of crude products. The initial approach used a commercially available hydrophobic membrane in a polypropylene cartridge to separate a chlorinated solvent from an aqueous phase. An alternative liquid solid extraction involving an absorbent packing has also been used by workers at Arris Pharmaceuticals in the preparation of triazine libraries [16].

A complementary protocol, appropriately termed the 'lollipop' method, for the separation involving solvent less dense than the aqueous phase was also disclosed [1]. The technique involves cooling the biphasic mixture in the presence of an array of pins, after the freezing process the solidified aqueous phase is removed attached to the pins.

There has been the anticipated increase in reports of the use of supported reagents to effect functional group transformations or remove excess reagents and by-products from crude solution phase reaction products. Combinatorial approaches using polymer supported reagents have been recently reviewed by Kaldor and Siegel [17].

Tartar and co-workers [18] reported the synthesis of polymer supported 1-hydroxybenzotriazole (Figure 7). Reaction of the reagent with a carboxylic acid in the presence of an activating agent afforded the polymer bound activated ester which was reacted with amines to liberate the amide in solution.

Supported electrophilic, nucleophilic or ionic reagents used to remove impurities from solution have been termed scavenger reagents, polymer supported quenching reagents (PSQ) or complementary molecular reactivity/molecular recognition polymer (CMR/R polymer). Use of such reagents provides a versatile counterpart to the approach described above. Booth and Hodges [19] utilised a high loading amine resin derived from chloromethylpolystyrene and tris(2-aminoethyl)amine in the preparation of ureas, thioureas, sulphonamides and amides.

Gayo and Suto [20] employed basic and acidic ion-exchange resins to remove excess reagents in the formation of amides and ureas, respectively. The combination of the ion exchange resin and solvent were optimised simultaneously in a parallel approach, demonstrating the further important application to reaction development.

An extension of the approach to additional chemical reactions has been reported by other groups. Flynn and co-workers [21] have developed some protocols for the Moffatt oxidation of secondary alcohol to ketones (Figure 8) and the addition of organometallic reagents to aldehydes. The same group has also developed the use of tetrafluorophthalic anhydride to assist in the purification of reaction mixtures [22].

Parlow et al. [23] used supported reagents and scavengers in two approaches to the optimisation of a lead pyrazole-5-carboxamide (Figure 9) investigating

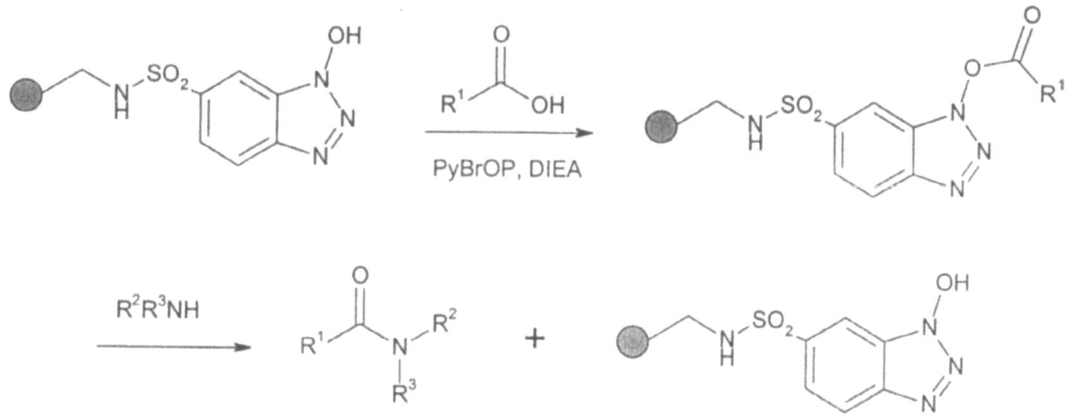

Figure 7. Use of polymer supported 1-hydroxybenzotriazole.

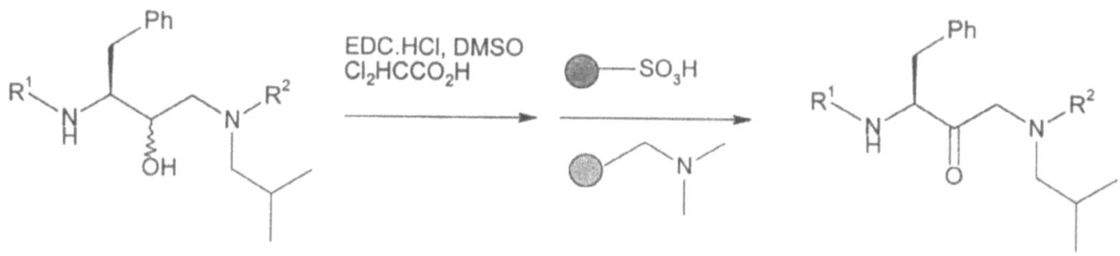

Figure 8. Use of amine encoded carbodiimide in Moffatt oxidations.

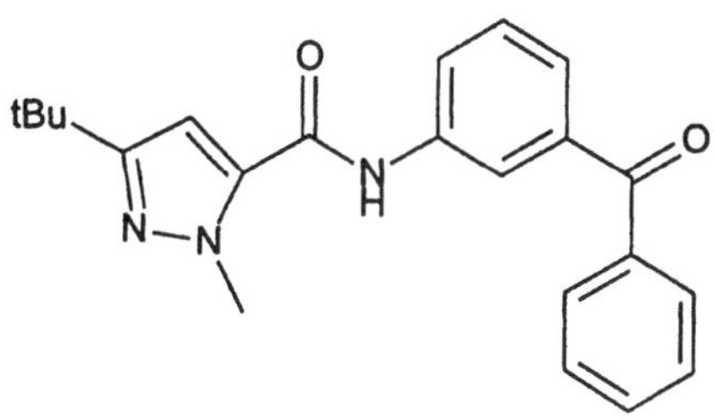

Figure 9. Pyrazole-5-carboxamide optimised by parallel synthesis.

variation in both the heterocyclic and aniline portions of the molecule.

Solid phase extraction (SPE) using a variety of absorbents is now a widely used technique allowing the rapid and facile purification of small molecules. Lawrence et al. [24] prepared an amide library of 225 discrete piperidine analogues using a carbodiimide mediated coupling followed by automated SPE extraction using a bonded silica functionalised with ethylbenzenesulfonic acid (SCX) (Figure 10). Two procedures were developed for the synthesis of neutral products: either a dual SPE protocol using an anionic exchange column followed by a cationic exchange column or modification of the carbodiimide reagents to one containing a basic centre.

A similar protocol was used by Siegel and co-workers [25] in the purification of products from reductive aminations, epoxide openings using amines and urea formation and by Chucholowski et al. [26] in the synthesis of thiazole libraries. A library of 48 ethanolamines in an 8×6 array prepared by the monoalkylation of amines with epoxides (Figure 11) was disclosed by Shuker et al. [27]. The synthetic protocol involved the in situ silylation of the amine with bis-(trimethylsilyl)acetamide, reaction with excess epoxide followed by SCX ion-exchange chromatography.

Fluorous synthesis

The potential of the application of fluorous synthesis as a means of purifying products via phase separation is now beginning to be realised. Curran and co-workers [28] have reported on a number of different fluorous strategies that could be used in solution phase work. In the strategy that has been exploited in the

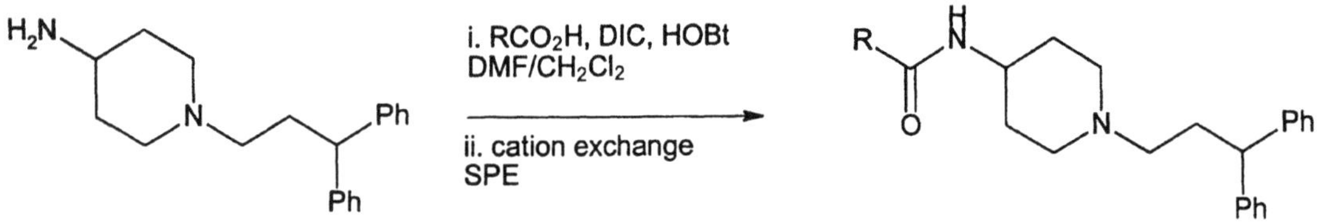

Figure 10. Use of solid phase extraction in purification of amides.

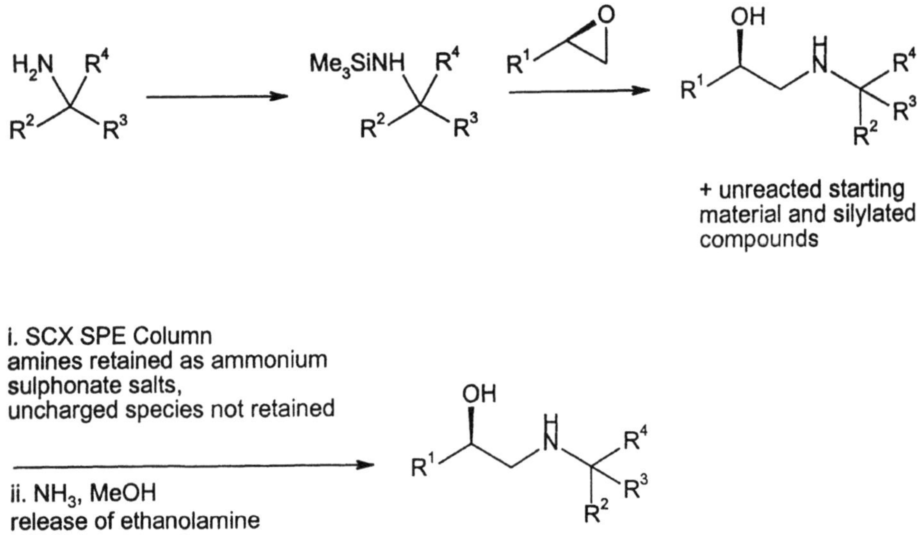

Figure 11. Synthesis of ethanolamines by epoxide opening.

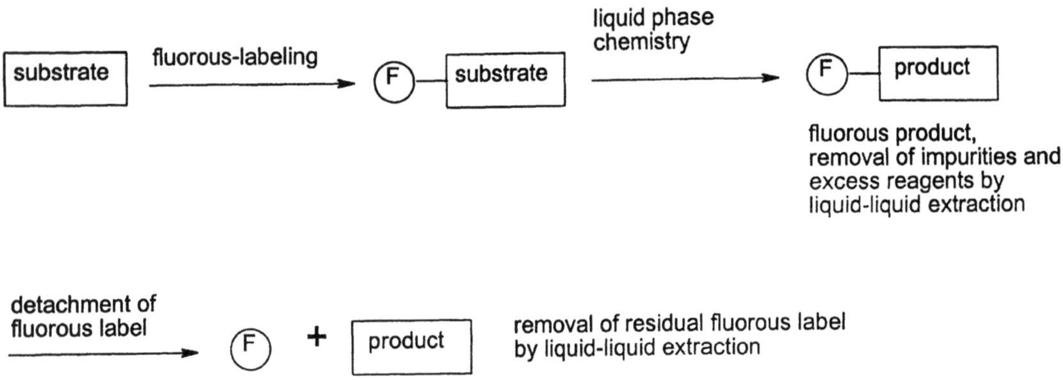

Figure 12. Fluorous synthesis.

preparation of combinatorial libraries an organic substrate has been converted into a fluorous compound by attachment of a 'fluorous label'. Reactions were then conducted and the product purified by a three-phase liquid extraction using fluorous solvents. On completion of the synthetic sequence the target molecule was released from the fluorous label (Figure 12).

Isoxazolines and isoxazoles have been prepared in good yield and excellent purity by reaction of excess nitrile oxide and fluorous labelled silyl ether and silyl propargyl ether respectively [29] (Figure 13). Fluorous variants of the Ugi and Biginelli reaction have also been developed [30].

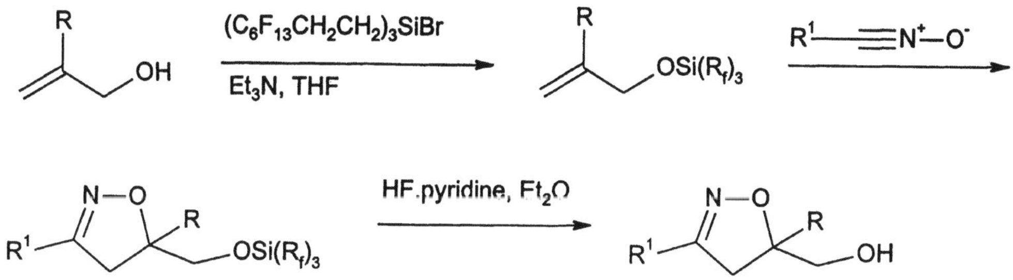

Figure 13. Synthesis of isoxazolines by fluorous synthesis.

Conclusions

Solution phase combinatorial chemistry continues to provide an important technique particularly to the medicinal chemist engaged in lead optimisation work. We anticipate that next year will see further development and application of purification technologies which will allow more complex chemistries to be employed. Although work on fluorous techniques has, currently, only been exploited by the original workers, the development of a solid phase extraction with fluorous reverse phase silica [31] and a soluble fluorous phase polymer support [32] indicates the opportunity for further innovative application of the strategy to solution phase approaches.

References

1. Bailey, N., Cooper, A.W.J., Deal, M.J., Dean, A.W., Gore, A.L., Hawes, M.C., Judd, D.B., Merritt, A.T., Storer, R., Travers, S. and Watson, S.P., *Solution-phase combinatorial chemistry in lead discovery*, Chimia, 51 (1997) 832–837.
2. Garr, C.D., Peterson, J.R., Schultz, L., Oliver, A.R., Underiner, T.L., Cramer, R.D., Ferguson, A.M., Lawless, M.S. and Patterson, D.E., *Solution phase synthesis of chemical libraries for lead discovery*, J. Biomol. Screening, 1 (1996) 179–186.
3. Maehr, H. and Yang, R., *Structure optimization of a leukotriene D_4 antagonist by combinatorial chemistry in solution*, Bioorg. Med. Chem., 5 (1997) 493–496.
4. An, H., Cummins, L.L., Griffey, R.H., Bharadwaj, R., Haly, B.D., Fraser, A.S., Wilson-Lingardo, L., Risen, L.M., Wyatt, J.R. and Cook, P.D., *Solution phase combinatorial chemistry. Discovery of novel polyazapyridinophanes with potent antibacterial activity by a solution phase simultaneous addition of functionalities approach*, J. Am. Chem. Soc., 119 (1997) 3696–3708.
5. An, H., Haly, B.D., Fraser, A.S., Guinosso, C.J. and Cook, P.D., *Solution phase combinatorial chemistry. Synthesis of novel linear pyridinopolyamine libraries with potent antibacterial activity*, J. Org. Chem., 62 (1997) 5156–5164.
6. Falorni, M., Giacomelli, G., Nieddu, F. and Taddei, M., *A new diketopiperazine tetra-carboxylic acid as template for the homogeneous phase synthesis of chemical libraries*, Tetrahedron Lett., 38 (1997) 4663–4666.
7. Chng, B.L. and Ganesan, A., *Solution-phase synthesis of a β-amino alcohol combinatorial library*, Bioorg. Med. Chem. Lett., 7 (1997) 1511–1514.
8. Boger, D.L., Chai, W., Ozer, R.S. and Andersson, C.-M., *Solution-phase combinatorial synthesis via the olefin metathesis reaction*, Bioorg. Med. Chem. Lett., 7 (1997) 463–468.
9. Boger, D.L., Ozer, R.S. and Andersson, C.-M., *Generation of targeted C_2-symmetrical compound libraries by solution-phase combinatorial chemistry*, Bioorg. Med. Chem. Lett., 7 (1997) 1903–1908.
10. Neuville, L. and Zhu, J., *Solution phase combinatorial synthesis of arylpiperazines*, Tetrahedron Lett., 38 (1997) 4091–4094.
11. Kaldor, S.W., Fritz, J.E., Tang, J. and McKinney, E.R., *Discovery of antirhinoviral leads by screening a combinatorial library of ureas prepared using covalent scavengers*, Bioorg. Med. Chem. Lett., 6 (1996) 3041–3044.
12. Jarvest, R.L., Parratt, M.J., Debouck, C.M., Gorniak, J.G., Jennings, L.J., Serafinowska, H.T. and Strickler, J.E., *Inhibition of HSV-1 protease by benzoxazinones*, Bioorg. Med. Chem. Lett., 6 (1996) 2463–2466.
13. Thomas, J.B., Fall, M.J., Cooper, J.B., Burgess, J.P. and Carroll, F.I., *Rapid in-plate generation of benzimidazole libraries and amide formation using EEDQ*, Tetrahedron Lett., 38 (1997) 5099–5102.
14. Sim, M.M. and Ganesan, A., *Solution-phase synthesis of a combinatorial thiohydantoin library*, J. Org. Chem., 62 (1997) 3230–3235.
15. Bush, M., Hamilton, T., Robbins, J. and DeWitt, S., *Phase separation devices for high-throughput synthesis, isolation and purification*, CHI Exploiting Molecular Diversity: Small Molecule Libraries for Drug Discovery, Coronado, CA, February 3–5, 1997.
16. Plunkett, M., *Chemistry Development for Library Synthesis*, 3rd IBC 'Nuts and Bolts' Forum Combinatorial Technologies, Coronado, CA, October 23–24, 1997.
17. Kaldor, S.W. and Siegel, M.G., *Combinatorial chemistry using polymer-supported reagents*, Curr. Opin. Chem. Biol., 1 (1997) 101–106.
18. Pop, I.E., Deprez, B.P. and Tartar, A.L., *Versatile acylation of N-nucleophiles using a new polymer supported 1-hydroxybenzotriazole*, J. Org. Chem., 62 (1997) 2594–2603.
19. Booth, R.J. and Hodges, J.C., *Polymer-supported quenching reagents for parallel purification*, J. Am. Chem. Soc., 119 (1997) 4882–4886.
20. Gayo, L.M. and Suto, M.J., *Ion-exchange resins for solution-phase parallel synthesis of chemical libraries*, Tetrahedron Lett., 38 (1997) 513–516.
21. Flynn, D.L., Crich, J.Z., Devraj, R.V., Hockermann, S.L., Par-

low, J.L., South, M.S. and Woodward, S., *Chemical library purification strategies based on the principles of complementary molecular reactivity and molecular recognition*, J. Am. Chem. Soc., 119 (1997) 4874–4881.
22. Parlow, J.J., Naing, W., South, M.S. and Flynn, D.L., *In situ chemical tagging: Tetrafluorophthalic anhydride as a 'sequestration enabling reagent' (SER) in the purification of solution-phase combinatorial libraries*, Tetrahedron Lett., 38 (1997) 7959–7962.
23. Parlow, J.J., Mischke, D.A. and Woodward, S.S., *Utility of complementary molecular reactivity and molecular recognition (CMR/R) technology and polymer supported reagents in solution-phase synthesis of heterocyclic carboxamides*, J. Org. Chem., 62 (1997) 5908–5919.
24. Lawrence, R.M., Biller, S.A., Fryszman, O.M. and Poss, M.A., *Automated synthesis and purification of amides: Exploitation of automated solid phase extraction in organic synthesis*, Synthesis, (1997) 553–558.
25. Siegel, M.G., Hahn, P.J., Dressman, B.A., Fritz, J.E., Grunwell, J.R. and Kaldor, S.W., *Rapid purification of small molecules libraries by ion exchange chromatography*, Tetrahedron Lett., 38 (1997) 3357–3360.
26. Chucholowski, A., Masquelin, T., Obrecht, D., Stadlwieser, J. and Villalgordo, J.M., *Novel solution- and solid-phase strategies for the parallel and combinatorial synthesis of small molecular weight compound libraries*, Chimia, 50 (1996) 525–530.
27. Shuker, A.J., Siegel, M.G., Matthews, D.P. and Weigel, L.O., *The application of high-throughput synthesis and purification to the preparation of ethanolamines*, Tetrahedron Lett., 38 (1997) 6149–6152.
28. Studer, A., Hadida, S., Ferritto, R., Kim, S.-Y., Jeger, P., Wipf, P. and Curran, D.P., *Fluorous synthesis: A fluorous-phase strategy for improving separation efficiency in organic synthesis*, Science, 275 (1997) 823–826.
29. Studer, A. and Curran, D.P., *A strategic alternative to solid phase synthesis: Preparation of a small isoxazoline library by 'fluorous synthesis'*, Tetrahedron, 53 (1997) 6681–6696.
30. Studer, A., Jeger, P., Wipf, P. and Curran, D.P., *Fluorous synthesis: Fluorous protocols for the Ugi and Biginelli multicomponent condensations*, J. Org. Chem., 62 (1997) 2917–2924.
31. Curran, D.P., Hadida, S. and He, M., *Thermal allylation of aldehydes with fluorous allylstannane. Separation of organic and fluorous products by solid phase extraction with fluorous reverse phase silica gel*, J. Org. Chem., 62 (1997) 6714–6715.
32. Bergbreiter, D.E. and Franchina, J.G., *A soluble fluorous phase polymer support*, J. Chem. Soc., Chem. Commun., (1997) 1531–1532.

M.R. Pavia and W.H. Moos (eds.), Annual Reports in Combinatorial Chemistry and Molecular Diversity, Vol. 2, 9–14.
KLUWER/ESCOM

New developments in solid phase synthesis supports

Irving Sucholeiki
Solid Phase Sciences Corporation, 550 Boston Avenue, 2nd Floor, Medford, MA 02155, U.S.A.

Key words: by-products, PEG, reaction rate, review, solid phase, supports

Summary

Developments in solid phase synthesis supports for the end of 1996 and much of 1997 are reviewed. Issues regarding support induced impurities and enhancing solid phase reaction rates are also discussed.

Introduction

Solid phase organic synthesis continues to play a vital role in the production of combinatorial libraries. Whether used in a mix-and-split strategy or in a multiple parallel synthesis format, solid phase synthesis allows for the rapid separation of product from unreacted starting material. In addition to synthesis, solid supports are also starting to be used in the purification of combinatorial libraries by ion-exchange chromatography [1,2]. Although no accurate statistics exist as to what is the most preferred solid phase synthesis support, a scan of the combinatorial literature shows 1–2% cross-linked polystyrene to be one of the most popular supports. Its relatively high loading capacity (0.5–3 mmoles reactive group/gram) as well as its relatively low cost make it an attractive support for solid phase synthesis. Another series of popular supports used in combinatorial chemistry are those that combine polyethylene glycol (PEG) with 1–2% cross-linked polystyrene. These supports offer good reactivity in both polar organic solvents such as methylene chloride and dimethylformamide as well as polar, protic solvents such as alcohols, acids and water. This review will look at the most recent developments in PEG-polystyrene composites as well as their most recent applications. In addition, the review will look at various methods for enhancing solid phase reaction rates.

Information on the physical and/or chemical stability of a support when exposed to various reaction conditions has generally not been well reported in the combinatorial literature. This is an important issue since combinatorial chemistry utilizes reaction conditions that can test even the most robust supports, yet may not have been envisioned when the material was first developed. This review will briefly look at some recently published data concerning this issue. Finally, the review will look at other newly developed materials for solid phase organic synthesis.

Enhancing solid phase reaction rates

Generally speaking, the rate of a solid phase reaction will depend on the diffusion rate of reagents and reaction products that enter and leave the pores of the solid matrix. There is evidence to suggest that shaking a resin or perturbing it through gas bubbling has no affect on the rate of a solid phase reaction [3]. Vortexing and continuous flow systems have been found to enhance solid phase peptide coupling rates [3]. Variables such as particle size, pore size, solvation effects as well as the polarity of the solvent can have a large affect on the diffusion rate of solute molecules in a support's matrix and hence on the overall reaction rate. For example, the rates of 9-fluorenylmethoxycarbonyl (Fmoc) deprotection using 25 μm sized TentaGel particles were found to be more than three times faster than with 90 μm sized TentaGel particles [3]. This indicates that simply reducing the particle size of the support can dramatically enhance the reaction rate. What it also means is that under diffusion control, if one were to run a solid phase reaction using a support having a broad particle size range, the rate of the reaction would ultimately be limited by the diffusion rate of the largest particles used in the mixture.

One of the simplest means of enhancing solid phase reaction rates is through the application of heat. Other methods such as microwave irradiation [4] and high energy ultrasound [5], have recently been applied to the solid phase production of non-peptide combinatorial libraries. Although very little data have been published on the advantages of using microwave radiation over heating, there have been several early studies showing the marked advantage of using ultrasound in solid phase chemical reactions. Early work by Takahashi and Shimonishi [6] showed that ultrasound accelerated the rates of solid phase peptide couplings as well as aided in resin washing. There have been many other research groups that have since applied ultrasound to various aspects of solid phase peptide synthesis [7–9].

The attachment of a polyethylene glycol (PEG) spacer to the surface of 1–2% cross-linked polystyrene has been shown to enhance the overall yield and purity of solid phase reactions. One reason to account for this greater reactivity is that it gives the tethered molecule greater mobility. This greater mobility is thought to be a result of either the PEG spacer lowering steric interactions between the surface of the support and the bound molecules and/or lowering the internal viscosity of the support itself [10]. Rapp has shown that the addition of a PEG spacer enhances overall sorption. He showed that PEG grafted polystyrene particles displayed greater sorption for the dye Oracett blue 2 as compared to standard 1% cross-linked polystyrene [3]. Recently another study looked at the level of penetration of the enzyme trypsin (23.5 kDa) on TentaGel beads bound with a strong trypsin peptide antagonist [11]. Using confocal microscopy, the researchers showed that fluorescein labeled pancreatic trypsin was able to penetrate to the core of the TentaGel bead. Most surprisingly, the researchers observed that trypsin was able to diffuse preferentially between contacting TentaGel beads at their interface. The researchers concluded that the TentaGel beads were acting as a separate gel phase enhancing the diffusion rate of the large trypsin molecules, preferentially, between contacting beads.

Support stability and support induced by-products

Since the ultimate aim in using solid phase synthesis is to be able to more rapidly produce the desired products in high yield and purity, it would be counter productive to have the support itself leave behind impurities on the final product. Recently a group of researchers found that upon exposing a series of 1–2% cross-linked polystyrene supports to various strong acid and base cleavage conditions, resin induced impurities were formed [12]. The researchers exposed Merrifield, Wang and chlorotrityl chloride resin to various ratios of trifluoroacetic acid (TFA) in methylene chloride, as well as the aminomethyl resin to NH_3OH in dioxane at elevated temperatures. After concentrating the filtrates, they found resin induced impurities ranging from between 7–35% by weight. The researchers also found that prewashing the support did little to prevent the formation of these resin induced by-products. Fortunately, they were able to easily remove these resin induced by-products by utilizing solid phase extraction as a post cleavage purification method [13].

PEG grafted supports also suffer from similar instabilities. Exposure of 150 mg of a Fmoc deprotected TentaGel S RAM support (**1**, Figure 1) to 99% aq. trifluroacetic acid produced 2.8 mg of cleaved PEG [4]. Prolonged heating of the TentaGel support at elevated temperatures (110 °C) has also been reported to cause impurities [14]. Exposure of the PEG-PS support (**2**, Figure 1) to sodium in liquid ammonium has been found to cleave some of the PEG graft [15]. Recently, researchers at Argonaut Technologies reported some chemical and thermal stability data on their PEG grafted support, ArgoGel [16]. The researchers exposed ArgoGel-OH (**3**, Figure 1) and Argo-Gel-NH_2 (**4**, Figure 1) to 95/5 trifluoroacetic acid/water for 4 h and then concentrated the filtrates to afford cleaved linear PEGs at levels of $\leq$0.5 wt.-%. Researchers in Korea compared the acid stability of their β-hydroxy polystyrene based PS-g-PEG resin (**5**, Figure 1) to an early type of PEG grafted support (**6**, Figure 1) synthesized using Merrifield resin [17]. They exposed both supports to 50% trifluoroacetic acid in methylene chloride for 2 h. They found that while 20% of the PEG was cleaved in the Merrifield based support **6**, the β-hydroxy based PS-g-PEG support **5** retained most of the PEG.

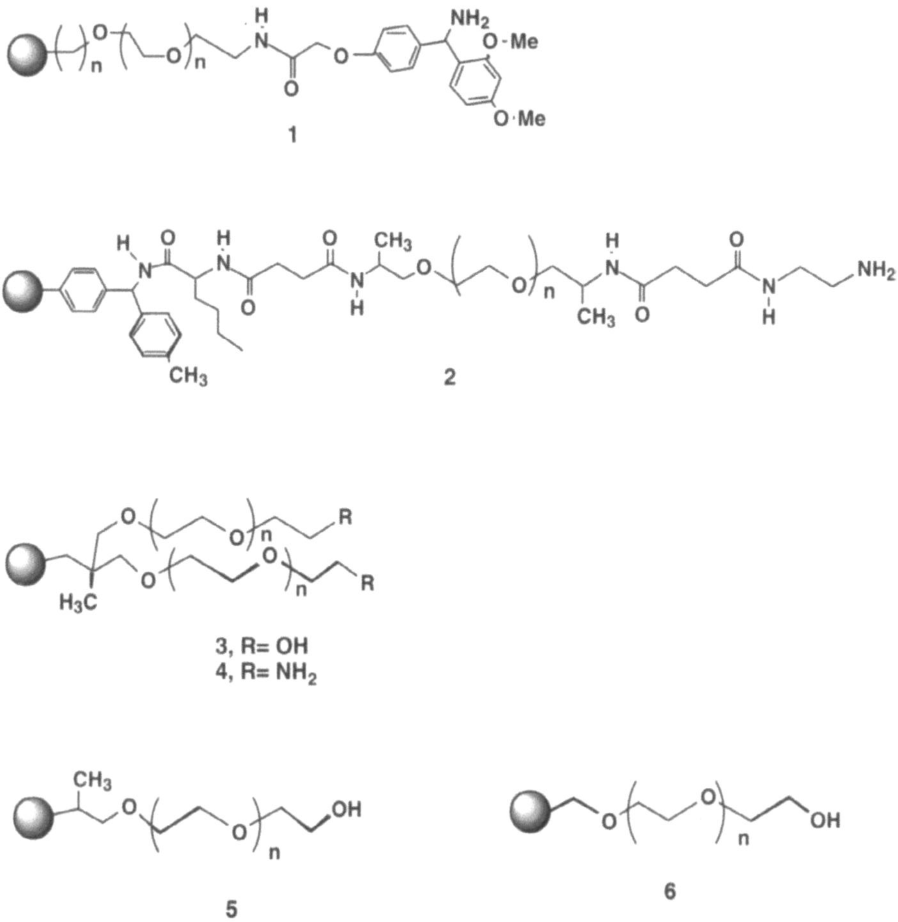

Figure 1. PEG grafted polystyrene supports – TentaGel S RAM resin **1**, PEG-PS resin **2**, Argo-Gel-OH **3**, ArgoGel-NH_2 **4**, β-hydroxy based PS-g-PEG resin **5** and Merrifield based PS-g-PEG resin **6**.

New solid phase synthesis supports

Since the last review on solid phase synthesis supports [10] several new supports have appeared that are based on well known materials such as polyethylene, polyethylene glycol and polymethacrylate. These new supports fall into two main categories: gelatinous, low cross-linked resins and macroporous supports. Gelatinous resins, such as Merrifield resin and TentaGel resin are characterized by solvent induced expansion of the support. Macroporous supports, on the other hand, exhibit large rigid pores that are not dramatically influenced by the solvent or its polarity. Control Pore Glass (CPG), Kieselgur and high cross-linked polystyrene are examples of macroporous supports.

New gelatinous supports

Cross-linked polyoxyethylene-polystyrene (POEPS) and polyoxyethylene-polyoxypropylene (POEPOP)

The POEPS (**7**, Figure 2) and POEPOP (**8**, Figure 2) supports consist of polystyrene or polyoxypropylene which has been cross-linked with PEG (MW = 1500). These supports have loading capacities in the range of 0.1 to 0.6 mmoles hydroxyl/g. Like most PEG based resins, these supports exhibit good swelling characteristics in both organic and aqueous solvents [18]. The authors demonstrated the use of these two supports through the solid phase synthesis and purification of the penta-peptide G-F-S-F-G-NH_2 in 85–93% isolated yields.

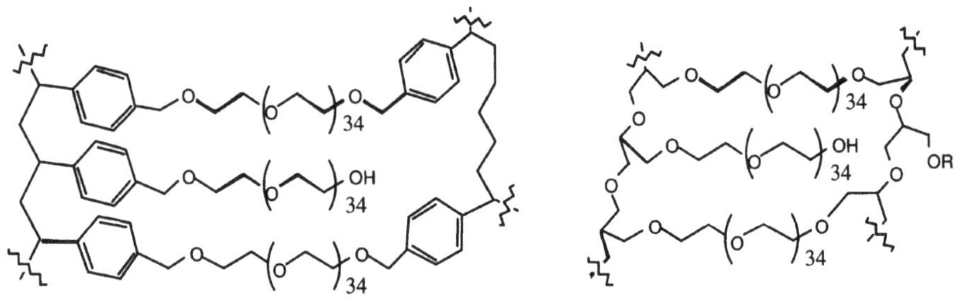

7 8

Figure 2. The POEPS **7** and POEPOP **8** polyoxyethylene based resins.

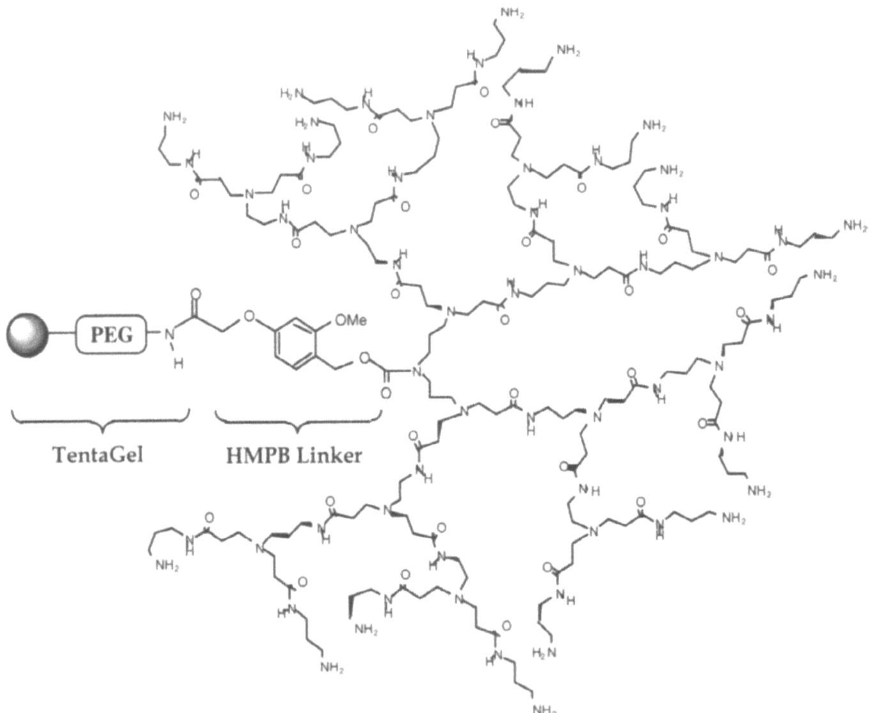

9

Figure 3. Super-high-loading TentaGel-dendrimer beads **9**.

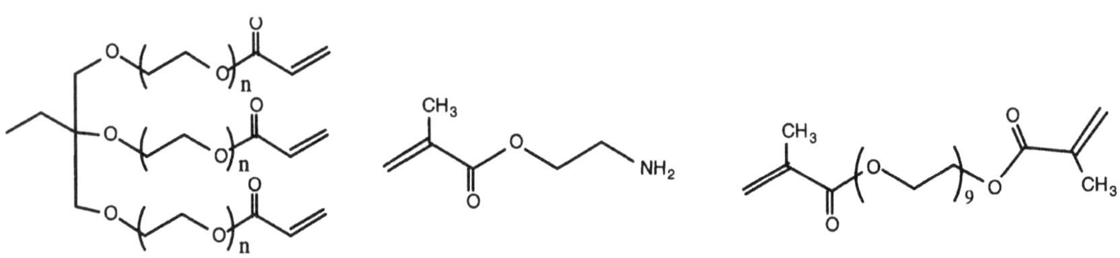

9 10 11

Figure 4. Components of CLEAR supports: trimethylolpropane ethoxylate triacrylate **9**, 2-aminoethyl methacrylate **10** and poly(ethylene glycol-400) dimethylacrylate **11**.

Polyamidoamine (PAMAM) dendrimers attached to TentaGel

Similar in concept to earlier branched lysine supports [10], researchers at the University of Southampton synthesized a series of polyamidoamine (PAMAM) dedrimers off a TentaGel support (**9**, Figure 3) [19]. Their intent was to not only create a support which exhibited a high loading capacity (around 2.3 mmoles amine/gram), but also to make purified dendrimers by subsequent cleavage of the dedrimer using 50% trifluoroacetic acid in methylene chloride. The researchers demonstrated the use of their support through the solid phase synthesis of a dipeptide in 44% overall yield starting from the underivatized TentaGel support [19].

TentaGel based macrobeads

Researchers at Rapp Polymere and at the University of Tübingen reported on the development and use of large diameter TentaGel beads (400–800 μm) which were used as 'microreactors' in the synthesis of peptide, non-peptide and oligonucleotide compounds. Each TentaGel macrobead was reported to have 2000–100 000 nmol reactive groups/bead as compared to 20–100 nmole/bead for standard sized TentaGel. The researchers synthesized several hydantoins off these TentaGel macrobeads, which were contained in glass capillary tubes [3, 20].

New macroporous supports

Highly cross-linked ethoxylate acrylate resins (CLEAR)

These are a series of supports which are composed of trimethylolpropane ethoxylate triacrylate (**10**, Figure 4) cross-linked to a series of different methacrylate monomers such as 2-aminoethylmethacrylate (**11**, Figure 4) and poly(ethylene glycol-400)dimethylmethacrylate (**12**, Figure 4) [21]. Although the supports are highly cross-linked, they exhibit swelling characteristics similar to hydrophilic gelatinous resins in both polar organic solvents and in water. Loading capacities for these supports are reported to be in the range of 0.13–0.29 mmoles amine/gram. The researchers demonstrated the use of these supports in both batchwise and continuous-flow solid phase synthesis of a series of peptides ranging in size from 9–17 amino acids long. The supports were found to be relatively resistant to strong acids such as neat trifluoroacetic acid, yet undergo ester base hydrolysis and ammonolysis with NaOH and NH_4OH, respectively.

Augmented surface polyethylene (ASPECT) particles

Researchers have found that upon exposing polyethylene particles to the oxidating conditions of oxygen with transition metals or chromic acid alone, a network of large pores can be created [22]. These pores range in size from 900–5000 Å. These supports which have a reported loading capacity in excess of 100 μmole of functional groups/g were shown to be suitable for both batchwise and continuous flow solid phase synthesis.

Summary and conclusions

This review has looked at various techniques that have been employed to enhance the reaction rate of solid phase reactions. In particular, we have seen that by simply reducing the particle size of a support, one can greatly enhance the reaction rate. The review has also touched upon the use of other methods such as the use of ultrasound to enhance not only the solid phase reaction rate, but also to enhance the efficiency of resin washing. Issues such as support stability and, more importantly, support induced by-products are beginning to take on greater importance. This review has looked at several cases of this phenomenon and, it is hoped, with more research new supports or techniques will be developed that will remove or at least reduce this problem. Lastly, we have looked at various new supports that have been developed for solid phase synthesis. Again we see that the areas of solid phase peptide and nucleotide synthesis are leading the way in the development of new solid phase synthesis supports. Although this is not a bad thing, it does limit the types of reactions and reaction conditions that a support is exposed to, thereby making it difficult to access the support's strengths and weakness in other areas of solid phase synthesis. On the other hand, nowhere can one find a greater wealth of knowledge concerning solid phase organic synthesis than in the peptide and nucleotide synthesis literature, a fact that should be emphasized to any one beginning their first solid phase production of a combinatorial library.

References

1. Gayo, L.M. and Suto, M.J., *Ion-exchange resins for solution phase parallel synthesis of chemical libraries*, Tetrahedron Lett., 38 (1997) 513–516.

2. Siegel, M.G., Hahn, P.J., Dressman, B.A., Fritz, J.E., Grunwell, J.R. and Kaldor, S.W., *Rapid purification of small molecule libraries by ion exchange chromatography*, Tetrahedron Lett., 38 (1997) 3357–3360.
3. Rapp, W.E., *Macro beads as microreactors: New solid-phase synthesis methodology*, In Wilson, S.R. and Czarnik, A.W. (Eds.) Combinatorial Chemistry – Synthesis and Application, Wiley, New York, NY, 1997, pp. 65–93.
4. Larhed, M., Lindeberg, G. and Halberg, A., *Rapid microwave-assisted Suzuki coupling on solid-phase*, Tetrahedron Lett., 37 (1997) 8213–8222.
5. Sucholeiki, I., *Solid-phase methods in combinatorial chemistry*, In Wilson, S.R. and Czarnik, A.W. (Eds.) Combinatorial Chemistry – Synthesis and Application, Wiley, New York, NY, 1997, pp. 119–133.
6. Takahashi, S. and Shimonishi, Y., *Solid phase peptide synthesis using ultrasonic waves*, Chem. Lett., (1974) 51–56.
7. Adamson, R., Gregson, S. and Shaw, G., *New application of sporopollenin as a solid phase support for peptide synthesis and the use of sonic agitation*, Int. J. Pept. Protein Res., 22 (1983) 560–564.
8. Krchnák, V. and Vágner, J., *Color-monitored solid-phase multiple peptide synthesis under low-pressure continuous-flow conditions*, Pept. Res., 3 (1990) 182–193.
9. Bláha, I., Nemec, J., Tözsér, J. and Oroszlan, S., *Synthesis of homologous peptides using fragment condensation: Analogs of an HIV proteinase substrate*, Int. J. Pept. Protein Res., 38 (1991) 453–458.
10. Sucholeiki, I., *Selection of supports for solid-phase organic synthesis*, In Moos, W.H., Pavia, M.R., Ellington, A.D. and Kay, B.K. (Eds.) Annual Reports in Combinatorial Chemistry and Molecular Diversity, Vol. 1, ESCOM, Leiden, 1997, pp. 41–49.
11. Quarrell, R., Claridge, T.D.W., Weaver, G.W. and Lowe, G., *Structure and properties of TentaGel resin beads – Implication for combinatorial library chemistry*, Mol. Div., 1 (1996) 223–232.
12. MacDonald, A.A., Dewitt, S.H., Ghosh, S., Hogan, E.M., Kieras, L., Czarnik, A.W. and Ramage, R., *The impact of polystyrene resins in solid-phase organic synthesis*, Mol. Div., 1 (1996) 183–186.
13. MacDonald, A.A., Dewitt, S.H., Hogan, E. and Ramage, R., In Epton, R. (Ed.) Innovation and Perspectives in Solid Phase Synthesis, Mayflower Scientific, Birmingham, 1996, pp. 115–116.
14. Hutchins, S.M. and Chapman, K.T., *Fisher indole synthesis on a solid support*, Tetrahedron Lett., 37 (1996) 4869–4872.
15. Sucholeiki, I., *The use of polymer-assisted organic synthesis for the production of small organic molecules*, In Epton, R. (Ed.) Innovation and Perspectives in Solid Phase Synthesis, Mayflower Scientific, Birmingham, 1996, pp. 107–110.
16. Labadie, J.W., Deegan, T.L., Gooding, O.W., Heisler, K., Newcomb, W.S., Porco, J.A., Jr., Tran, T.W. and van Eikeren, P., *New poly(styrene-oxyethylene) graft copolymers as supports for solid-phase organic synthesis*, Polym. Mater. Sci. Eng., 75 (1996) 389–390.
17. Park, B.-D., Lee, H.-I., Ryoo, S.-J. and Lee, Y.-S., *Convenient method for preparing polystyrene having β-hydroxy group: Its application to the synthesis of polyethylene glycol-grafted polystyrene resin*, Tetrahedron Lett., 38 (1997) 591–594.
18. Renil, M. and Meldal, M., *POEPOP and POEPS: Inert polyethylene glycol crosslinked polymeric supports for solid synthesis*, Tetrahedron Lett., 37 (1996) 6185–6188.
19. Swali, V., Wells, N.J., Langley, J.G. and Bradley, M., *Solid-phase dendrimer synthesis and the generation of super-high-loading resin beads for combinatorial chemistry*, J. Org. Chem., 62 (1997) 4902–4903.
20. Rapp, W., Nicholson, G., Maier, M., Schlotterbeck, G., Pursch, M. and Albert, K., *Miniaturization in chemistry: Chemical possibilities and physicochemical properties of polymeric microreactors*, In Epton, R. (Ed.) Innovation and Perspectives in Solid Phase Synthesis, Mayflower Scientific, Birmingham, 1996, pp. 97–100.
21. Kempe, M. and Barany, G., *CLEAR: A novel family of highly cross-linked polymeric supports for solid-phase peptide synthesis*, J. Am. Chem. Soc., 118 (1996) 7083–7093.
22. Cook, R.M. and Hudson, D., *The synthesis and separation of peptides, DNA and hybrid molecules*, In Epton, R. (Ed.) Innovation and Perspectives in Solid Phase Synthesis, Mayflower Scientific, Birmingham, 1996, pp. 19–26.

M.R. Pavia and W.H. Moos (eds.), Annual Reports in Combinatorial Chemistry and Molecular Diversity, Vol. 2, 15–26.
KLUWER/ESCOM

Recent advances in solid phase synthesis

Steven E. Hall
Sphinx Pharmaceuticals, A Division of Eli Lilly, P.O. Box 13951, Research Triangle Park, NC 27709, U.S.A.

Key words: bond forming reactions, function group conversion, heterocyclic synthesis, linker methodology, solid phase synthesis

Summary

The use of solid phase synthesis continues to expand as chemists identify methodology that enables complex reactions. Recent efforts in this area have focused on new carbon-carbon bond forming reactions as well as a variety of heterocyclic systems. These examples are described along with updates on new linking strategies for solid phase synthesis.

Introduction

Interest in solid phase synthesis continues to increase as articles which expand the scope of non-oligomer organic synthesis are published and combinatorial chemistry is applied to an increasing variety of problems. A special thematic issue of *Chemical Reviews* on combinatorial chemistry [1] has been published which includes reviews of solid-phase chemistry as applied to the one bead–one compound approach [2], heterocycle synthesis [3], spatially addressable libraries [4], as well as soluble resin-based synthesis [5]. This review focuses on work published from June 1996 through September 1997. A discussion of solid-phase based reagents is not included as this topic is covered elsewhere.

Linker and cleavage methodology

Publications continue to appear that deal with new applications of known linkers as well as novel linkers to expand the scope of functionality which can serve as a handle for solid-phase synthesis. Prominent among recent manuscripts are linkers for a variety of nitrogen-based functional groups such as amines, sulfonamides, amidines, and hydroxamic acids (Figure 1).

Secondary amines were prepared by the Mitsunobu reaction of a resin bound sulfonamide with alcohols. The presence of the *ortho*-nitro group allowed nucleophilic cleavage to liberate the desired amine [6]. An amidine linker [7,8] provided a resin bound amine that was stable to oxidative, reductive, and alkylative conditions. The use of 2-chlorotrityl resin proved to be useful for the preparation of a secondary amine library [9] as well as hydroxamic acids [10]. Tertiary amines were prepared by 1,4 addition to a resin bound acrylate followed quaternization and a retro-Michael reaction [11]. The *p*-nitrophenyl carbonate linker has been shown to be useful for the immobilization of benzamidines [12] and for the synthesis of sulfonamides [13]. Sulfonamides and amides were also prepared using a halogenated Wang resin [14].

Roush has described an alkylsulfonyl chloride resin which was prepared and used for the solid-phase synthesis of 6-deoxyoligo-saccharides [15]. Advances in the use of silyl linkers for the synthesis of aromatic products continue to be described including a full paper on the use of silyl and germanium linkers by Plunkett and Ellman [16]. Ellman et al. [17] also described a trialkylsilyl chloride linker for the direct loading of aromatic compounds as shown in Figure 2. Darling et al. described a similar resin prepared by the hydrosilylation of residual vinyl groups in a divinylbenzene copolymer [18]. A dimethylsilyl-propionic acid linker was successfully employed by Newlander et al. [19] under a variety of reaction conditions including alkylation, acylation, and Mitsunobu reac-

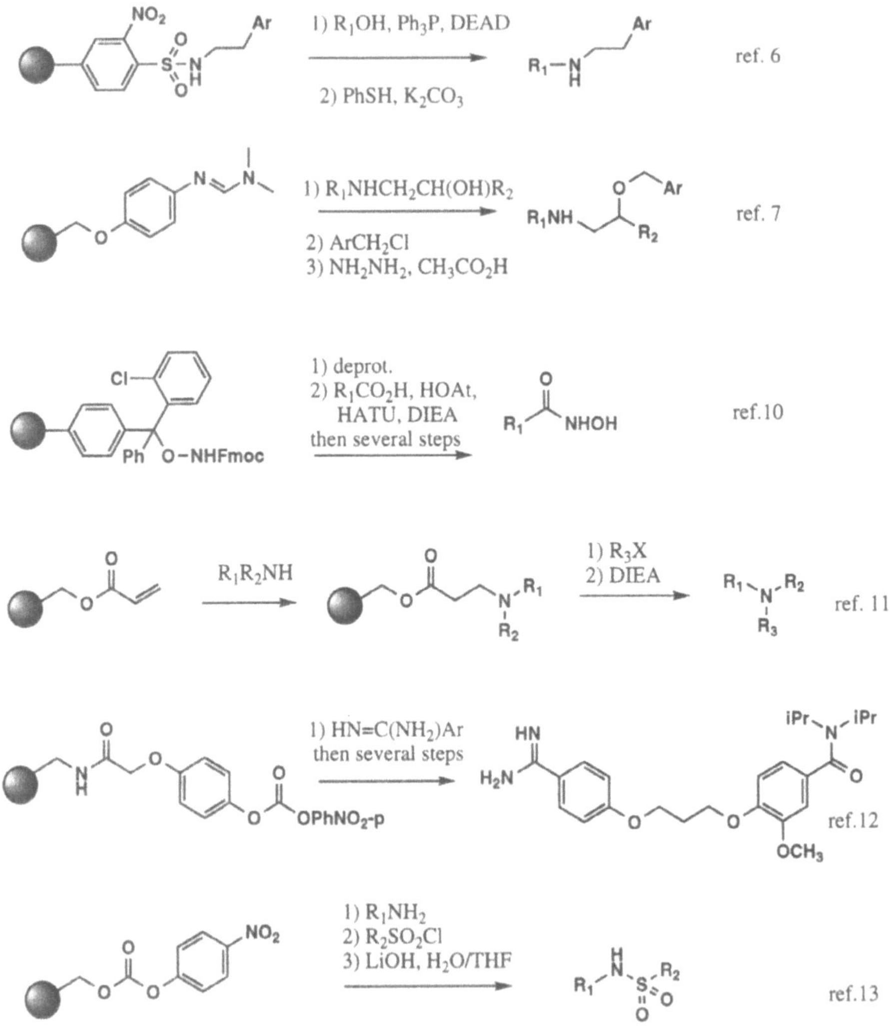

Figure 1. Linker and cleavage methodology.

tions. A dithiane protected benzoin has been reported by Balasubramanian [20] which serves as a photolabile safety catch linker. Other reports include the use of halogenated Wang resins for sulfonamide synthesis [21] and the attachment of HMB linker to continuous cellulose membranes [22] for parallel spot synthesis. Ho and Kukla [23] have described the synthesis of *N*-methyl amines by reductive cleavage of secondary carbamates. Latent secondary amines have been released by the cleavage of tertiary benzylamines using α-chloroethyl chloroformate [24].

Functional group conversion

Interconversion of functional groups is fundamental in organic synthesis. A summary of recently described conversions is shown in Figure 3. Palladium catalyzed arylation of secondary amines has been reported by groups at both Boehringer Ingelheim [25] and Merck [26]. Preparation of tertiary amines by reductive amination proceeds smoothly using borane-pyridine complex as the reducing agent [27]. Borane–THF complex was used to produce secondary amines from the corresponding amides which were unaccessible by reductive amination [28]. Resin bound acetylenes, in the

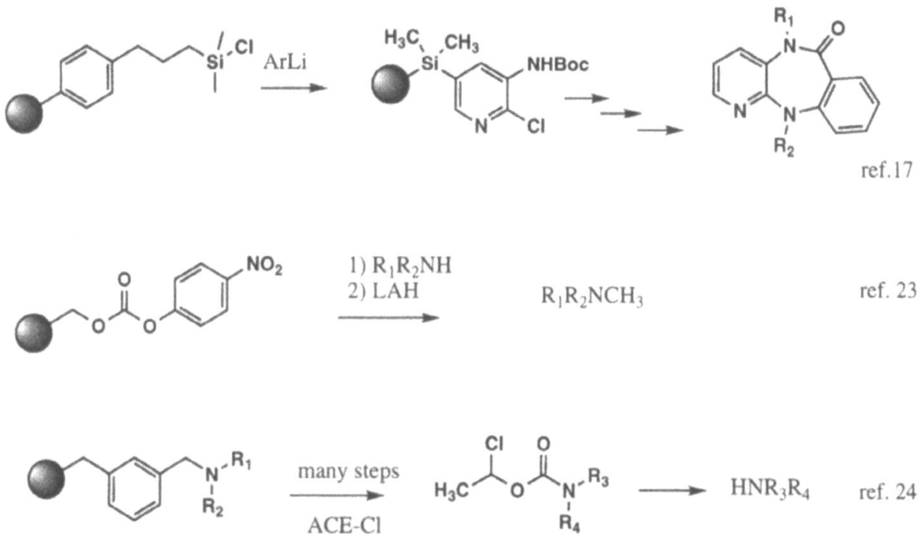

Figure 2. Traceless linker methodology.

presence of Cu(I), undergo the Mannich reaction to produce propargyl amines [29]. Complex ureas have been prepared by reaction of resin bound secondary amines with phosgene to afford the intermediate carbamoyl chlorides, followed by addition of secondary amines [30].

Peptide hydroxamic acids were prepared by condensation of the peptide acid with H_2NOBn [31]. The benzyl protecting group was removed during the final HF cleavage. A method for preparing hydroxamides as the linking group has also been published [32].

Trisubstituted guanidines have been generated [33] by reaction of a resin bound benzylazide with isocyanates. The intermediate diimide was then condensed with secondary amines affording the guanidine. Methods for the synthesis of bisureas using diisocyanates [34] and oligomeric ureas using *p*-nitrophenylcarbamates of aminoethylazides [35] have also been described. The synthesis of extremely hindered peptide fragments has been accomplished through the use of azido acids [36]. For example, fragments containing up to four diphenyl glycine residues could be assembled. Wallace reported the synthesis of ketones from esters by reduction of in situ formed *N*-methoxy-*N*-methylamides [37]. Several ether forming reactions have been published. Tortolani and Biller described the iodoetherification of styrenes followed by nucleophilic displacement to afford miconazole analogs [38]. Fancelli et al. prepared 2-substituted benzofurans through the copper/palladium heteroannulation of terminal acetylenes with resin bound aryl iodides [39]. Other ether producing reactions include the synthesis of diarylethers in cyclic peptides [40] and the synthesis of oligosaccharides [41,42].

Reactions: Carbon-carbon bond formation

Carbon-carbon bond formation continues to challenge those involved in solid-phase synthesis. Nonetheless, researchers continue to develop high yield variants of this ever important bond in organic synthesis. Recent advances in the area are summarized in Figures 4 and 5.

Diaryl coupling and olefination

Diaryl coupling reactions continue to attract interest given the importance of this structural sub-unit. Several groups have described their efforts using variants of the Suzuki coupling [43–46]. Hallberg et al. found that microwave irradiation accelerated the Suzuki reaction providing high yields in only several minutes [43]. Marquais and Arlt demonstrated successful cross-coupling of resin bound aryl bromides with arylzinc reagents [47].

Tempest and Armstrong were successful in coupling squaric acid derivatives to aryl halides using either a Stille coupling or direct metallation of the resin bound aryl halide and subsequent addition to the

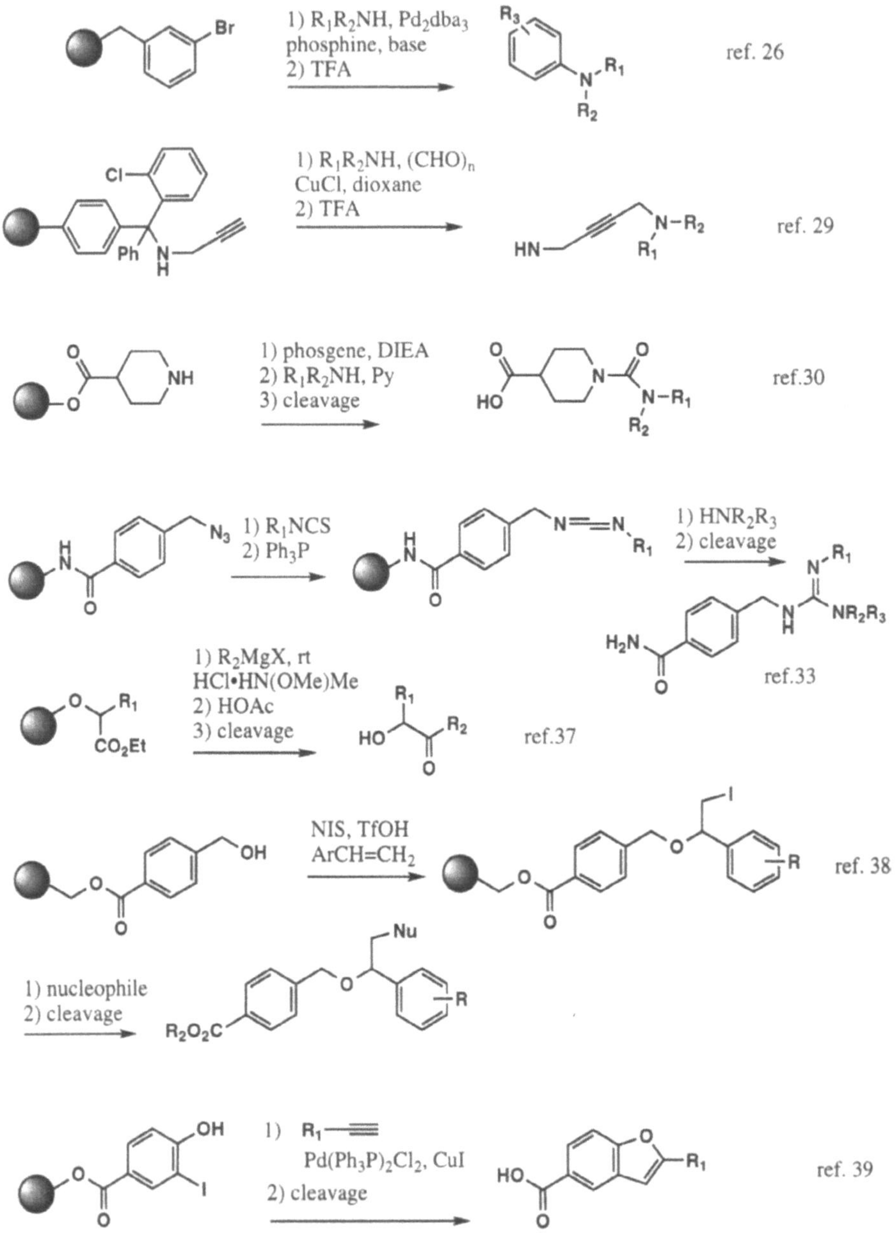

Figure 3. Functional group conversions.

diisopropyl squarate [48]. Tetrasubstituted ethylenes were prepared by capture of a vinyl boronate with a resin bound aryl iodide [49]. A novel approach to the synthesis of seven-membered lactams involved a ring-closing metathesis reaction [50]. A ring opening metathesis reaction was reported by Cuny et al. to afford highly substituted cyclopentanes [51] which were then further elaborated. Wipf and Henniger [52] prepared alkene peptidomimetics by the ring opening of resin bound *N*-nosyl alkeneaziridines.

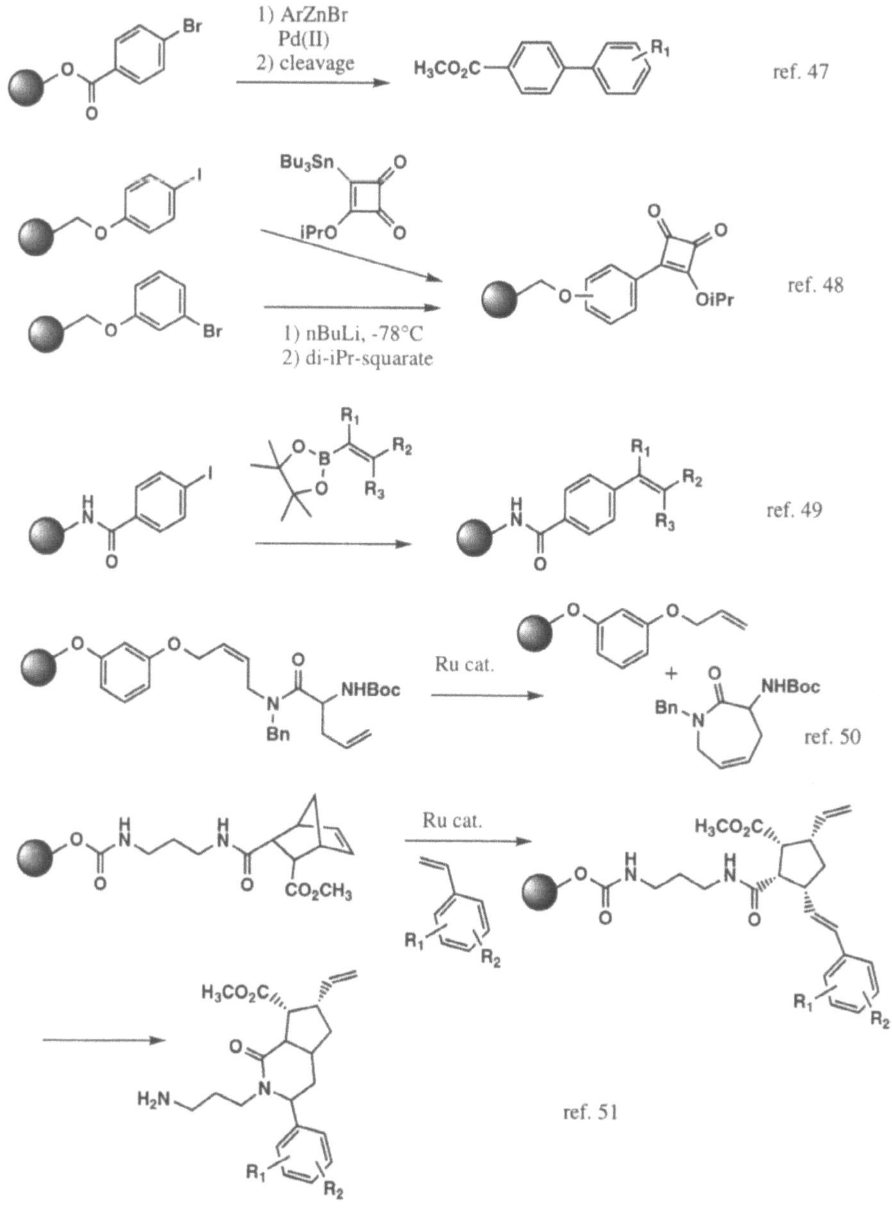

Figure 4. Carbon-carbon bond forming reactions.

Aldol reactions and miscellaneous

Several groups have described aldol reactions on the solid phase in which either the aldehyde or the 'enolate' are resin bound (Figure 5). Kobayashi et al. reported the scandium triflate catalyzed condensation of thiosilylenol ethers with aryl aldehydes [53]. His group extended this chemistry demonstrating the condensation of the same resin bound enolates with imines to afford, after reductive cleavage, 1,3-aminoalcohols [54]. Reggelin and Brenig [55] devised an iterative process using resin bound aldehydes with Evans boron enolates [56].

Katritzky showed that primary amines could be

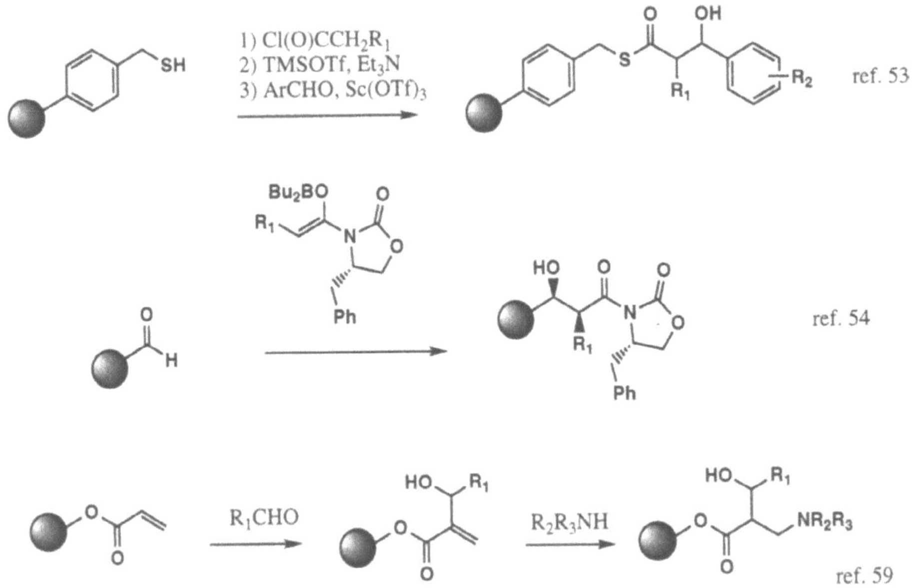

Figure 5. Carbon-carbon bond forming reactions.

synthesized via the nucleophilic addition of Grignard and organolithium reagents to resin bound aldimines [57]. A 20-membered cyclic tetrapeptide derivative was prepared by an intramolecular Heck reaction wherein the solid-phase environment accelerated the cyclization relative to the cyclization in solution [58]. Kunzer et al. prepared a series of 3-hydroxypropionamides using a Bayliss–Hillman reaction [59].

Reactions: Heterocycle synthesis

Most prominent in the recent literature are papers describing the solid-phase synthesis of heterocyclic products. A review of solid-phase heterocycle synthesis has been published, focusing on the work from the Chiron group [60]. This section has been organized based on the size of the ring constructed.

Five-membered ring heterocycles

A variety of five-membered heterocycles have been synthesized on solid phase and have been summarized in Figures 6 and 7. Substituted indoles have been prepared by copper/palladium catalyzed coupling of terminal acetylenes with resin bound aryl iodides [61]. Tetramethylguanidine (TMG), which was used as a solvent, catalyzed both the coupling and cyclization reactions. Other groups have described solid phase indole syntheses using similar conditions with internal alkynes [62] or alternatively, using the Heck reaction [63] or Fisher indole conditions [64]. Benzoxazoles were constructed in a straightforward manner by Mitsunobu reactions of the precursor *o*-hydroxyanilides [65]. Benzimidazoles have been assembled by nucleophilic displacement of aryl fluorides with amines, followed by reduction and condensation with imidates [66].

Condensation of resin bound ketones with hydrazines followed by cyclizative cleavage was used in a synthesis of pyrazolinones [67]. Samarium iodide induced radical cyclization of allyl aryl ethers led to benzofuran derivatives [68]. Kurth et al. have reported a two reaction iterative process for the construction of polyisoxazolines [69] and polyisoxazoles [70]. An intramolecular Heck reaction has been employed to prepare 2-oxindole derivatives [71].

Gowravaram and Gallop have developed a traceless synthesis of substituted furans using 1,3-dipolar cycloadditions of isomunchones [72]. Zaragoza published the synthesis of 1,2,3-triazoles [73] and aminothiophenes [74] wherein the final targets were monoamides of piperazine. Hydantoins continue to be a favorite target for combinatorial chemists. Matthews and Rivero [75] have described a modification of

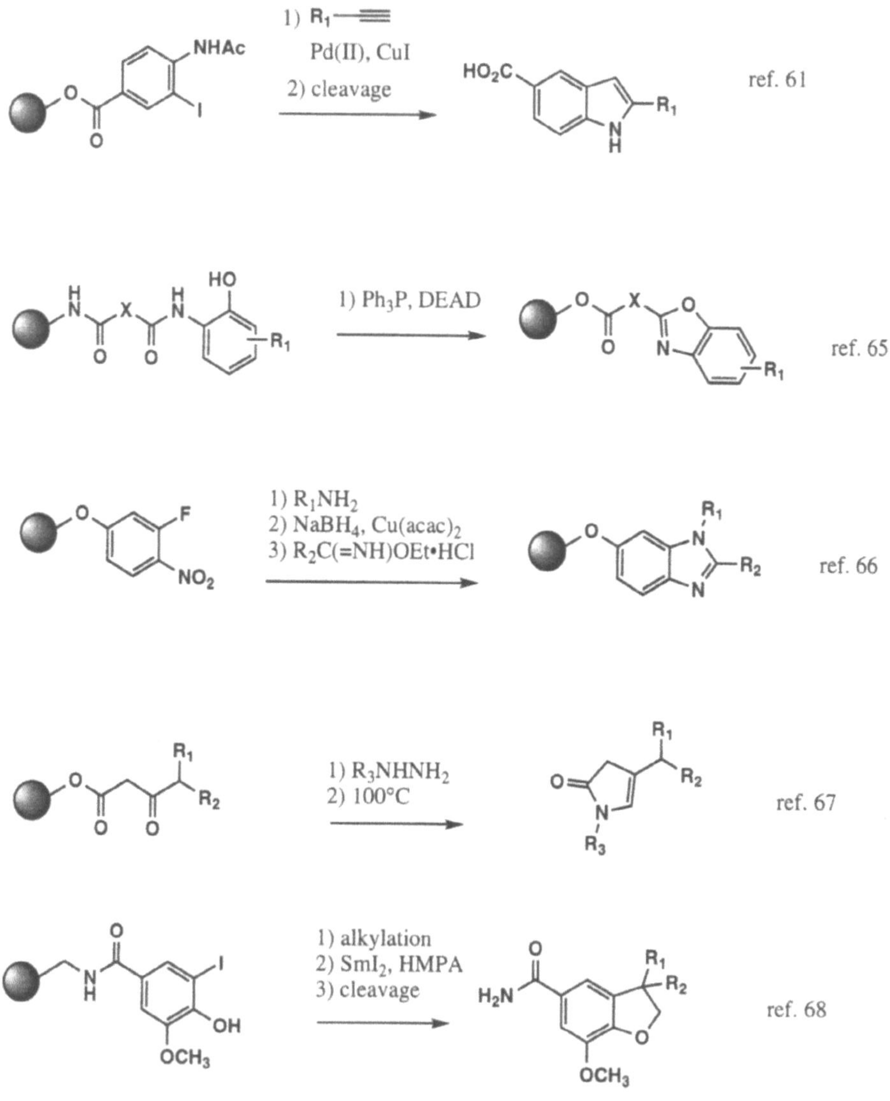

Figure 6. Five-membered ring heterocycle synthesis.

prior methods which expand the diversity of the final products. Mjalli et al. described the synthesis of hydantoin-4-imides [76] and Hanessian reported on the synthesis of 5-alkoxyhydantoins [77]. Procedures for the construction of ureas [78], spiroindolines [79], and isoquinolines [80] have also been reported. Finally, two additional papers have appeared describing the use of azomethine ylide cycloadditions to prepare polyfunctionalized pyrrolidines [81,82].

Six-membered ring heterocycles

A three component condensation involving aryl amines, aldehydes, and olefins was used to assemble a variety of tetrahydroquinolines [83]. Dihydropyrimidine-2,4-diones were prepared from condensing amines with resin bound acrylates followed by treatment with isocyanates [84]. Two additional groups have published their work using the Pictet-Spengler reaction to prepare tetrahydro-β-carbolines [85], tetrahydroisoquinolines [86] and

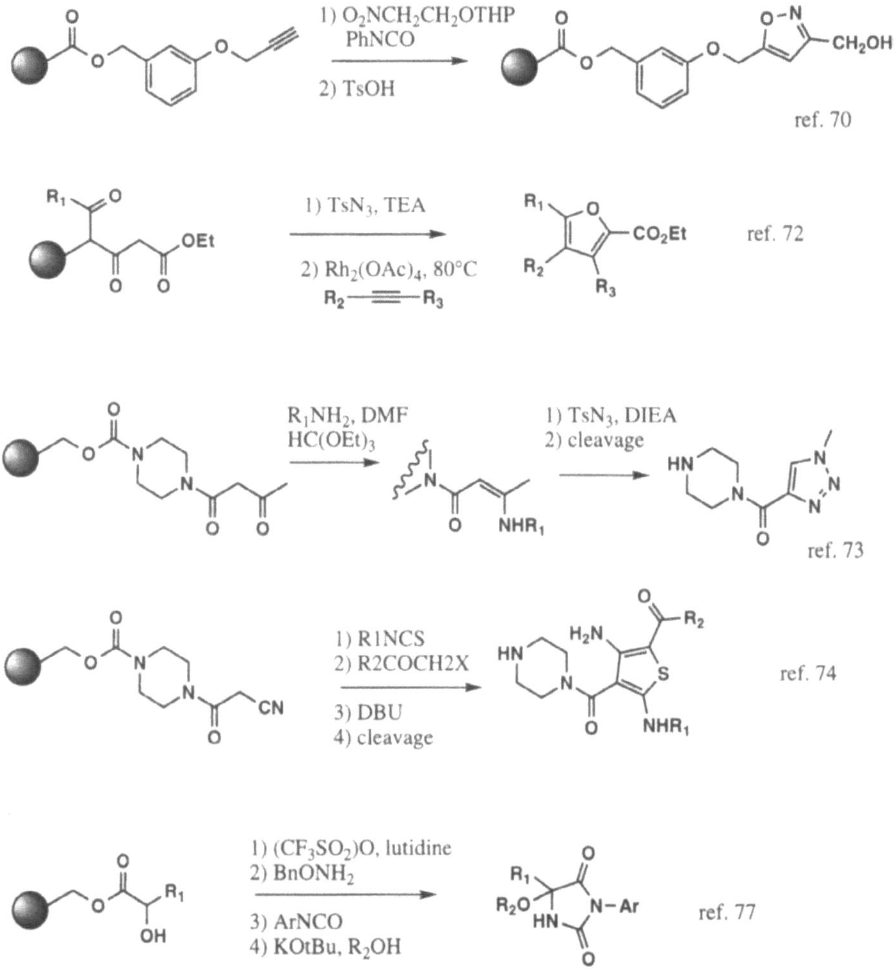

Figure 7. Five-membered ring heterocycle synthesis.

tetrahydroimidazopyridines [86]. Outlined in Figure 8 is one of the three reports on the synthesis of quinazoline-2,4-diones. The Versicor procedure [87] provides products with diversity derived from amino acids, with the other reports using amines [88] and isocyanates [89]. Aryl quinoline-4-carboxylic acids were prepared using the Doebner quinoline synthesis [90]. Panek and Zhu employed an inverse electron demand Diels-Alder reaction of resin bound tetrazines to synthesize 1,2-diazines [91]. Other six-membered heterocycles which have been reported include 3,4-disubstituted tetrahydroquinoxalin-2-ones [92], pyridines [93], and pyrido[2,3-*d*]pyrimidines [93].

Larger ring heterocycles

Seven-membered ring targets such as benzodiazepines and benzodiazepinones have been popular targets for solid-phase synthesis. Mayer et al. described a modified approach to benzodiazepin-diones which maximizes the diversity of the products [94]. The related perhydroazepin-diones [95] have been prepared starting from resin bound aspartic acid. Nine- and ten-membered ring β-turn mimetics were prepared by intramolecular nucleophilic displacement of an α-bromo amide by a mercaptan [96]. Several groups have reported on the synthesis of large ring modified peptides including protein loop mimetics [97], oscillamide Y [98], and artificial β-sheets [99].

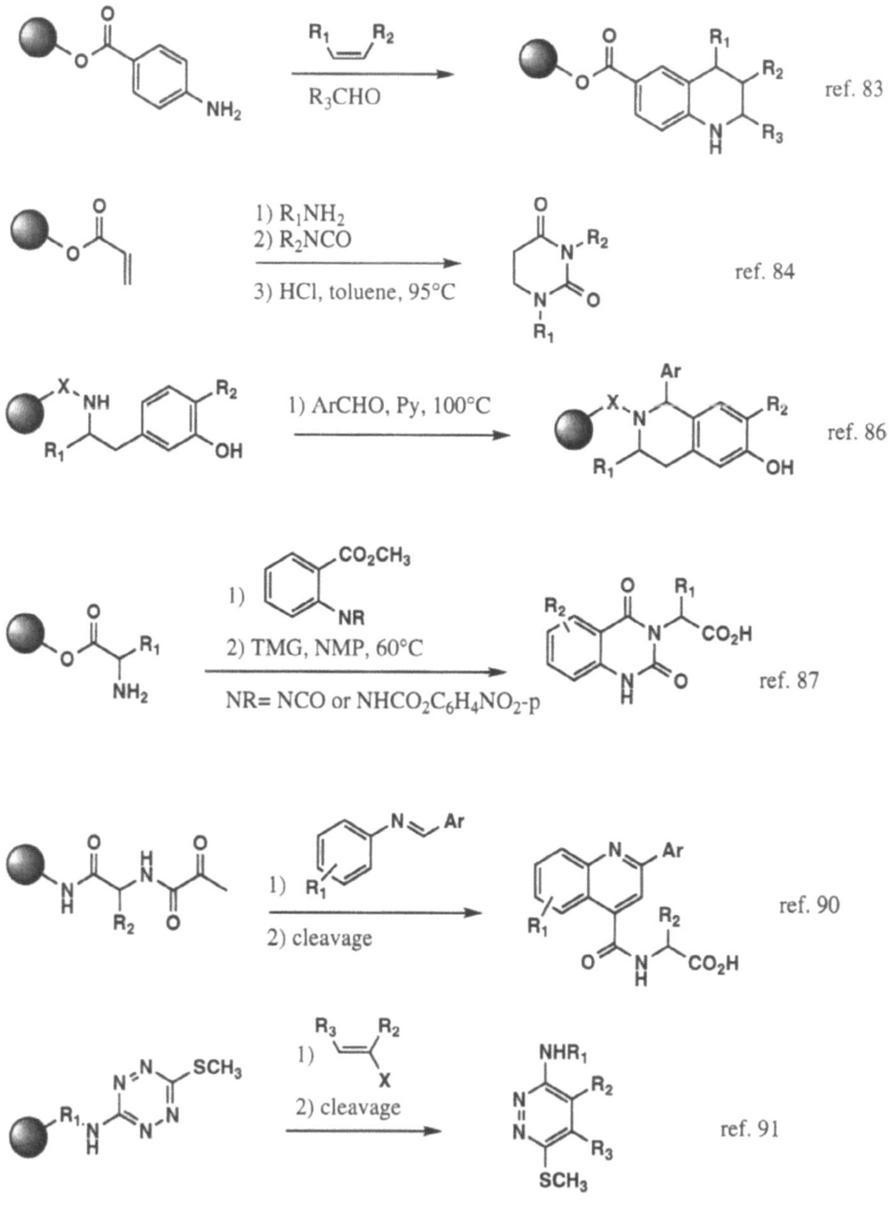

Figure 8. Six-membered ring heterocycle synthesis.

Reactions: Total synthesis of complex compounds

Over the past year, solid-phase synthesis approaches have been employed to tackle the preparation of a number of complex natural and synthetic targets. Fraser-Reid, Geysen et al. reported the solid-phase synthesis of oligosaccharides [100], which adds to the efforts reported previously by Kahne et al. [101]. Workers at IRORI have published their approach to a Taxoid library in which three modification sites were used, introducing them as esters and amides [102]. Armstrong et al. [103] have described the application of the four-component Ugi reaction in which the aldehyde and/or the carboxylic acid were derived from

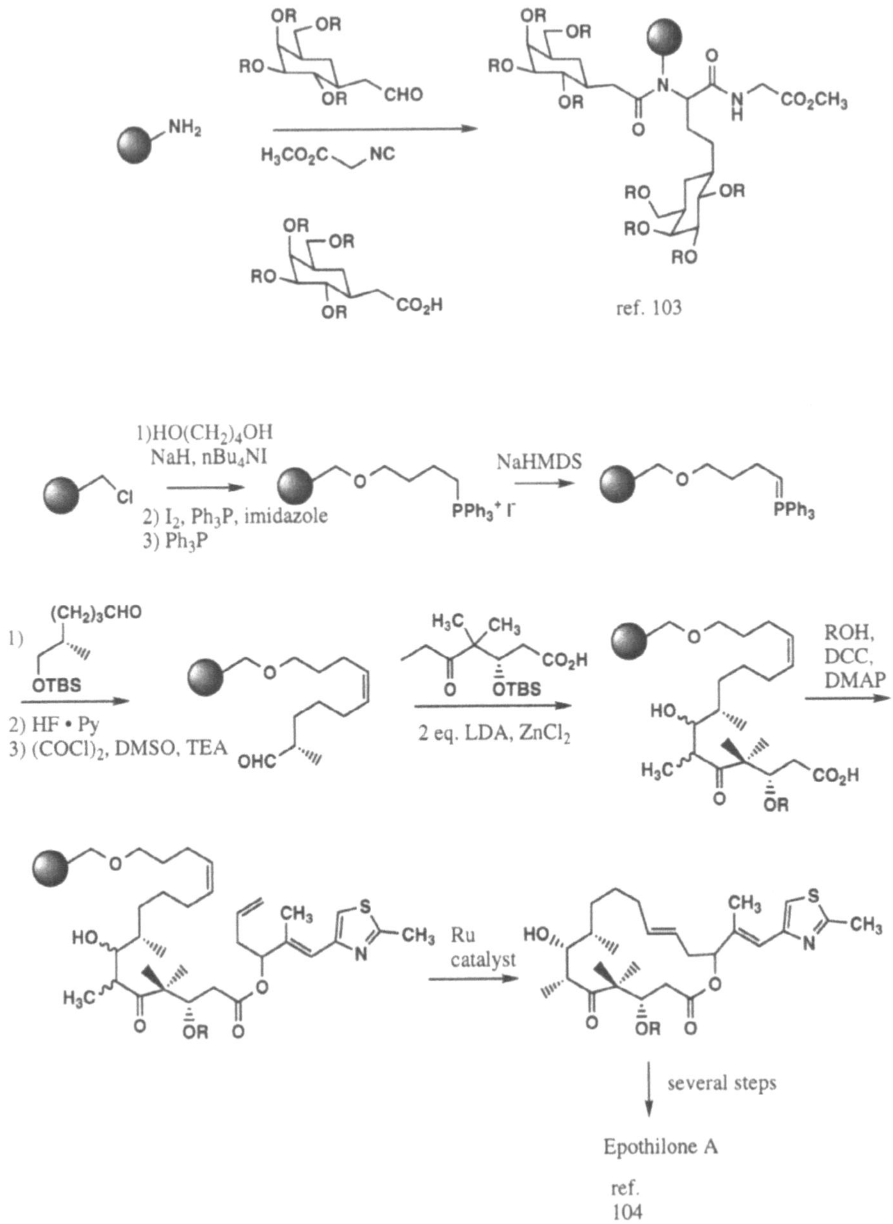

Figure 9. Solid-phase synthesis of complex compounds.

carbohydrates. This allowed for the one step construction of highly functionalized products as shown in Figure 9. Finally, Nicolaou et al. published the synthesis of epithilones A and B [104]. Key reactions in the construction of this macrocycle included the ring closing metathesis reaction.

Conclusions

The number and complexity of solid-phase syntheses continues to increase. Despite the advances in resin bound reagents, solid-phase synthesis, wherein the growing product of interest is attached to the resin, will continue to be a useful paradigm in organic synthesis. The increased number and variation of linking technologies will also encourage those chemists currently not taking advantage of solid-phase methods to give it serious consideration.

References

1. Szostak, J. (Ed.) Chem. Rev., 97 (1997) 347–510.
2. Lam, K.S., Lebl, M. and Krchnak, V., Chem. Rev., 97 (1997) 411.
3. Nefzi, A., Ostresh, J.M. and Houghten, R.A., Chem. Rev., 97 (1997) 449.
4. Pirrung, M.C., Chem. Rev., 97 (1997) 473.
5. Gravert, C.J. and Janda, K.D., Chem. Rev., 97 (1997) 489.
6. Kay, C., Murray, P.J., Sandow, L. and Holmes, A.B., Tetrahedron Lett., 38 (1997) 6941.
7. Furth, P.S., Reitman, M.S. and Cook, A.F., Tetrahedron Lett., 38 (1997) 5403.
8. Furth, P.S., Reitman, M.S., Gentles, R. and Cook, A.F., Tetrahedron Lett., 38 (1997) 6643.
9. Hoekstra, W.J., Greco, M.N., Yabut, S.C., Hulshizer, B.L. and Maryanoff, B.E., Tetrahedron Lett., 38 (1997) 2629.
10. Mellor, S.L., McGuire, C. and Chan, W.C., Tetrahedron Lett., 38 (1997) 3311.
11. Brown, A.R., Rees, D.C., Rankovic, Z. and Morphy, J.R., J. Am. Chem. Soc., 119 (1997) 3288.
12. Roussel, P., Bradley, M., Matthews, I. and Kane, P., Tetrahedron Lett., 38 (1997) 4861.
13. Raju, B. and Kogan, T.P., Tetrahedron Lett., 38 (1997) 3373.
14. Raju, B. and Kogan, T.P., Tetrahedron Lett., 38 (1997) 4965.
15. Hunt, J.A. and Roush, W.R., J. Am. Chem. Soc., 118 (1996) 9998.
16. Plunkett, M.J. and Ellman, J.A., J. Org. Chem., 62 (1997) 2885.
17. Woolard, F.X., Paetsch, J. and Ellman, J.A., J. Org. Chem., 62 (1997) 6102.
18. Stranix, B.R., Liu, H.Q. and Darling, G.D., J. Org. Chem., 62 (1997) 6183.
19. Newlander, K.A., Chenera, B., Veber, D.F., Yim, N.C.F. and Moore, M.L., J. Org. Chem., 62 (1997) 6726.
20. Routledge, A., Abell, C. and Balasubramanian, S., Tetrahedron Lett., 38 (1997) 1227.
21. Ngu, K. and Patel, D.V., Tetrahedron Lett., 38 (1997) 973.
22. Volkmer-Engert, R., Hoffmann, B. and Schneider-Mergener, J., Tetrahedron Lett., 38 (1997) 1029.
23. Ho, C.Y. and Kukla, M.J., Tetrahedron Lett., 38 (1997) 2799.
24. Conti, P., Demont, D., Cals, J., Ottengeijm, H.C.J. and Leysen, D., Tetrahedron Lett., 38 (1997) 2915.
25. Ward, Y.D. and Farina, V., Tetrahedron Lett., 37 (1996) 6993.
26. Willoughby, C.A. and Chapman, K.T., Tetrahedron Lett., 37 (1996) 7181.
27. Khan, N.M., Arumugam, V. and Balasubramanian, S., Tetrahedron Lett., 37 (1996) 4819.
28. Brown, P.J., Hurley, K.P., Stuart, L.W. and Willson, T.M., Synthesis, (1997) 778.
29. Youngman, M.A. and Dax, S.L., Tetrahedron Lett., 38 (1997) 6347.
30. Wang, G.T., Chen, Y., Wang, S., Sciotti, R. and Sowin, T., Tetrahedron Lett., 38 (1997) 1895.
31. Chen, J.J. and Spatola, A.F., Tetrahedron Lett., 38 (1997) 1511.
32. Floyd, C.D., Lewis, C.N., Patel, S.R. and Whittaker, M., Tetrahedron Lett., 37 (1996) 8045.
33. Drewry, D.H., Gerritz, S.W. and Linn, J.A., Tetrahedron Lett., 38 (1997) 3377.
34. Scialdone, M.A., Tetrahedron Lett., 37 (1996) 8141.
35. Kim, J.-M., Bi, Y., Paikoff, S.J. and Schultz, P.G., Tetrahedron Lett., 37 (1996) 5305.
36. Meldal, M., Juliano, M.A. and Jansson, A.M., Tetrahedron Lett., 38 (1997) 2531.
37. Wallace, O.B., Tetrahedron Lett., 38 (1997) 4939.
38. Tortolani, D.R. and Biller, S.A., Tetrahedron Lett., 37 (1996) 5687.
39. Fancellli, D., Fagnola, M.C., Severino, D. and Bedeschi, A., Tetrahedron Lett., 38 (1997) 2311.
40. Burgess, K., Lim, D., Bois-Choussy, M. and Zhu, J., Tetrahedron Lett., 38 (1997) 3345.
41. Wang, Z.-G., Douglas, S.P. and Krepinsky, J.J., Tetrahedron Lett., 37 (1996) 6985.
42. Adinolfi, M., Barone, G., De Napoli, L., Iadonisi, A. and Piccialli, G., Tetrahedron Lett., 37 (1996) 5007.
43. Larhed, M., Lindeberg, G. and Hallberg, A., Tetrahedron Lett., 37 (1996) 8219.
44. Guiles, J.W., Johnson, S.G. and Murray, W.V., J. Org. Chem., 61 (1996) 5169.
45. Yoo, S., Seo, J., Yi, K. and Gong, Y., Tetrahedron Lett., 38 (1997) 1203.
46. Piettre, S.R. and Baltzer, S., Tetrahedron Lett., 38 (1997) 1197.
47. Marquais, S. and Arlt, M., Tetrahedron Lett., 37 (1996) 5491.
48. Tempest, P.A. and Armstrong, R.W., J. Am. Chem. Soc., 119 (1997) 7607.
49. Brown, S.D. and Armstrong, R.W., J. Am. Chem. Soc., 118 (1996) 6331.
50. Van Maarseveen, J.H., den Hartog, J.A.J., Engelen, V., Finner, E., Visser, G. and Kruse, C.G., Tetrahedron Lett., 37 (1996) 8249.
51. Cuny, G.D., Cao, J. and Hauske, J.R., Tetrahedron Lett., 38 (1997) 5237.
52. Wipf, P. and Henninger, T.C., J. Org. Chem., 62 (1997) 1586.
53. Kobayashi, S., Hachiya, I. and Yasuda, M., Tetrahedron Lett., 37 (1996) 5569.
54. Koybayashi, S., Moriwaki, M., Akiyama, R., Suzuki, S. and Hachiya, I., Tetrahedron Lett., 37 (1996) 7783.
55. Reggelin, M. and Brenig, V., Tetrahedron Lett., 37 (1996) 6851.
56. Evans, D.A., Bartroli, J.A. and Shis, T.L., J. Am. Chem. Soc., 103 (1981) 2127.
57. Katritzky, A.R., Xie, L. and Zhang, G., Tetrahedron Lett., 38 (1997) 7011.
58. Akaji, K. and Kiso, Y., Tetrahedron Lett., 38 (1997) 5185.
59. Prien, O., Rolfing, K., Thiel, M. and Kunzer, H., Synlett., (1997) 325.
60. Nuss, J.M., Desai, M.C., Zuchermann, R.N., Singh, R., Renhowe, P.A., Goff, D.A., Chinn, J.P., Wang, L., Dorr, H., Brown, E.G. and Subramanian, S., Pure Appl. Chem., 69 (1997) 447.
61. Fagnola, M.C., Candiani, I., Visentin, G., Cabri, W., Zarini, F., Mongelli, N. and Bedeschi, A., Tetrahedron Lett., 38 (1997) 2307.
62. Zhang, H.-C., Brumfield, K.K. and Maryanoff, B.E., Tetrahedron Lett., 38 (1997) 2439.
63. Yun, W. and Mohan, R., Tetrahedron Lett., 37 (1996) 7189.

64. Hutchins, S.M. and Chapman, K.T., Tetrahedron Lett., 37 (1996) 4869.
65. Wang, F. and Hauske, J.R., Tetrahedron Lett., 38 (1997) 6529.
66. Phillips, G.B. and Wei, G.P., Tetrahedron Lett., 37 (1996) 4887.
67. Tietze, L.F. and Steinmetz, A., Synlett., (1996) 667.
68. Du, X. and Armstrong, R.W., J. Org. Chem., 62 (1997) 5678.
69. Kurth, M.J., Ahlberg-Randall, L.A. and Takenouchi, K., J. Org. Chem., 61 (1996) 8755.
70. Kantorowski, E.J. and Kurth, M.J., J. Org. Chem., 62 (1997) 6797.
71. Arumugam, V., Routledge, A., Abell, C. and Balasubramanian, S., Tetrahedron Lett., 38 (1997) 6473.
72. Gowravaram, M.R. and Gallop, M.A., Tetrahedron Lett., 38 (1997) 6973.
73. Zaragoza, F. and Petersen, S.V., Tetrahedron, 52 (1996) 10823.
74. Zaragoza, F., Tetrahedron Lett., 37 (1996) 6213.
75. Matthews, J. and Rivero, R.A., J. Org. Chem., 62 (1997) 6090.
76. Short, K.M., Ching, B.W. and Mjalli, A.M.M., Tetrahedron Lett., 37 (1996) 7489.
77. Hanessian, S. and Yang, R.-Y., Tetrahedron Lett., 37 (1996) 5835.
78. Xiao, X., Ngu, K., Chao, C. and Patel, D.V., J. Org. Chem., 62 (1997) 6968.
79. Cheng, Y. and Chapman, K.T., Tetrahedron Lett., 38 (1997) 1497.
80. Lorsbach, B.A., Miller, R.B. and Kurth, M.J., J. Org. Chem., 61 (1996) 8716.
81. Marx, M.A., Grillot, A.-L., Louer, C.T., Beaver, K.A. and Bartlett, P.A., J. Am. Chem. Soc., 119 (1997) 6153.
82. Bicknell, A.J. and Hird, N.W., Bioorg. Med. Chem. Lett., 6 (1996) 2441.
83. Kiselyov, A.S. and Armstrong, R.W., Tetrahedron Lett., 38 (1997) 6163.
84. Kolodziej, S.A. and Hamper, B.C., Tetrahedron Lett., 37 (1996) 5277.
85. Mohan, R., Chou, Y.-L. and Morrissey, M.M., Tetrahedron Lett., 37 (1996) 3963.
86. Hutchins, S.M. and Chapman, K.T., Tetrahedron Lett., 37 (1996) 4865.
87. Gordeev, M.F., Hui, H.C., Gordon, E.M. and Patel, D.V., Tetrahedron Lett., 38 (1997) 1729.
88. Gouilleux, L., Fehrentz, J.-A., Winternitz, F. and Martinez, J., Tetrahedron Lett., 37 (1996) 7031.
89. Buckman, B.O. and Mohan, R., Tetrahedron Lett., 37 (1996) 4439.
90. Gopalsamy, A. and Pallai, P.V., Tetrahedron Lett., 38 (1997) 907.
91. Panek, J.S. and Zhu, B., Tetrahedron Lett., 37 (1996) 8151.
92. Lee, J., Murray, W.V. and Rivero, R.A., J. Org. Chem., 62 (1997) 3874.
93. Gordeev, M.F., Patel, D.V., Wu, J. and Gordon, E.M., Tetrahedron Lett., 37 (1996) 4643.
94. Mayer, J.P., Zhang, J., Bjergarde, K., Lenz, D.M. and Gaudino, J.J., Tetrahedron Lett., 37 (1996) 8081.
95. Nefzi, A., Ostresh, J.M. and Houghten, R.A., Tetrahedron Lett., 38 (1997) 4943.
96. Virgilio, A.A., Schurer, S.C. and Ellman, J.A., Tetrahedron Lett., 37 (1996) 6961.
97. Emery, F., Bisang, C., Favre, M., Jiang, L. and Robinson, J.A., J. Chem. Soc., Chem. Commun., (1996) 2155.
98. Marsh, I.R. and Bradley, M., J. Org. Chem., 62 (1997) 6199.
99. Holmes, D.L., Smith, E.M. and Nowick, J.S., J. Am. Chem. Soc., 119 (1997) 7665.
100. Rodebaugh, R., Joshi, S., Fraser-Reid, B. and Geysen, H.M., J. Org. Chem., 62 (1997) 5660.
101. Liang, R., Yan, L., Loebach, J., Ge, M., Uozumi, Y., Sekanina, K., Horan, N., Gildersleeve, J., Thompson, C., Smith, A., Biswas, K., Still, W.C. and Kahne, D.E., Science, 274 (1996) 1520.
102. Xiao, X.-Y., Parandoosh, Z. and Nova, M.P., J. Org. Chem., 62 (1997) 6029.
103. Sutherlin, D.P., Stark, T.M., Hughes, R. and Armstrong, R.W., J. Org. Chem., 61 (1996) 8350.
104. Nicolaou, K.C., Winssinger, N., Pastor, J., Ninlovic, S., Sarabia, F., He, Y., Vourloumis, D., Yang, Z., Li, T. and Giannakakou, P., Nature, 387 (1997) 268.

M.R. Pavia and W.H. Moos (eds.), Annual Reports in Combinatorial Chemistry and Molecular Diversity, Vol. 2, 27–32.
KLUWER/ESCOM

Purification of combinatorial libraries

Harold N. Weller
Bristol-Myers Squibb Pharmaceutical Research Institute, P.O. Box 4000, Princeton, NJ 08543, U.S.A.

Key words: chromatography, HPLC, library, purification, SPE

Summary

In the early days of combinatorial chemistry, much attention focused on preparation of large libraries for lead discovery. Recently, though, the focus has shifted toward smaller, more focused libraries for lead optimization. These focused libraries generally consist of individual discrete compounds. Biological assay requirements often require compounds of high purity, thus development of automated high throughput purification methods has received new attention in the past several years. This paper covers automated high throughput purification methods that have been applied to libraries of discrete compounds. Literature published through February 1998 is included. Purification methods discussed include extraction methods, scavenger methods, solid phase extraction, and preparative HPLC.

Introduction

The drug discovery process is a continuum beginning with lead generation, proceeding through lead optimization, and resulting in drug candidate selection. Most early work in the emerging field of combinatorial chemistry focused on synthesis of large compound libraries for lead generation. In the lead generation phase, the objective is only to discover some biological activity in a compound which will serve as a starting point for further lead optimization. As a result, compound purity is not generally an important factor and, indeed, synthesis of mixtures is a common practice. Once projects progress into the lead optimization phase, however, compound purity becomes more important as comparisons among compounds are made and biological assays become more time and labor intensive. In the last few years it has become apparent that if combinatorial chemistry is to have as great an impact on late stage lead optimization as it has had on early stage lead generation then high throughput automated analysis and purification methods must be developed. This paper will review purification methods that have been developed and applied to the purification of discrete compounds resulting from parallel solution or solid phase synthesis.

Extractive procedures

The purification challenge for products of parallel solution phase synthesis is primarily one of reagent removal. For compounds synthesized conventionally, reagent removal generally consists of an organic/aqueous extraction procedure. While liquid-liquid extraction has been applied to work-up of parallel synthesis libraries [1] it is generally difficult to automate. A recent report has described an automated device for liquid/liquid extraction. The device reportedly uses a hydrophobic membrane to separate solutions of methylene chloride and water, but full details are lacking [2]. A preliminary report of a related extraction procedure recently appeared from the Arris group [3]. This procedure uses commercially available diatomaceous earth (Varian ChemElute^{TM}) to remove water and water soluble impurities. Aqueous buffer (for example 1 N hydrochloric acid to remove amines, or potassium hydroxide solution to remove acids) is added to the diatomaceous earth. The reaction mixture in organic solvent is then added, and the products are extracted by filtration of the organic solvent through the aqueous/earth filter pad. While this procedure may eventually have widespread impact, full experimental details have not yet been described.

An alternative to liquid/liquid extraction is to use solid supported electrophilic or nucleophilic scavengers to remove excess reagents based on molecular reactivity. This procedure has been reported by several groups. Kaldor and co-workers [4] used polymer supported aminomethyl resin to scavenge excess isocyanate, acid chloride, or sulfonyl chloride reagents. The resulting acylated resin was removed by filtration, leaving highly purified neutral products in solution. Alternatively, polymer supported isocyanate, aldehyde, and benzoyl chloride groups were used to scavenge excess amine reagents. The method has been applied to synthesis of secondary amines (via reductive amination), amides, ureas, thioureas, sulfonamides, and carbamates. The Searle group reported nearly identical results, but extended their work to include trapping the carbodiimide and urea by-products of Moffat oxidation using dimethylaminopropyl ethyl carbodiimide [5]. A subsequent report by the same group described Mitsunobu reactions using phosphines containing masked carboxylates which could be removed by sequestration with amine resin [6]. Gayo and Suto [7] reported use of anion exchange resins to remove carboxylic acid by-products, and use of cation exchange resin to remove excess amines. These authors describe a series of experiments designed to optimize the nature of the resin, solvent, and addition method. The Ellman group [8] described the use of a unique guanidine-containing scavenger resin, while the Argonaut group [9] described a scavenger for DDQ. The Armstrong group [10] described product capture by covalent linking to resin; this method makes products directly available for further elaboration by solid phase synthesis.

In general, liquid/solid extraction methods offer the simplicity of solution phase synthesis and the ease of work-up by simple filtration normally offered by solid phase synthesis. As such, their use will almost certainly become much more widespread in the future.

Solid phase extraction

Solid phase extraction (SPE) is a method originally developed for concentration of biological and environmental samples prior to analysis by conventional methods. The method is essentially a short column chromatography through sorbent packed in disposable polypropylene syringe-type cartridges of various sizes. A variety of commercially available devices can be used to automate the process. Such devices range from simple vacuum boxes to elaborate robotic systems such as the Zymark Benchmate™, Gilson Aspec™, and Hamilton Microlab™ instruments. A customized robotic device for high speed parallel solid phase extraction has also recently been reported [11], as has a device for performing SPE in standard 96-well format [12]. Because of the low cost and availability of standard automation, solid phase extraction has become popular as a method for performing automated chromatography of products from parallel synthesis.

Automated solid phase extraction systems are generally capable of collecting only a limited number of fractions per sample, and do not offer real time on line detection. Consequently, automated SPE methods are only successful when the elution profile of products and by-products is known with certainty. For that reason, ion-exchange methods have been far more successful than partition chromatography methods when applied to purification of diverse libraries from parallel synthesis. For example, the Bristol-Myers Squibb group reported purification of a library of 225 tertiary amine containing amides by SPE cation exchange [13]. Amine (**1**) was allowed to react with a variety of carboxylic acids (**2**) using diisopropylcarbodiimide and 1-hydroxybenzotriazole as coupling agents. The crude reaction mixtures were loaded onto SCX cation exchange cartridges, washed to remove neutral by-products, then the products were eluted with a solution of ammonia in methanol (Figure 1). Average purity of the resulting library was 90%.

In a variation on the theme, the same authors reported purification of neutral products resulting from addition of an excess of various amines to a constant *p*-nitrophenyl ester (Figure 2). Sequential anion exchange and cation exchange removed charged by-products, resulting in pure (>90%) neutral products.

The Lilly group recently reported similar use of ion exchange SPE to separate (basic) amine products (**7**) from (neutral) excess aldehyde (**6**) in a reductive amination reaction (Figure 3) [14].

The Lilly group also prepared a series of neutral ureas (**10**) by reaction of amines (**8**) with excess isocyanate (**9**). In this case, both the products (**10**) and by-products (excess **9**) were neutral. Excess isocyanate was scavenged by reaction with a tertiary amine containing nucleophile to provide basic ureas (**11**) which were separated from products (**10**) by cation exchange SPE (Figure 4). The use of scavenger reagents thus extends the utility of ion exchange SPE.

While there have been anecdotal reports of partition chromatography of parallel synthesis products

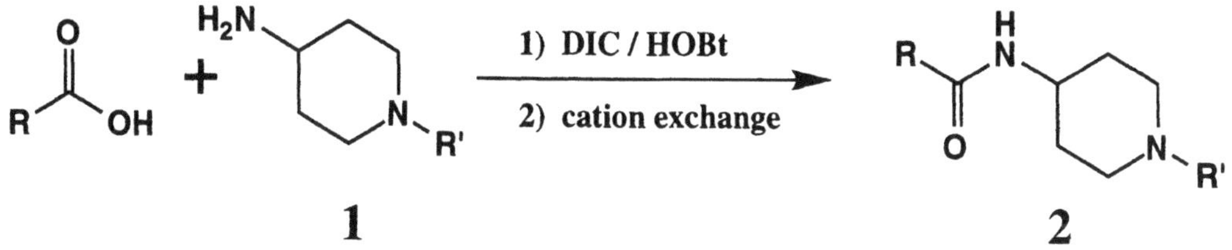

Figure 1. Purification of amines on SCX resin.

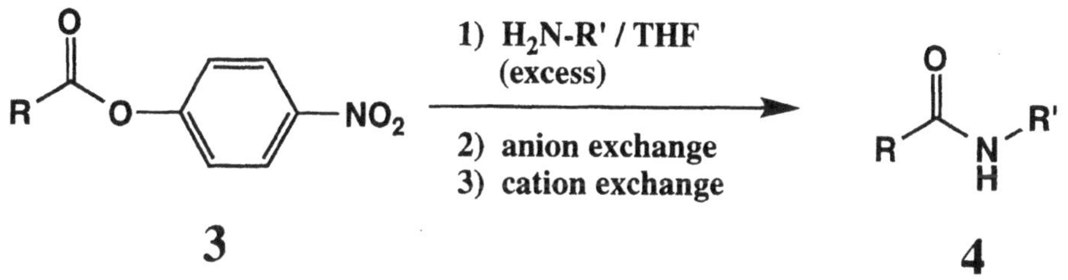

Figure 2. Purification of neutral amides by sequential ion exchange.

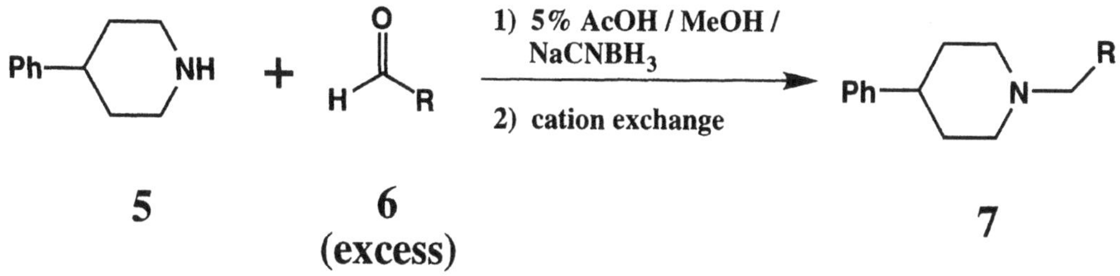

Figure 3. Purification of amines on SCX resin.

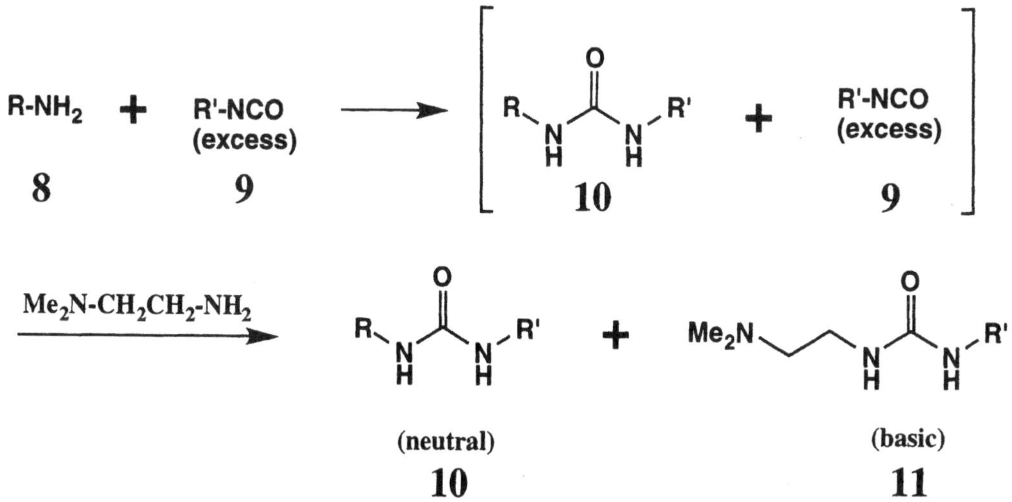

Figure 4. Trapping of isocyanates with amines.

in SPE cartridges, no well described written reports have appeared. Sorbent based SPE of a combinatorial library is generally difficult because of the lack of real time detection. By very definition, products

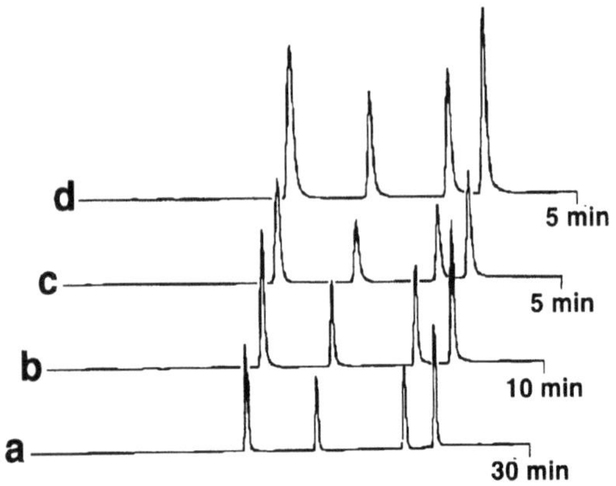

Figure 5. Effect of column length and flow rate on preparative HPLC separation. Gradient test mixture chromatograms using linear gradient elution from 10% to 90% aqueous methanol containing 0.1% TFA. UV absorbance monitored at 220 nm. (a) 20 × 250 mm column, 10 mL min^{-1} flow rate, 30 min gradient time. (b) 20 × 100 mm column, 20 mL min^{-1} flow rate, 10 min gradient time. (c) 20 × 100 mm column, 40 mL min^{-1} flow rate, 5 min gradient time. (d) 20 × 50 mm column, 20 mL min^{-1} flow rate, 5 min gradient time.

of parallel synthesis have diverse physical properties and thus diverse chromatographic elution profiles. Partition chromatography without real time detection is thus difficult and tedious. The development of rapid automated preparative HPLC methods with real time detection has made partition chromatography by SPE an unattractive option.

Preparative HPLC

Preparative HPLC has been widely used for purification of peptides and small molecules for over a decade. Its inherent serial nature, lack of automation, and severe throughput limits, would seem to make it unsuitable for purification of large and diverse compound libraries. Recent advances in automation, detection, and method development, however, now make it possible to purify hundreds of compounds per day using a single instrument. Further throughput improvements may come from parallel HPLC systems which have been recently reported.

The Bristol-Myers Squibb group has reported an automated preparative HPLC system capable of processing dozens to hundreds of samples in unattended mode [15]. The system requires customized hardware and software and is not commercially available. The same group reported very rapid gradient reverse phase methods for purifying up to 200 mg of sample in as little as 5 min. Very short columns and high flow rates are used to achieve rapid separation (Figure 5) and excellent loading (Figure 6) [16]. A full report of this work has appeared [17]. The YMC group, working in collaboration with Bristol-Myers Squibb, has separately reported on rapid separation methods [18]. The combination of full automation, rapid methods, and real time on-line detection makes it possible to routinely provide products of high purity from automated parallel synthesis.

Another proprietary fully automated semi-preparative HPLC system was described by the Parke-Davis group [19]. This system relies heavily on equivalence between an analytical scale 'scouting' chromatogram and the corresponding preparative chromatogram to set an appropriate fraction collection window. The system has been used to purify up to 10 mg of product per run, using conventional HPLC methods. While this system shows promise, its use in true high throughput mode has not yet been reported. The Abbott group made a preliminary report of their efforts to construct a preparative HPLC system [20]. Final system development and application to purification of products from parallel synthesis have not yet been reported.

The systems reported above all rely on detection by ultraviolet absorbance, and require some separate means of identifying the desired product among the various components (and thus collected fractions) of the reaction mixture. Efficiency and throughput can be considerably enhanced by real-time detection *and identification* of products. Two groups have now reported preparative HPLC systems where fractionation decisions are based upon output from a mass spectrometer detector. The first preliminary report of such a system came from a collaboration between CombiChem and Sciex [21]. This system bases fractionation decisions on the output of a single ion mass chromatogram for the predicted molecular ion of the desired product. Reverse phase preparative HPLC is used in conjunction with electrospray ionization mass detection. A full report of this work has appeared in early 1998 [22]. The second report of such a system came from a collaboration between Pfizer and Micromass [23]. This system uses a flexible combination of UV and/or ion chromatograms to control fractionation. Unlike other systems, fractionation parameters are set by the mass spectrometer control software. Variants of both of the above systems will probably become commercially available in late 1997 [24].

The above preparative LC-MS systems both require a stream splitting device to divert a small portion

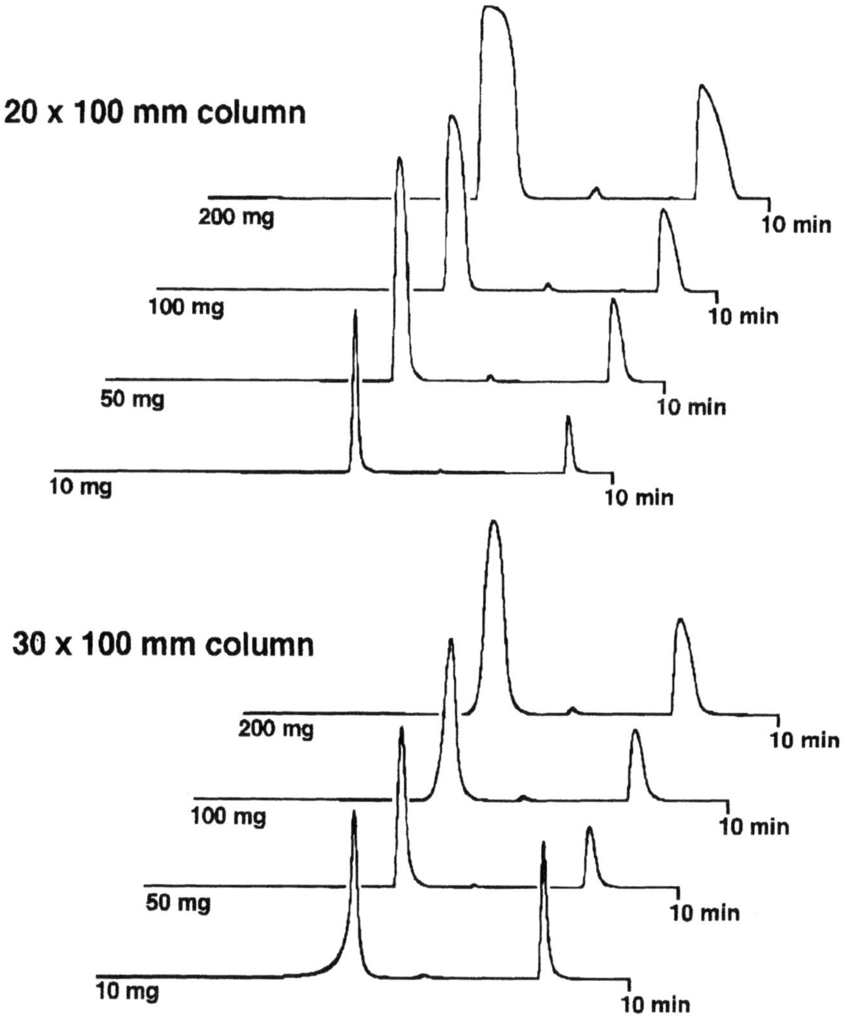

Figure 6. Effect of column diameter on loading. C18 reverse phase 20 × 100 mm and 30 × 100 mm columns. Operated in gradient elution mode from 10% to 90% aqueous methanol over 10 min.

of the eluant stream (0.1–1.0 mL min^{-1}) to the mass spectrometer inlet while allowing the balance (~20–50 mL min^{-1}) to pass to the fraction collection device. CombiChem recently reported a variant of their system that uses APCI rather than electrospray mass detection [25]. Because of the increased sensitivity of the APCI method relative to electrospray, an additional stream dilutor is necessary to avoid saturation of the mass detector. In this system, then, the stream is split to divert 0.1 mL min^{-1} to the mass detector. An additional solvent pump then dilutes this stream with an additional 1.0 mL min^{-1} of solvent before introducing the combined stream into the mass detector at 1.1 mL min^{-1}. The resulting 10-fold dilution of the eluant stream ensures that the mass detector is not saturated.

The preparative HPLC systems described above offer rapid, high throughput purification. They remain, however, essentially serial processing devices which process samples one at a time. The first report of a parallel processing preparative HPLC system has now appeared [26]. This system processes up to four samples through four columns simultaneously, thus offering true parallel sample processing. The system uses a single solvent pump outfitted with four pump heads (one pump head per flow channel), a single autosampler with four separate injection loops, a single detector with four separate flow cells, and multiple fraction collectors. Companion software operates the entire system and tracks product fractions. This system is now commercially available [27]. This is the first example of a true parallel processing preparative HPLC system. Since automated synthesis is now routinely achieved in a parallel sense, it will undoubtedly follow that parallel purification will also become routine. Competition from internal development groups and vendor companies will no doubt improve the availability and efficiency of these systems in the future.

Concluding remarks

During the last several years it has been recognized that automated parallel synthesis methods can be extremely powerful tools for lead optimization in late stage drug discovery projects. These techniques complement methods for synthesis of very large libraries which are most powerful in the early lead discovery phase of the overall drug discovery process. Sample purity may be an important issue in late stage drug discovery as compounds find their way into multiple assays for intrinsic activity, pharmacokinetics, metabolism, etc. Two broad categories of automated purification methods have been most widely applied to parallel synthesis libraries: purification by ion-exchange or ion-entrapment for solution phase synthesis libraries, and purification by automated preparative reverse phase HPLC for all types of libraries. These methods have proven to be very powerful, and libraries of many hundreds of compounds can now be routinely purified in this way. In spite of this success, though, severe limitations still exist. Most of these methods are intrinsically serial in nature, while library synthesis is done in parallel. The first early reports of parallel purification described in this manuscript will no doubt be followed by many more in the years to come. Solvent consumption, evaporation, and disposal are significant problems that must be addressed if these methods are to be extended to very large libraries. Most of the methods that have been described for library purification are based on conventional chromatography. Alternate methods such as electrophoresis and supercritical fluid chromatography may offer advantages over conventional chromatography methods for purification of very large libraries, and we look forward to reports of use of these methods in the future.

References

1. Cheng, S., Comer, D.D., Williams, J.P., Meyers, P.L. and Boger, D.L., J. Am. Chem. Soc., 118 (1996) 2567–2573.
2. DeWitt, S.H., Solid and Solution Phase Combinatorial Synthesis, New Orleans, LA, 1997.
3. Breitenbucher, J.G., Solid and Solution Phase Combinatorial Synthesis, Princeton, NJ, 1997.
4. Kaldor, S.W., Siegel, M.G., Fritz, J.E., Dressman, B.A. and Hahn, P.J., Tetrahedron Lett., 37 (1996) 7193–7196.
5. Flynn, D.L., Crich, J.Z., Devraj, R.V., Hockerman, S.L., Parlow, J.J., South, M.S. and Woodard, S., J. Am. Chem. Soc., 119 (1997) 4874–4881.
6. Devraj, R., Solid and Solution Phase Combinatorial Synthesis, Princeton, NJ, 1997.
7. Gayo, L.M. and Suto, M.J., Tetrahedron Lett., 38 (1997) 513–516.
8. Virgilio, A.A., Schurer, S.C. and Ellman, J.A., Tetrahedron Lett., 37 (1996) 6961–6964.
9. Deegan, T.L., Gooding, O.W., Baudart, S. and Porco, J.A., Tetrahedron Lett., 38 (1997) 4973–4976.
10. Keating, T.A. and Armstrong, R.W., J. Am. Chem. Soc., 118 (1996) 2574–2583.
11. Young, M.G., Solid and Solution Phase Combinatorial Synthesis, New Orleans, LA, 1997.
12. Kaye, B., Herron, W.J., Macrae, P.V., Robinson, S., Stopher, D.S., Venn, R.F. and Wild, William, Anal. Chem., 68 (1996) 1658–1660.
13. Lawrence, R.M., Biller, S.A., Fryszman, O.M. and Poss, M.A., Synthesis, (1997) 553–558.
14. Siegel, M.G., Hahn, P.J., Dressman, B.A., Fritz, J.E., Grunwell, J.R. and Kaldor, S.W., Tetrahedron Lett., 38 (1997) 3357–3360.
15. Weller, H.N., IBC Interactive Forum on Molecular Diversity, San Diego, CA, January 1996.
16. Weller, H.N., ComTech '97, Frieberg, February 1997.
17. Weller, H.N., Young, M.G., Reitnauer, G.H., Michalczyk, S.J., Cooley, R.S., Rahn, P.J., Loyd, D.J., Fiore, D., and Fischman, S.J., Mol. Div., 3 (1997) 61–70.
18. Rahn, P.C. and Bickler, J.R., Pittcon '97, Atlanta, March 1997.
19. Kibbey, C.E., Robertson, G.A., MacDonald, A. and DeWitt, S.H., In Proceedings of the International Society of Lab Automation and Robotics, 1996, pp. 336–348.
20. Routburg, M., Swenson, R., Schmitt, B., Washington, A., Mueller, S., Hochlowski, J., Masiana, G., Minin, B., Matuszak, K., Searle, P. and Pan, J., Proceedings of the International Society of Lab Automation and Robotics, 1997, pp. 360–374.
21. Kassel, D.B., ComTech '97, Frieberg, February 1997.
22. Zeng, L., Burton, L., Yung, K., Shushan, B. and Kassel, D.B., J. Chromatogr. A, 794, (1998) 3–13.
23. Kiplinger, J.P., Ware, R.S., Roskamp, E., Dorf, P., Cole, R.G., Robinson, S., Brailsford, A., Raso, R. and Frey, C., Pittcon '97, Atlanta, March 1997.
24. a. Perkin-Elmer Sciex Instruments, 71 Four Valley Drive, Concord, ON, Canada.
    b. Micromass UK Limited, Altrincham, Cheshire, U.K.
25. Kassel, D.B., American Society of Mass Spectrometry fall workshop on combinatorial chemistry, San Diego, CA, October 1997.
26. Coffey, P., Lab. Automation News, 2 (1997) 7–13.
27. Biotage division of Dyax Corp., Charlottesville, VA, U.S.A.

M.R. Pavia and W.H. Moos (eds.), Annual Reports in Combinatorial Chemistry and Molecular Diversity, Vol. 2, 33–39.
KLUWER/ESCOM

Analytical methods for quality control of combinatorial libraries

William L. Fitch
Affymax Research Institute, 3410 Central Expressway, Santa Clara, CA 95051, U.S.A.

Key words: analytical chemistry, combinatorial chemistry, infrared spectroscopy, magic angle spinning, mass spectrometry, nuclear magnetic resonance spectroscopy, solid phase organic chemistry

Summary

This literature review covers the applications of analytical techniques to solid phase organic chemistry and combinatorial chemistry published between June 96 and September 1997. Highlighted are mass spectrometry, NMR, IR and chromatographic analyses of solid phase synthesis reactions and combinatorial libraries.

Introduction

This literature review covers the applications of analytical techniques to solid phase organic chemistry and combinatorial chemistry. It updates a similar review completed in June 96 [1] and covers the literature through September 1997.

Several reviews have appeared which deal in part or exclusively with solid phase analytical methods [2–6]. It is heartening to see that documentation of solid phase chemistry is improving in the literature. A comprehensive review of the solid phase reaction literature before November, 1995 [7] reports that only 69/169 (41%) of the reactions included yield information. In a review of the subsequent year [8] the number had increased to 198/270 (73%). We can hope for a continued increased stringency to the reporting of solid phase reaction data.

Infrared spectroscopy

Fourier transform infrared spectroscopy (FTIR) is a useful way to confirm the presence of a desired functional transformation in a solid phase reaction. A high throughput diffuse reflectance infrared FT (DRIFT) technique allows spectra of beads to be collected without the need for KBr pelleting [9] and is reported to yield better data in the 'OH' spectral region [10]. But a similar improvement in quality of spectra for KBr pellets can be obtained by removing bound water at 110 °C for 2–3 h. [11]. Spectral enhancements have also been obtained by using photoacoustic FTIR methods, where IR radiation absorbed by analyte is detected as an acoustic wave generated in an inert gas in contact with the sample. PA-FTIR spectra of resin beads were found to be free of baseline artifacts caused by light scattering and sample inhomogeniety in typical FTIR experiments [12].

The Novartis group, continuing their exploitation of single bead IR [13–17] have shown that recording transmission spectra on flattened resin particles significantly enhances the spectral resolution by providing a more uniform pathlength for the IR irradiation. The sensitivity of FTIR microscopy in these single bead experiments is impressive: by using an attenuated total reflectance (ATR) microscope objective, it is possible to collect IR absorption data from just a fraction of a bead's surface. This approach has been used to monitor the kinetics of esterification of hydroxyl groups on TentaGel and Wang resins, and to compare the acylation rates of surface hydroxyls with OH groups embedded in the bead.

Russell and coworkers have demonstrated that IR absorbances from deuterated protecting groups can provide a quantitative measurement of deuterium content in an immobilized substrate, and that this can be used to monitor reaction progress and yields with such molecules [18]. The carbon-deuterium stretching absorbance falls in a largely interference-free spectral region (2300–2200 cm^{-1}) and for appropriate groups (e.g. Boc-d_9) the molar absorptivity of the C-D stretch was found to be independent of the chemical environment of the deuterated moiety. This research group has also developed a specialized flow-through infrared cell to permit real-time *in situ* infrared analysis of solid-phase organic reactions [19]. In contrast to the single-bead reaction monitoring studies described earlier, this experimental set-up permits on-line analysis of reaction kinetics and has provided a unique insight [20] into the rates of single bead photocleavage of **1**.

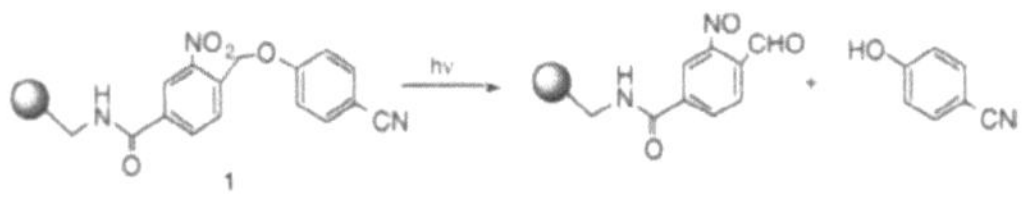

1

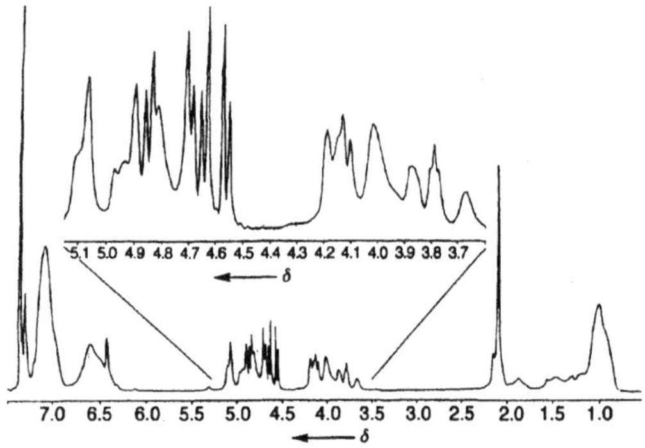

Figure 1. ^1H MAS NMR spectrum of **2**, reprinted from [32].

NMR spectroscopy

Solid phase NMR techniques have advanced to the point where there are routine techniques for following reactions and confirming structures on all common resin materials. Natural abundance gel phase ^{13}C and ^{19}F NMR are used for assessing intermediate structures in solid phase synthesis [21–23]. ^{13}C NMR has proved useful for determining anomeric configuration in a solid phase glycosidation reaction [24].

Obtaining high-resolution ^1H NMR spectra for compounds bound to TentaGel™ using the MAS technique has become routine [25, 26]. Useful data has recently been obtained from single beads of a macro TentaGel™-type resin (400–750 mm diameter, 10–65 nanomoles sample) where direct ^1H observation in a CP/MAS probe gave clear assignable resonances, and enabled intermediates in the synthesis of a hydantoin moiety to be readily characterized [27].

But increasingly, MAS is used to gain information from the less flexible polystyrene resins [28]. Spectra for the pure polystyrene resins (e.g. Wang, Merrifield) can benefit from use of a spin echo pulse sequence with appropriate t values which preferentially suppresses the signals from the rotationally restricted polymer atoms compared to the ligand atoms. New NMR experimental techniques applied to improving the information content of ^1H spectra taken on Merrifield resin include gradient field technology [29], J-resolved 2D spectra [30] and reduction of inhomogeneous line broadening [31]. A particularly excellent illustration [32] of the quality of spectrum which can be obtained for a complex structure on Merrifield resin is shown as Figure 1.

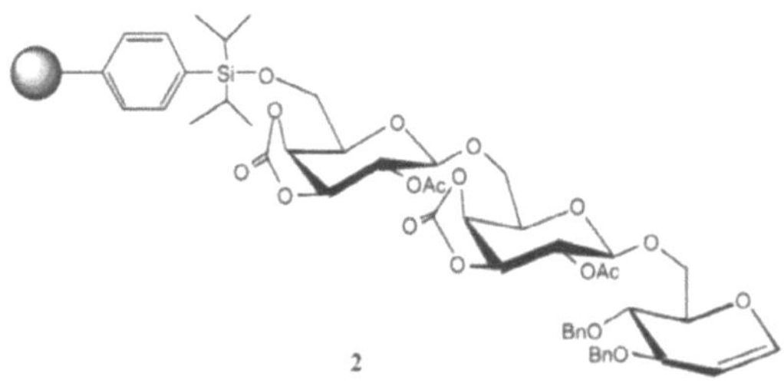

2

MAS NMR can also be productively applied to compounds grafted onto polypropylene crowns. Good signals could be obtained for the aldehyde and olefin protons which indicated the success of this Wittig reaction [33]. Broadened signals for the more rigid portions of the structure were removed using a spin echo pulse sequence.

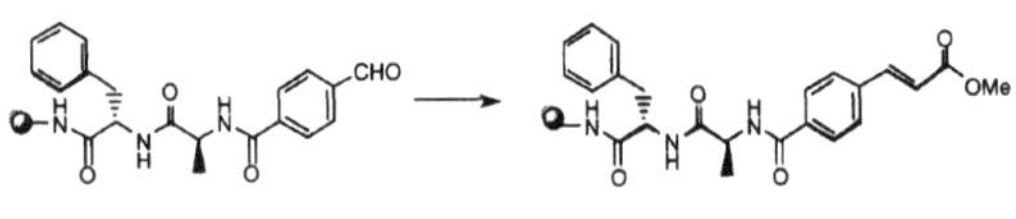

Shapiro has explored the use of pulsed-field gradient technology together with the TOCSY experiment as a means of obtaining useful NMR from solution mixtures of several compounds (e.g. a small combinatorial compound pool) [34]. 2D experiments have also been used for quantifying the members of a small library of peptides on solid support [35]. It is worth remembering that NMR can be a very quantitative technique with external standardization. New flow probes and gradient technology make NMR more applicable in a high-throughput laboratory.

Chromatographic methods

Considerable attention has been paid to the evaporative light-scattering detector (ELSD) [36] as a more universal HPLC detector than UV; in general it detects any analyte above a molecular weight of about 200 amu. However, the ELSD response is not linear and varies considerably between analogs of similar structure. These issues limit the usefulness of the ELSD for quantitative assessment of purity and yield.

Another new development is the chemiluminescent nitrogen detector (CLND) for HPLC [37]. This destructive detector, which oxidizes nitrogen compounds to NO, converts the NO to excited NO_2 by reaction with ozone and detects the emitted light in a photomultiplier, was reported to give equivalent response/mole of N for all nitrogen compounds. In practice, we have found that the CLND detector as currently manufactured has severe limitations. The data in Table 1 were collected by flow injection/CLND of pure compounds. While many compounds give excellent, predictable response, a few significant outliers are observed. It was also observed that the relative response of glycine to nitrobenzene varies as a function of mobile phase composition (Figure 2), minimizing the value of this detector for gradient HPLC. More recently, the manufacturer reports that a reengineered detector has solved these limitations [38]. It remains to be seen how robust the CLND detector is in regular practice, but it still offers the tantalizing promise of quantitative organic analysis of nitrogen containing compounds without purified reference samples for each analyte.

It is not usually possible to monitor solid phase reactions with the standard solution phase chromatographic techniques (TLC, HPLC) because the reaction components are either tied up by the solid phase or present in large excess. However, in a special case, Xiao et al. were able to follow the consumption of a precious solution phase taxoid intermediate as it reacted ('resin-trapped') with excess resin [39].

Table 1. Relative response of organic compounds tested by flow injection CLND

Compound	Relative response
Nitrobenzene	1.00
2,5-Di-t-butylnitrobenzene	0.96
Benzamide	0.95
1-Nitronaphthalene	0.91
NMP	0.96
Nitromethane	0.97
4-Cyanophenyl 4-heptylbenzoate	0.99
Nitropropane	1.00
Benzonitrile	0.96
DMF	0.99
Acetonitrile	1.04
Uracil	0.38
Benzotriazole	0.29
Triethylamine	1.08
Benzylamine	0.95
Pyridine	0.95

Each compound was tested by flow injection in 100% methanol and reported relative to nitrobenzene.

Mass spectrometry

Mass spectrometry is by far the major analytical technique used in library characterization. But the literature does not reflect this preponderance. The mass spectroscopists may be too busy running samples to publish!

Several groups engaged in large scale parallel synthesis programs have described the integration of robotic synthesis systems with MS or LC/MS autosampling interfaces that permit MS data to be acquired on some or all of the compounds prepared in a synthesis run [40]. LC/MS/NMR systems are gaining in popularity for quickly assessing the quality of libraries [41].

Two groups have reported on the advantages of high resolution Fourier transform mass spectrometry for characterizing libraries [42, 43]. This technique can determine the accurate mass of a compound and thus distinguish between isobars which have the same nominal mass. However, high resolution MS does not solve the portion of the mass redundancy in libraries which is due to true structural isomers.

Single bead mass spectrometry can be used to readily measure the molecular weight of the component from each bead of a split/pool library. GC/MS is exquisitely sensitive for those limited molecules which are volatile. Chen et al. report the use of single bead GC/MS for quality control of a small library [11]. Figure 3 shows the GC/MS total ion chromatogram and spectra for the material from a single polystyrene resin bead. GC/MS is an underutilized technique for characterization of libraries. It can be a very universal detector for small molecules up to roughly 600 amu (especially if carboxyl and hydroxyl groups are silylated). GC/MS response factors can be estimated

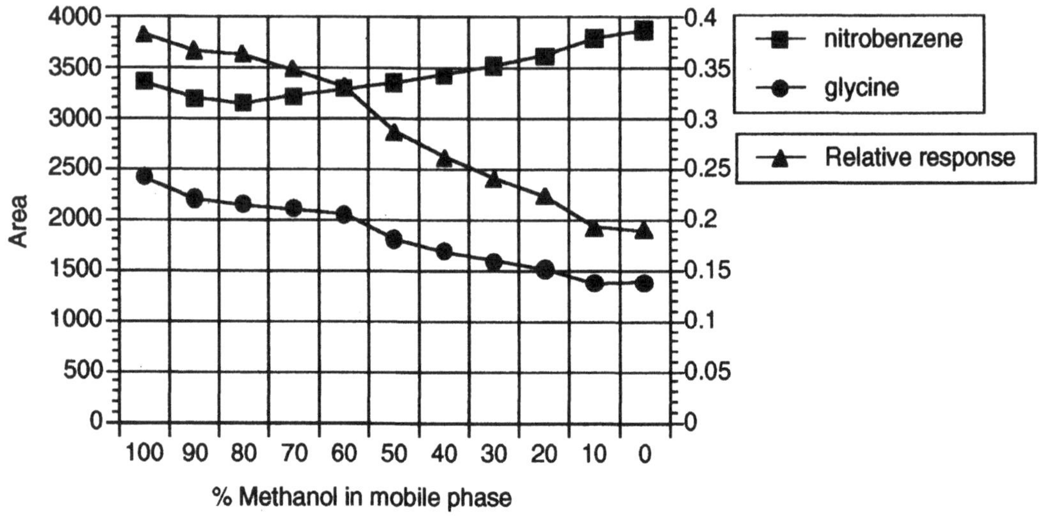

Figure 2. Effect of mobile phase composition on relative CLND response of glycine and nitrobenzene.

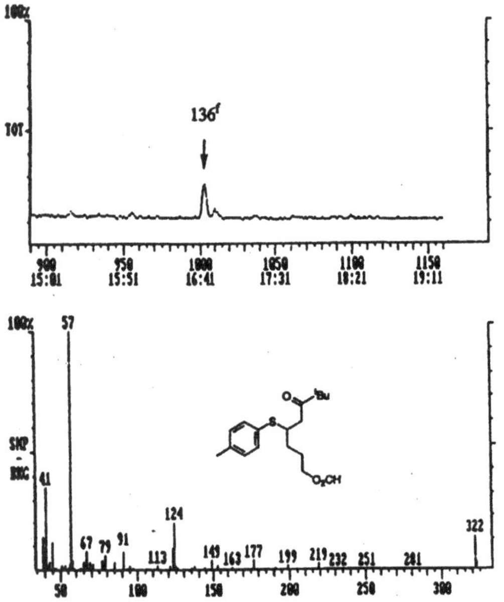

Figure 3. GC/MS analysis of a single bead, holding the indicated structure. Top: the total ion current trace; bottom: the spectrum of the indicated peak; reprinted from [11].

from structure for semi-quantitative analysis of novel structures.

Electrospray MS has sufficient sensitivity for routine MW determination of the material cleaved from a single bead [44] without the volatility limitations of GC/MS. However, MS data alone are not sufficient for quality control of a split/pool synthetic library. The mass spectra of a sampling of beads can be compared to the theoretical distribution of molecular weights [45, 46] but this will be of little value. At Affymax, we prefer to compare the mass spectrum of a single bead's cleavage product with the molecular weight predicted by the expected structure as determined from 'reading' the code on the same bead. This process normally uses flow injection electrospray MS. But we have recently demonstrated the ability to do single bead, capillary LC/UV/MS (Figure 4) to generate confidence in both the structure and purity of the material on each bead [47]. These data were collected from a thiazolidinone library; the selected bead contained structure **3**.

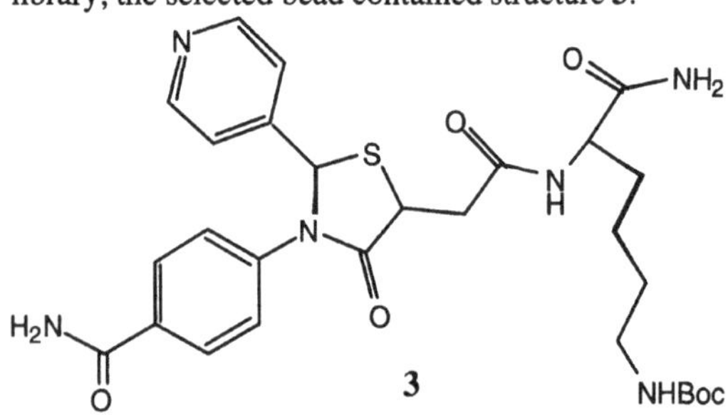

3

Berlin et al. [48] report the use of MALDI of mixtures as a method for screening of libraries. This intriguing method holds promise for quickly characterizing behavior of mixtures where the assay does not introduce contaminating ions into the mixture. The au-

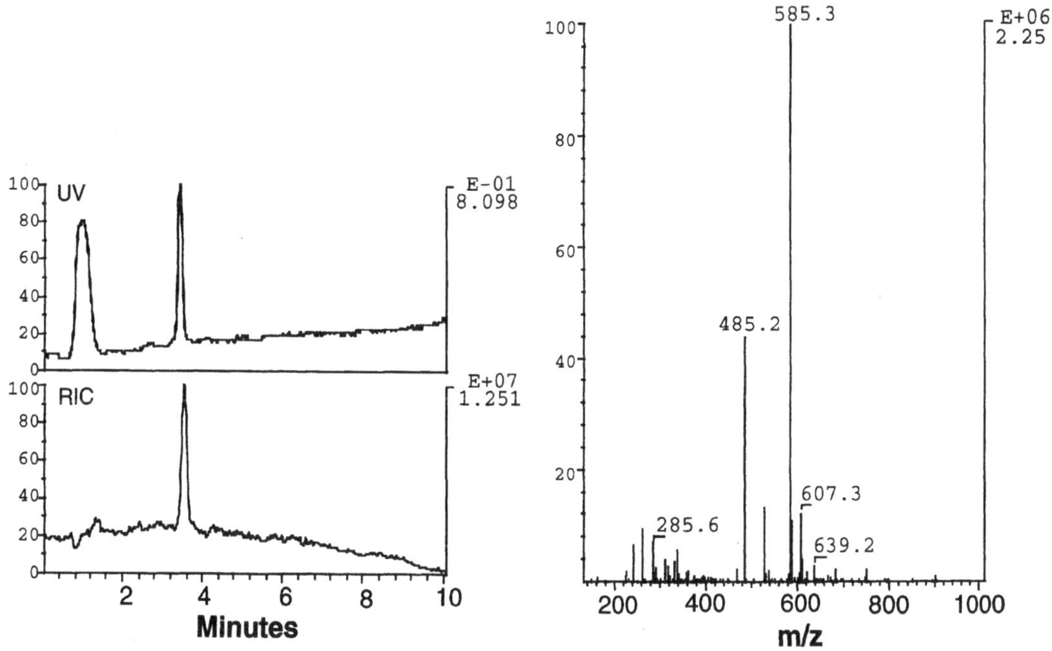

Figure 4. LC/UV/MS of compound **3**, cleaved from a single bead. The top trace is the UV chromatogram; the bottom trace is the total ion current chromatogram.

thors do not contend that all members of the mixture give equal response. So this technique is not intended for quality control of the library. In addition, real libraries are far too contaminated with small molecule reaction byproducts and resin cleavage impurities [49] to make mass spectra of mixtures useful. Attempts have been made to equalize MALDI response using ionization enhancement linkers in combinatorial synthesis [50], but extra complexity in synthesis will not be a popular approach.

Other techniques

Combustion (or cleavage followed by) elemental analysis remains a favorite method for measuring loading of relatively pure or unstable intermediates in solid phase synthesis [11, 21, 51]. Another common technique is to derivatize a reactive intermediate to add an elemental 'tag' solely for the purpose of quantifying it [51, 52]. An interesting fluorescence depletion method for measuring loading has been developed by Li et al. Solid phase reactions are normally performed with a large excess of reagent, which will barely be depleted during incorporation. But these authors have found that a highly reactive aldehyde group can be quantified by measuring the decrease of fluorescence of a dansylhydrazine solution, present in only two-fold excess [53]. A similar method for on-resin aldehyde measurement [54] uses the reaction with O-tritylhydroxylamine followed by UV assay for trityl [55]. On-bead fluorescence has been reviewed but does not appear to have any applications in structure determination or quantitative reaction assessment [56].

Conclusions

We have available a powerful collection of methods for determining structures on solid phase (IR, MAS NMR) and for quantifying yield in solid phase reactions (combustion analysis, gravimetric analysis of resin weight gain or cleaved product, colorimetric assay (FMOC)). NMR and LC with MS or other new detectors provide useful tools for assessing the quality of parallel and split pool libraries in a high throughput analytical environment.

The general analytical chemistry community has begun to take notice of the challenges provided by combinatorial chemistry [57]. New, parallel approaches to analysis and new methods for solids should be forthcoming.

The modern, high throughput laboratory can generate a lot of data in a short time. The next major hurdle for the analytical laboratory which supports combinatorial chemistry is to find ways to automate the reduction of this data into useful information.

Acknowledgments

Thanks to Chris Holmes for helpful suggestions in the reading of this review.

References

1. Fitch, W.L., *Analytical methods for quality control of combinatorial libraries*, In Moos, W.H., Pavia, M.R., Kay, B.K. and Ellington, A.D. (Eds.), Annual Reports in Combinatorial Chemistry and Molecular Diversity, Vol. 1., ESCOM, Leiden, 1997, pp. 59–68.
2. Balkenhohl, F., von dem Bussche-Hunnefeld, C., Lansky, A. and Zechel, C., *Combinatorial synthesis of small molecules*, Angew Chem. Int. Ed. Engl., 35 (1996) 2288–2337.
3. Egner, B.J. and Bradley, M., *Analytical techniques for solid-phase organic and combinatorial synthesis*, Drug Dev. Technol., 2 (1997) 102–109.
4. Gallop, M.A. and Fitch, W.L., *New methods for analyzing compounds on polymeric supports*, Cur. Opin. Chem. Biol., 1 (1997) 94–100.
5. von dem Bussche-Hunnefeld, C., Balkenhohl, F., Lansky, A. and Zechel, C., *Combinatorial chemistry*, Fresenius J. Anal. Chem., 359 (1997) 3–9.
6. Schlick, S., Bortel, E. and Dyrek, K., *Catalysis on polymer supports*, Acta Polymer., 47 (1996) 1–15.
7. Hermkens, P.H.H., Ottenheijm, H.C.J. and Rees, D., *Solid phase organic reactions: A review of the literature*, Tetrahedron, 52 (1996) 4527–4554.
8. Hermkens, P.H.H., Ottenheijm, H.C.J. and Rees, D.C., *Solid-phase organic reactions II: A review of the literature Nov 95–Nov 96*, Tetrahedron, 53 (1997) 5643–5678.
9. Chan, T.Y., Chen, R., Sofia, M.J., Smith, B.C. and Glennon, D., *High throughput on-bead monitoring of solid phase reactions by diffuse reflectance infrared Fourier transform spectroscopy (DRIFTS)*, Tetrahedron Lett., 38 (1997) 2821–2824.
10. Deben, I.E.A., van Wijk, T.H.M., van Doornum, E.M., Hermens, P.H.H. and Kellenbach, E.R., *Characterization of reaction products on solid supports by FT-IR*, Mikrochim. Acta, 14 (1997) 245–246.
11. Chen, C., Ahlberg Randall, L.A., Miller, R.B., Jones, A.D. and Kurth, M.J., *The solid-phase combinatorial synthesis of β-thioketones*, Tetrahedron, 53 (1997) 6595–6609.
12. Gosselin, F., Di Renzo, M., Ellis, T.H. and Lubell, W.D., *Photoacoustic FTIR spectroscopy, a nondestructive method for sensitive analysis of solid-phase organic chemistry*, J. Org. Chem., 61 (1996) 7980–7981.
13. Li, W. and Yan, B., *A direct comparison of the mixing efficiency in solid-phase synthesis by single bead IR and fluorescence spectroscopy*, Tetrahedron Lett., 37 (1997) 6485–6488.
14. Marti, R.E., Yan, B. and Jarosinski, M.A., *Solid-phase synthesis via 5-oxazolidinones. Ring opening reactions with amines and reaction monitoring by single-bead FTIR microspectroscopy*, J. Org. Chem., 62 (1997) 5615–5618.
15. Yan, B., Fell, J.B. and Kumaravel, G., *Progression of organic reactions on resin supports monitored by single bead FTIR microspectroscopy*, J. Org. Chem., 61 (1996) 7467–7472.
16. Yan, B. and Gstach, H., *An indazole synthesis on solid support monitored by single bead FTIR microspectrocopy*, Tetrahedron Lett., 37 (1996) 8325–8328.
17. Yan, B., Sun, Q., Wareing, J.R. and Jewell, C.F., *Real-time monitoring of the catalytic oxidation of alcohols to aldehydes and ketones on resin support by single-bead Fourier Transform Infrared Microspectroscopy*, J. Org. Chem., 61 (1996) 8765–8770.
18. Russell, K., Cole, D.C., McLaren, F.M. and Pivonka, D.E., *Analytical techniques for combinatorial chemistry: Quantitative infrared spectroscopic measurements of deuterium-labeled protecting groups*, J. Am. Chem. Soc., 118 (1996) 7941–7945.
19. Pivonka, D.E., Russell, K. and Gero, T., *Tools for combinatorial chemistry: In situ infrared analysis of solid-phase organic reactions*, Appl. Spectrosc., 50 (1996) 1471–1478.
20. Pivonka, D.E. and Simpson, T.R., *Tools for combinatorial chemistry: Real-time single-bead infrared analysis of a resin-bound photocleavage reaction*, Anal. Chem., 69 (1997) 3851–3853.
21. Nugiel, D.A., Wacker, D.A. and Nemeth, G.A., *Facile preparation of chloromethylaryl solid supports using methanesulfonyl chloride and Hunig's base*, Tetrahedron Lett., 38 (1997) 5789–5790.
22. Shapiro, M.J., Kumaravel, G., Petter, R.C. and Beveridge, R., *^{19}F Monitoring of a SnAr reaction on solid support*, Tetrahedron Lett., 37 (1996) 4671–4674.
23. Svensson, A., Fex, T. and Kihlberg, J., *Use of ^{19}F NMR spectroscopy to evaluate reactions in solid phase organic synthesis*, Tetrahedron Lett., 37 (1996) 7649–7652.
24. Rodebaugh, R., Joshi, S., Fraser-Reid, B. and Geysen, H.M., *Polymer supported oligosaccharides via n-pentenyl glycosides: Methodology for a carbohydrate library*, J. Org. Chem., 62 (1997) 5660–5661.
25. Pop, I.E., Dhalluin, C.F., Deprez, B.P., Melnyk, P.C., Lippens, G.M. and Tartar, A.L., *Monitoring of a three step solid phase synthesis reaction involving a Heck reaction using magic angle spinning NMR spectroscopy*, Tetrahedron, 52 (1996) 12209–12222.
26. Wehler, T. and Westman, J., *Magic angle spinning NMR: A versatile tool for monitoring the progress of solid phase synthesis*, Tetrahedron Lett., 37 (1996) 4771–4774.
27. Pursch, M., Schlotterbeck, G., Tseng, L., Albert, K. and Rapp, W., *Monitoring the reaction progress in combinatorial chemistry: ^{1}H MAS NMR investigations on single macro beads in the suspended state*, Angew. Chem. Int. Ed. Engl., 35 (1996) 2867–2869.
28. van Maarseveen, J.H., den Hartog, J.A.J., Engelen, V., Finner, E., Visser, G. and Kruse, C.G., *Solid phase ring-closing metathesis: Cyclization/cleavage approach towards a seven membered cycloolefin*, Tetrahedron Lett., 37 (1997) 8249–8252.
29. Maas, W.E., Lankien, F.H. and Cory, D.G., *Gradient, high-resolution, magic angle spinning NMR*, J. Am. Chem. Soc.,

118 (1996) 13085–13086.
30. Shapiro, M.K., Chin, J., Marti, R.E. and Jarosinski, M.A., *Enhanced resolution in MAS NMR for combinatorial chemistry*, Tetrahedron Lett., 38 (1997) 1333–1336.
31. Meissner, A., Bloch, P., Humpfer, E., Spraul, M. and Sørensen, O.W., *Reduction of inhomogeneous line broadening in two-dimensional high-resolution MAS NMR spectra of molecules attached to swelled resins in solid-phase synthesis*, J. Am. Chem. Soc., 119 (1997) 1787–1788.
32. Seeberger, P.H., Beebe, X., Sukenick, G.D., Pochapsky, S. and Danishefsky, S.J., *Monitoring the progress of solid-phase oligosaccharide synthesis by high-resolution magic angle spinning NMR: Observations of enhanced selectivity for β-glycoside formation from α-1,2-anhydrosugar donors in solid-phase couplings*, Angew. Chem. Int. Ed. Engl., 36 (1997) 491–493.
33. Chin, J., Fell, B., Shapiro, M.J., Tomesch, J., Wareing, J.R. and Bray, A.M., *Magic angle spinning NMR for reaction monitoring and structure determination of molecules attached to multipin crowns*, J. Org. Chem., 62 (1997) 538–539.
34. Lin, M.S. and Shapiro, M.J., *Mixture analysis in combinatorial chemistry. Application of diffusion-resolved NMR spectroscopy*, J. Org. Chem., 61 (1996) 7617–7619.
35. Fauchère, J.L., Henlin, J.M. and Boutin, J.A., *Peptide and nonpeptide lead discovery using robotically synthesized soluble libraries*, Can. J. Physiol. Pharmacol., 75 (1997) 683–689.
36. Kibbey, C.E., *Quantitation of combinatorial libraries of small organic molecules by normal phase HPLC with evaporative light scattering detection*, Mol. Div., 1 (1996) 247–258.
37. Fitch, W.L., Szardenings, A.K. and Fujinari, E.M., *Chemiluminescent nitrogen detection for HPLC: An important new tool in organic analytical chemistry*, Tetrahedron Lett., 38 (1997) 1689–1692.
38. Homan, G., unpublished results, 1997.
39. Xiao, X.-Y., Parandoosh, Z. and Nova, M.P., *Design and synthesis of a taxoid library using radiofrequency encoded combinatorial chemistry*, J. Org. Chem., 62 (1997) 6029–6033.
40. Hegy, G., Gorlach, E., Richmond, R. and Bitsch, F., *High throughput electrospray mass spectrometry of combinatorial chemistry racks with automated contamination surveillance and results reporting*, Rapid Commun. Mass Spectrom., 10 (1996) 1894–1900.
41. Holt, R.M., Newman, M.J., Pullen, F.S., Richards, D.S. and Swanson, A.G., *High-performance liquid chromatography/NMR spectroscopy/mass spectrometry: Further advances in hyphenated technology*, J. Mass Spectrom., 32 (1997) 64–70.
42. Winger, B.E. and Campana, J.E., *Characterization of combinatorial peptide libraries by electrospray ionization Fourier transform mass spectrometry*, Rapid Commun. Mass Spectrom., 10 (1996) 1811–1813.
43. Nawrocki, J.P., Wigger, M., Watson, C.H., Hayes, T.W., Senko, M.W., Benner, S.A. and Eyler, J.R., *Analysis of combinatorial libraries using electrospray Fourier transform ion cyclotron resonance mass spectrometry*, Rapid Commun. Mass Spectrom., 10 (1996) 1860–1864.
44. Brummel, C.L., Vickerman, J.C., Carr, S.A., Hemling, M.E., Roberts, G.D., Johnson, W., Weinstock, J., Gaitanopoulos, D., Benkovic, S.J. and Winograd, N., *Evaluation of mass spectrometric methods applicable to the direct analysis of non-peptide bead-bound combinatorial libraries*, Anal. Chem., 68 (1996) 237–242.
45. Demirev, P.A. and Zubarev, R.A., *Probing combinatorial library diversity by mass spectrometry*, 69 (1997) 2893–2900.
46. Steinbeck, C., Berlin, K. and Richert, C., *MASP-A program predicting mass spectra of combinatorial libraries*, J. Chem. Inf. Comput. Sci., 37 (1997) 449–457.
47. Lewis, K.C., Fitch, W.L. and Maclean, D., *Characterization of a split-pool combinatorial library*, LC.6C., 16 (1998) 644–649.
48. Berlin, K., Jain, R.K., Tetzlaff, C., Steinbeck, C. and Richert, C., *Spectrometrically monitored selection experiments: Quantitative laser desorption mass spectrometry of small chemical libraries*, Chem. Biol., 4 (1997) 63–77.
49. Burgess, K., Martinez, C.I., Russell, D.H., Shin, H. and Zhang, A.J., *Photolyic mass sladdering for fast characterization of oligomers on single resin beads*, J. Org. Chem., 62 (1997) 5662–5663.
50. Carrasco, M.R., Fitzgerald, M.C., Oda, Y. and Kent, S.B.H., *Direct monitoring of organic reactions on polymeric supports*, Tetrahedron Lett., 38 (1997) 6331–6334.
51. Booth, R.J. and Hodges, J.C., *Polymer-supported quenching reagents for parallel purification*, J. Am. Chem. Soc., 119 (1997) 4882–4886.
52. Tietze, L.F. and Steinmetz, A., *Stereoselective solid phase synthesis of cyclopentane and cyclohexane derivatives by two component domino reactions: Generation of combinatorial libraries*, Angew. Chem. Int. Ed. Engl., 35 (1996) 651–652.
53. Yan, B. and Li, W., *Rapid fluorescence determination of the absolute amount of aldehyde and ketone groups on resin supports*, J. Org. Chem., 62 (1998) 9354–9357.
54. Holmes, C.P., unpublished results, 1997.
55. Pon, R.T., *Tips for oligonucleotide synthesis*, A.B.R.F. News, 5 (1994) 14–18.
56. Scott, R.H. and Balasubramanian, S., *Properties of fluorophores on solid phase resins; Implications for screening, encoding and reaction monitoring*, Bioorg. Med. Chem. Lett., 7 (1997) 1567–1572.
57. Murray, R., *Analytical chemistry in parallel*, Anal. Chem., 69 (1997) 575A.

M.R. Pavia and W.H. Moos (eds.), Annual Reports in Combinatorial Chemistry and Molecular Diversity, Vol. 2, 41–52.
KLUWER/ESCOM

Accessing carbohydrate-based combinatorial libraries through solid phase and solution phase approaches

Michael J. Sofia
Intercardia Inc., 8 Cedar Brook Drive, Cranbury, NJ 08512, U.S.A.

Introduction

The pursuit of chemical diversity for the purpose of drug discovery has focused primarily on the generation of peptide and heterocyclic small molecule combinatorial libraries [1–3]. With genomics initiatives beginning to dramatically increase the number of interesting biomolecular targets, a great need has arisen for expanding the universe of molecular structures available for screening. Although carbohydrates play fundamental roles in many biological processes [4,5] and are found as structural units in many drugs [6,7], they are poorly represented in the historical structurally biased chemical sample collections being used by most drug discovery organizations in their screening programs. However, it is because of their biological importance and unique chemical characteristics that carbohydrates are valuable in generating leads through drug discovery screening efforts.

Through the description of actual library syntheses, this review shows how individual investigators have attempted to address the construction of combinatorial libraries of oligosaccharides and glycoconjugates. Glycosidic bond construction in combinatorial synthesis is described. Library synthesis strategies developed to address the regiochemical selectivity issues associated with the presence of multiple reactive sites on a sugar nucleus are outlined. Deconvolution strategies used when constructing carbohydrate libraries are presented. Chemical strategies that maximize the chemical diversity accessible from a carbohydrate library are also discussed.

Oligosaccharide libraries

In exploring oligosaccharide molecular diversity, the glycosidic bond plays a central role. Because the glycosidic bond is the dominant construct by which sugars are attached to one another, much of the work in solution and solid phase carbohydrate chemistry and therefore in library synthesis has focused on how to construct glycosidic linkages efficiently and stereoselectively. In conjunction with the focus on glycosidic bond construction, most of the oligosaccharide library work has concentrated on introducing simple monomer building block diversity. However, through the use of derivatizable monomers, a few strategies have been able to extend the accessible diversity beyond the simple sugar monomers themselves.

Solid phase approaches

The construction of oligosaccharides on the solid phase has now been successfully demonstrated using a number of glycosylation methods. These methods include the sulfoxide glycosyl donor chemistry [8], the glycal-based glycosylation technology [9,10], the trichloroacetimidate-based glycosylation chemistry [11,12], thioglycoside donor chemistry [13,14], pentenyl-glycoside technology [15], glycosyl halide [16–18] and enzyme based strategies [19–21]. In each of these solid phase methods, glycosylation was shown to be compatible with a wide range of solid supports and chemically cleavable linkers.

Even with the vast amount of work now reported on the development of solid phase oligosaccharide synthesis, there has been only one reported solid phase construction of an oligosaccharide library. This library reported by Kahne and co-workers used the sulfoxide glycosylation method to construct each of the glycosidic linkages in a 1300 compound 1→3-linked di- and trisaccharide library (see Figure 1) [22]. The library diversity was achieved by stereoselectively constructing both α- and β-linked oligosaccharides, by exploring the unit diversity of the sugar building blocks and by derivatizing one of the substituents appended to the oligosaccharide products. This library

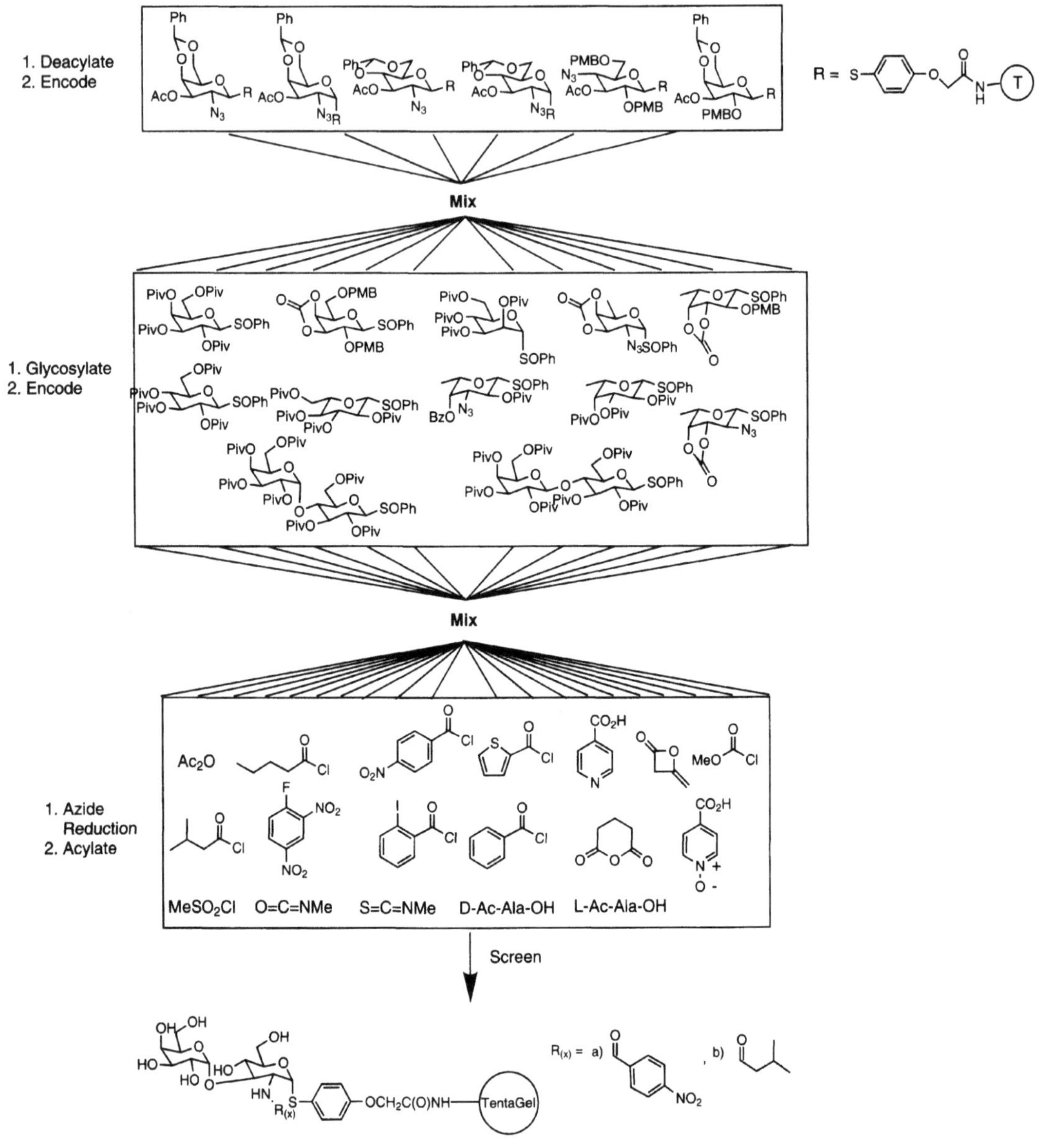

1

Figure 1. The solid phase construction of a 1300 compound di- and trisaccharide library using the mix and split combinatorial strategy. Anomeric linkage diversity, sugar building block diversity and substituent diversity were achieved. Chemical encoding was used to identify the structure of the active compounds **1a** and **1b** which bound most tightly to the *Bauhinia purpurea* lectin.

of 1300 compounds was constructed by first glycosylating six distinct monosaccharide acceptors bound to TentaGel. The diversity around the resulting 72 different di- and trisaccharides was subsequently expanded to 1300 by reducing an azido group contained in these oligosaccharides to an amine and acylating it with 18 different acylating agents. The library was completed by removal of the sugar protecting groups. Polyvalent presentation of these modified oligosaccharides on the polymer support was used to screen the library products as ligands for the *Bauhinia purpurea* lectin. From the library, two novel high affinity disaccharide

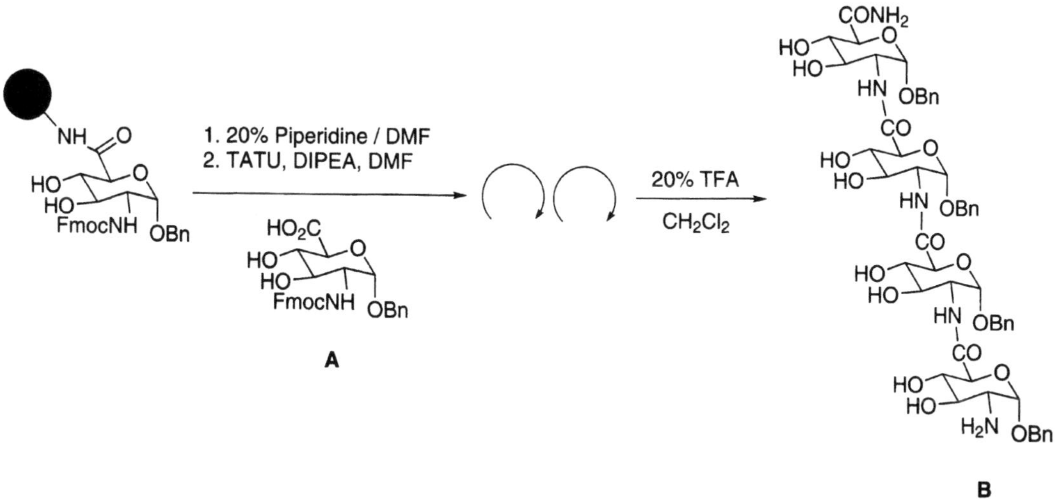

Figure 2. The solid phase chemistry developed for the construction of an oligosaccharide mimetic library is shown. Glucosaminouronic acid building blocks **A** are linked using standard peptide coupling chemistry producing the homotetramer **B**.

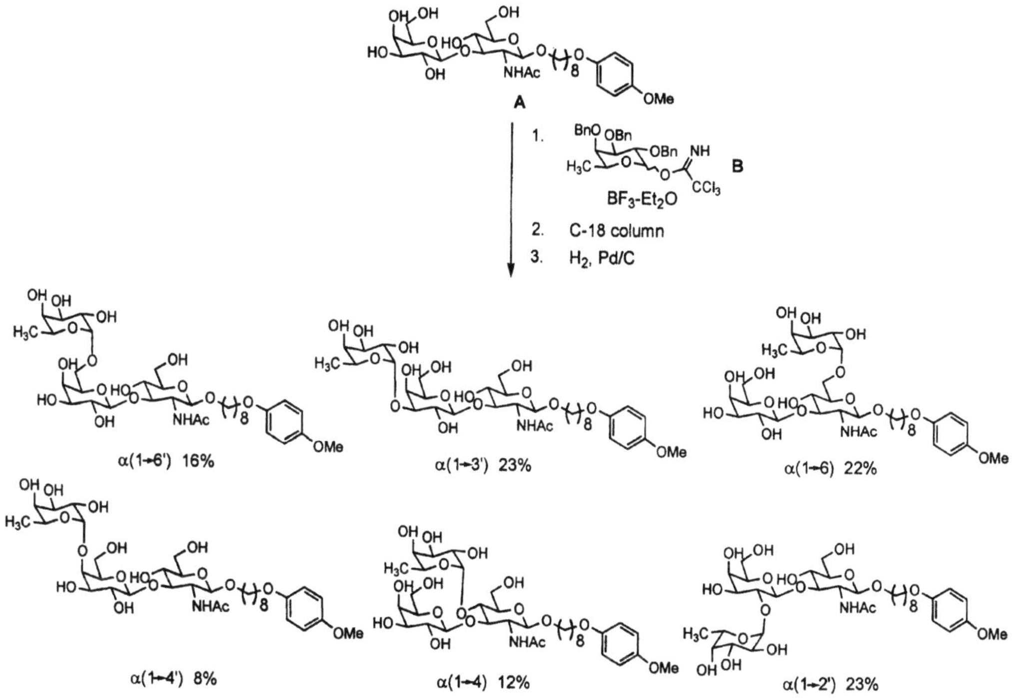

Figure 3. The solution phase random glycosylation of the unprotected disaccharide **A** with the trichloroacetimidate donor **B** produced six trisaccharides in nearly equal amounts.

ligands were identified. To identify the structure of the active oligosaccharides, the encoding technology developed by W. Clark Still was used [23].

In attempts to eliminate the need to construct the glycosidic linkage on the solid phase, some reports have described strategies which link sugars using other

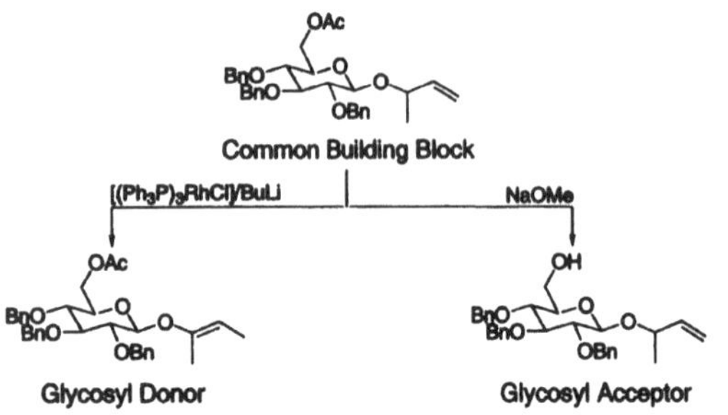

Figure 4. The preparation of a glycosyl donor and glycosyl acceptor from a common building block. This conversion is accomplished by isomerization of the allylic glycoside to the vinyl glycoside. The glycosyl acceptor can be thought of as a latent glycosyl donor thus providing the potential for subsequent homologations through glycosylation.

types of chemical bonds. The solid phase construction of oligosaccharide mimetic libraries was proposed using glucosaminuronic acids [24]. In this strategy, libraries of oligosaccharides can be produced by constructing amide bonds between amino and carboxylic acid functionalities of glucosaminuronic acid building blocks (see Figure 2). The solid phase chemistry needed to prepare such a mimetic library was demonstrated by preparing a tetrasaccharide on the solid phase. The sugars were linked by a 2,6-amide linkage. The growing oligosaccharide was attached to benzhydrylamine polystyrene resin functionalized with a *p*-[(*R*,*S*)-α-1-(9*H*-fluoren-9-yl)methoxyformamido-2,4-dimethoxybenzyl]phenoxyacetic acid linker, and the coupling of the building blocks was accomplished with the activating agent *O*-(7-azabenzotriaxol-1-yl)-1,1,3,3-tetramethyluronium tetrafluoroborate (TATU). The product tetrasaccharide was cleaved from the resin with 50% TFA and showed only a minor amount of contamination with dimer and trimer products. Other oligosaccharide mimetic strategies have been proposed and have the potential for constructing libraries [25,26].

Solution phase approaches

The advantage of solution phase approaches for constructing combinatorial libraries is the direct translation of known solution chemistry to library synthesis, versus the need to reengineer these chemistries to be compatible with the solid phase environment. Three approaches have been reported for generating solution phase oligosaccharide libraries [27–30]. The first reported strategy attempted to address one of the inherent challenges and costs associated with constructing oligosaccharide libraries: the need to construct sugar building blocks with orthogonal protecting group schemes to ensure regioselective glycosylation. Using a random glycosylation strategy, Hindsgaul and co-workers demonstrated that a nearly statistical distribution of di- or trisaccharides could be produced from sugar building blocks where the acceptor sugar was protected only at the anomeric hydroxyl group, thus leaving the other four hydroxyl groups free for glycosylation (see Figure 3) [27,28,31]. The glycosylation method used in this solution strategy was the trichloroacetimidate chemistry in which the donor sugar was fully protected. A demonstration of the random glycosylation strategy for library construction produced a six compound trisaccharide library where βGal(1$\rightarrow$3)βGlcNAc-O-$(CH_2)_8$O-*p*MeOPh was fucosylated to produce a mixture of fucosyltrisaccharides in the distribution of 12% $\alpha(1\rightarrow4)$, 22% $\alpha(1\rightarrow6)$, 19% $\alpha(1\rightarrow2')$, 23% $\alpha(1\rightarrow3')$, 8% $\alpha(1\rightarrow4')$ and 16% $\alpha(1\rightarrow6')$. Identification of the library products was accomplished by reverse phase HPLC separation of the library constituents followed by spectral characterization. The non-removable *p*-methoxyphenoxyoctyl substituent appended to the C-1 carbon of the acceptor disaccharide was critical for HPLC separation in this deconvolution approach.

Validation of the random glycosylation method for producing oligosaccharide libraries was provided by demonstrating that specific disaccharides generated in two libraries were substrates for selective glycosyltransferases. One example demonstrated the incorporation of [^{3}H]-Fuc into the library constituents β-Gal(1$\rightarrow$3)GlcNAc-OR and β-Gal(1$\rightarrow$4)GlcNAc-OR in the presence of human milk fucosyltransferase [27]. Another example showed that [^{3}H]-GalNAc was incorporated into a fucosylated-LacNAc library using human serum from a blood group A donor [28].

Another demonstration of oligosaccharide library synthesis by a solution phase approach was provided by Boons and co-workers [29]. In their approach, they chose to pursue regioselective glycosylation with conventionally protected sugar building blocks. Their approach used a common precursor building block to prepare both a glycosyl donor and acceptor (see Figure 4). The anomeric substituent of this common precursor, when converted to the reactive glycosyl donor by isomerization to the vinyl glycoside and activation with TMSOTf, glycosylates a free hydroxyl on the acceptor. The resulting oligosaccharide can be further homologated by now activating the anomeric

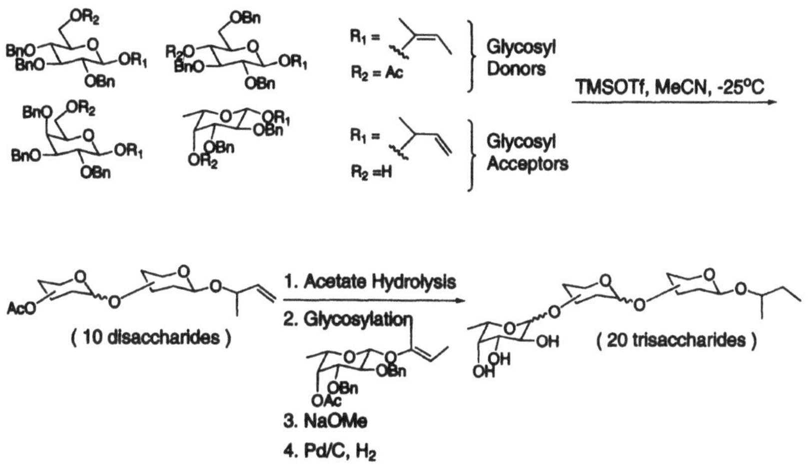

Figure 5. The solution phase construction of a 20 member trisaccharide library based on a latent active glycosylation strategy. This strategy relies on the latent glycosyl donor characteristic of the acceptor building blocks permitting the addition of another acceptor building block to the reducing end of the growing oligomer. The 20 oligosaccharide library products are a mixture of anomers.

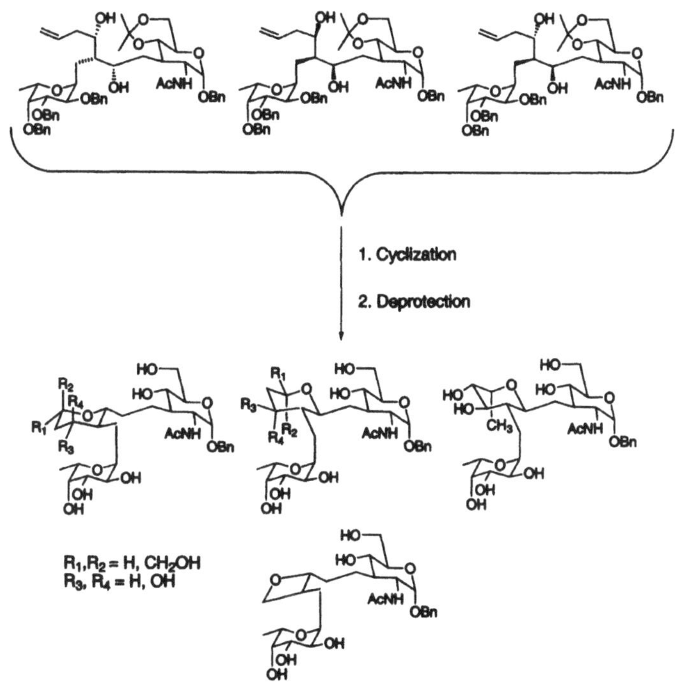

Figure 6. The solution phase construction of a library of C-trisaccharides using three building blocks that have sugars fixed to the termini of a functionalized chain is shown. The internal ring systems are constructed through intramolecular cyclizations of acyclic precursors. The chemical diversity in the library resides in the stereochemical orientation of the substituents on the central sugar of the Fuc-α(1→2)-hexose-(1→3)-GlcNAc saccharide products.

center of the reducing sugar by isomerization and treatment with TMSOTf, in the presence of another acceptor. Exploiting this strategy, a twenty member trisaccharide library was generated by first preparing five discrete disaccharides as (1:1) α/β mixtures (see Figure 5). After these ten disaccharides were

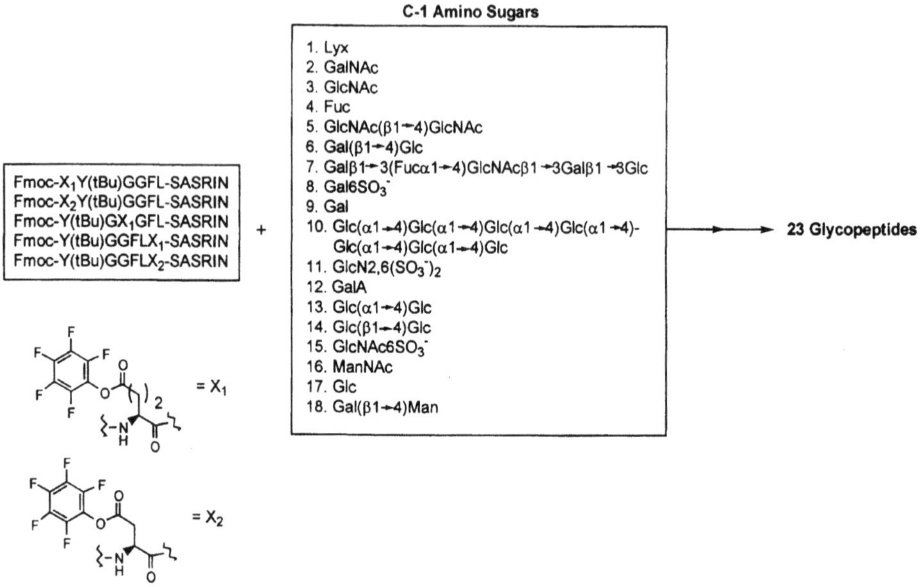

Figure 7. A library of 23 N-linked glycopeptides was constructed using the convergent approach. In constructing this library, 18 1-amino-1-deoxysaccharides were coupled to either the aspartic acid or glutamic acid residues of 5 peptides bound to a SASRIN resin. The conjugation was accomplished without the need for protecting groups on the sugar building blocks (from Reference 31).

combined, an acetate protecting group was removed and a second non-stereoselective glycosylation provided the mixture of 20 trisaccharides. The resulting protected trisaccharide library was purified by size exclusion chromatography to remove unreacted glycosyl donor. The remaining protecting groups were removed to give the final product mixture. Mass spectrometry and carbohydrate compositional analysis were used to validate the existence of the proposed library products.

A solution phase approach directed toward the construction of diverse sets of C-trisaccharides was described by Sutherlin and Armstrong [30]. Their strategy used a core building block fixing two pendent sugars, one at each end of a four carbon spacer (see Figure 6). The acyclic spacer was designed to permit the construction of a diverse set of deoxy C-glycosides between the existing two sugars. Diastereomeric mixtures of C-trisaccharides resulted from this approach. To address the deconvolution of such a diastereomeric mixture of products, these workers proposed the use of a recursive stereochemical deconvolution (RSD) strategy. Based on the use of archived intermediates, structure elucidation of an active compound may be achieved by employing conventional analytical methods or by executing a new set of more stereoselective chemical steps. In exemplifying this library strategy, Sutherlin and Armstrong prepared a set of 12 compounds of which 11 were C-trisaccharides of the general structure Fuc-$\alpha(1\rightarrow2)$-hexose-$(1\rightarrow3)$-GlcNAc and one was a compound containing a central oxetane unit. The resulting central hexose units were all deoxysugar derivatives. These compounds were prepared as potential inhibitors for the cell surface protein of the bacterium *Helicobactor pylori*.

Glycoconjugate libraries

Glycoconjugate systems such as glycopeptides and glycolipids are ubiquitous in biology where they play important regulatory roles [4]. In addition, many known drugs and therapeutically interesting natural products are glycoconjugates [6,7]. The introduction of combinatorial techniques to explore carbohydrate diversity around glycoconjugate systems has provided a powerful new tool for uncovering the significance of the sugar units in the activity of these important classes of molecules and for identifying molecules with enhanced biological activity.

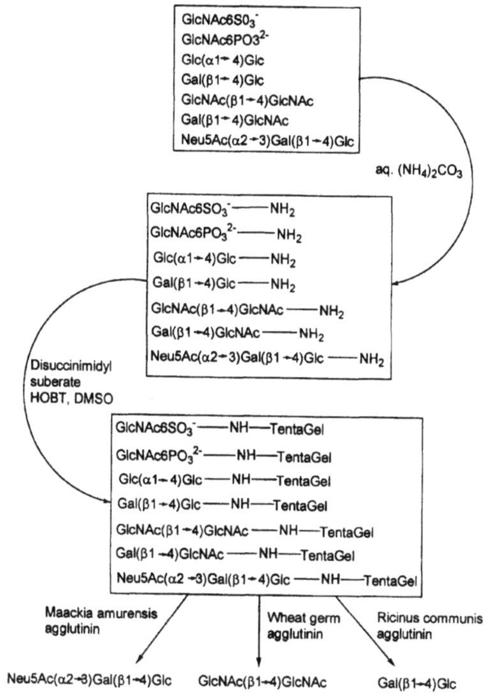

Figure 8. A solid phase glycoconjugate library was constructed from seven 1-amino-1-deoxysaccharides by linking to TentaGel using disuccinimidyl suberate as the linking reagent. The resin bound glycoconjugate library was screened against three lectins: *Maackia amurensis* agglutinin, wheat germ agglutinin and *Ricinus communis* agglutinin. The saccharides known to be specific for the individual lectins were identified by on-bead assays using either antidigoxigenin/peroxidase conjugate with precipitative stain formation (*Maackia amurensis*) or a fluorescently labeled lectin (wheat germ and *Ricinus communis*).

Solid phase approaches

The combinatorial construction of glycopeptide libraries dominates the work reported in the area of solid phase glycoconjugate library construction. In all of the reported glycopeptide libraries, preformed saccharides are appended to the peptide unit by amide bond construction via either a 'building blocks' or 'convergent' strategy [32–34]. Such approaches eliminate the need to construct a glycosidic bond on the solid phase; however, they also limit the accessible diversity.

Using 1-amino-1-deoxy saccharides, Vetter and coworkers constructed a small *N*-linked glycopeptide library (23 members) [35,36]. The carboxylic acid group of glutamic acid or aspartic acid residues of polymer bound peptides were activated by preparation of the pentafluorophenyl (Pfp) ester derivative. Eighteen commercially available 1-amino-1-deoxy saccharides were then coupled to the solid supported peptides by an amide bond construction (see Figure 7). Amide bond formation with fully unprotected charged and uncharged oligosaccharides was accomplished. Coupling yields of 50–80% were observed with uncharged oligosaccharides and yields of 30–59% were observed with charged oligosaccharides. Chemoselective amide bond formation eliminated the need to use protecting groups on the saccharide building blocks. Following cleavage of the products from the solid support with 50% TFA, the desired glycopeptides were obtained in good purity.

In another use of unprotected 1-amino-1-deoxy saccharides, Vetter and Gallop constructed a glycoconjugate library with the potential for taking advantage of on-bead encoding and screening strategies [37]. The resulting solution conjugates made from the 1-amino-1-deoxy saccharides of maltose, chitobiose, lactose, sialyllactose, LacNAc, GlcNAc6SO_3^-, and GlcNAc6PO_3^{2-} and disuccinyl suberate (DDS) were immobilized onto amino-functionalized ethyleneglycol-grafted polystyrene resin. Using two different on-bead screening protocols, this small library of saccharide conjugates was probed for lectin binding to *Maackia amurensis* agglutinin, wheat germ agglutinin and *Ricinus communis* agglutinin (see Figure 8). Saccharides selective for the individual lectins were identified in this library, thus indicating the viability of this type of conjugate screening strategy for studying carbohydrate molecular recognition processes.

The construction of another solid phase glycoconjugate library was described for the preparation of peptidoglycan monomer analogs [38,39]. These analogs were intended for screening as inhibitors of bacterial peptidoglycan biosynthesis. As in other convergent approaches, preformed saccharides were coupled to resin bound peptides through the formation of an amide linkage. However, unlike typical *N*-linked glycopeptides, the carboxylic acid group involved in amide bond construction did not derive from an aspartic acid or glutamic acid amino acid residue, but from a sugar muramic acid moiety, and the link to the peptide occurred at the peptide terminal amino group (see Figure 9). A series of 350 peptidoglycan monomer analogs were constructed using a spatial array format from twelve pre-constructed mono- and disaccharide building blocks and 22 peptides linked to polystyrene through a chlorotrityl linker. Solid phase coupling condi-

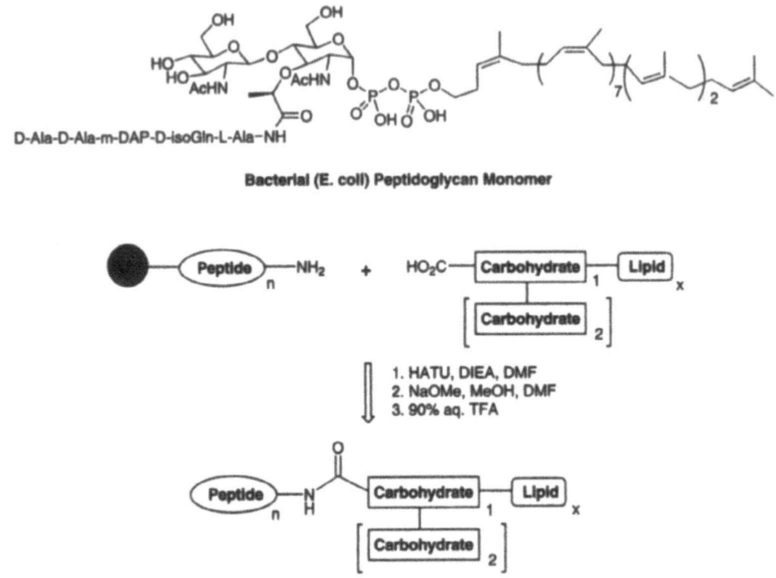

Figure 9. The solid phase library strategy for constructing a peptidoglycan monomer analog library. In this library construction, 350 peptidoglycan monomer analogs were prepared by coupling twelve pre-constructed carboxylic acid containing mono- and disaccharide building blocks to the amino terminus of 22 peptides linked to polystyrene through a chlorotrityl linker. Cleavage of the products from the solid phase under acidic conditions provided the products in solution (from Reference 31).

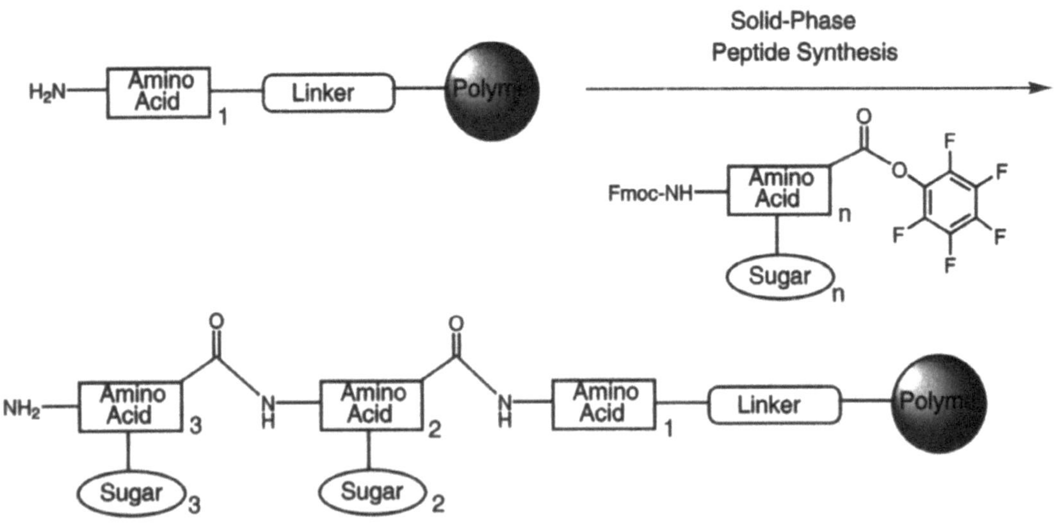

Figure 10. The well established 'building blocks' approach for the solid phase construction of glycopeptides uses glycosylated amino acid building blocks to introduce carbohydrate diversity into a peptide conjugate. The glycosylated amino acid building blocks use pentafluorophenyl activation of the carboxylate, Fmoc protection of the α-amino group and generally, base labile acetate protecting groups on the sugar moieties.

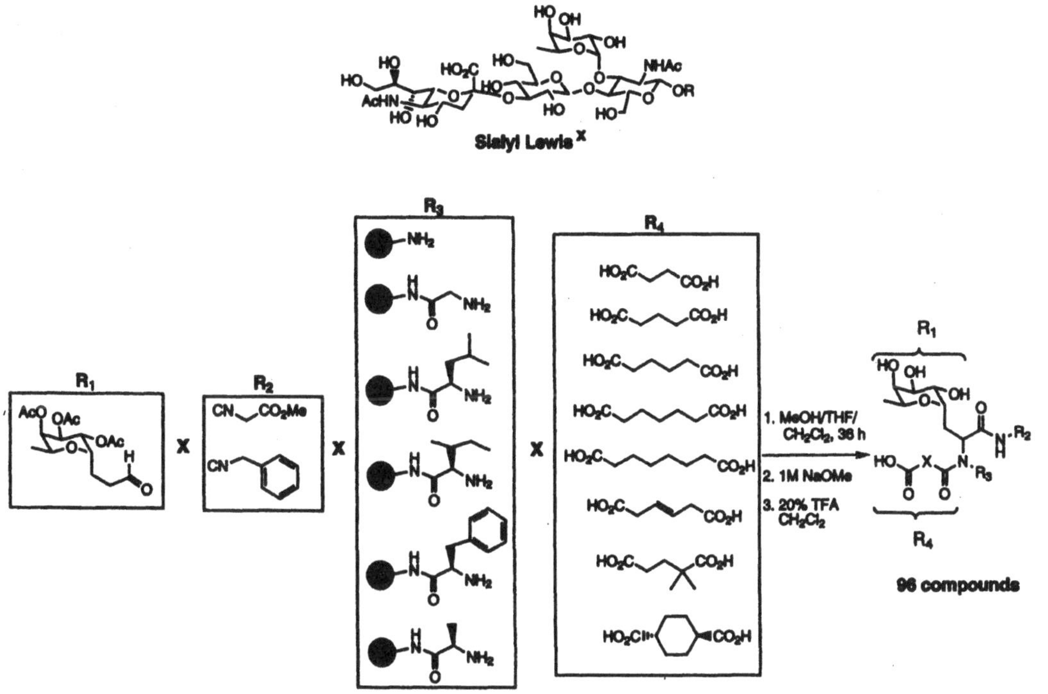

Figure 11. An Ugi multicomponent condensation strategy was used to construct a 96 compound library of C-glycoside mimetics of sialyl Lewis^x. This solid phase library construction used a single C-glycoside as the aldehyde component relying on the isocyanate, amine and acid components to introduce most of the molecular diversity (from Reference 31).

tions employed HATU ([*O*-(7-azabenzotriazol-1-yl)-1,1,3,3-tetramethyluronium hexafluorophosphate]) as the activating reagent and provided conjugate products in 89% yield and high purity as determined by HPLC. The library products were eventually obtained after base hydrolysis of the acetate protecting groups and product cleavage from the solid support with TFA.

An alternative and extensively developed strategy for constructing glycopeptides has relied on using pre-formed glycosylated amino acid building blocks and linking them by using established peptide solid phase synthesis methods (see Figure 10) [40–54]. This solid phase synthesis approach has produced libraries of glycopeptides where the diversity is achieved by either varying the sequence in which the glycosylated amino acids are fed into the synthesis or by introducing a different building block at a particular position within the sequence. The glycosylated amino acid building blocks typically contain an *N*-fluoren-9-ylmethoxycarbonyl (Fmoc) protecting group on the amino functionality of the amino acid, and activation of the amino acid carboxylate group is accomplished through the pentafluorophenyl (Pfp) ester. Since attachment to the solid support is usually accomplished through acid labile linkers such as either the 5-(4-aminoethyl-3,5-dimethoxyphenoxy)valeric acid (PAL) or *p*-(α-amino-2,4-dimethoxybenzyl)phenoxyacetic acid (Rink) linker, base labile protecting groups are used on the sugar hydroxyls. With the development of a wide variety of N-, aliphatic O- and aromatic O-linked glycosylated amino acid building blocks, access to an extensive array of glycopeptide systems is possible [40–54].

Using pre-formed glycosylated amino acid building blocks, libraries of glycopeptides were prepared on a multiple column solid phase synthesis apparatus [55,56]. Examples of glycopeptide library construction using these methods include the preparation of variants of the mucin-type *O*-glycan core structure and the synthesis of Tn-antigenic glycopeptide analogs [57,58].

The preparation of a C-glycoside glycoconjugate library was first described by Armstrong and coworkers [59]. Applying the Ugi four component con-

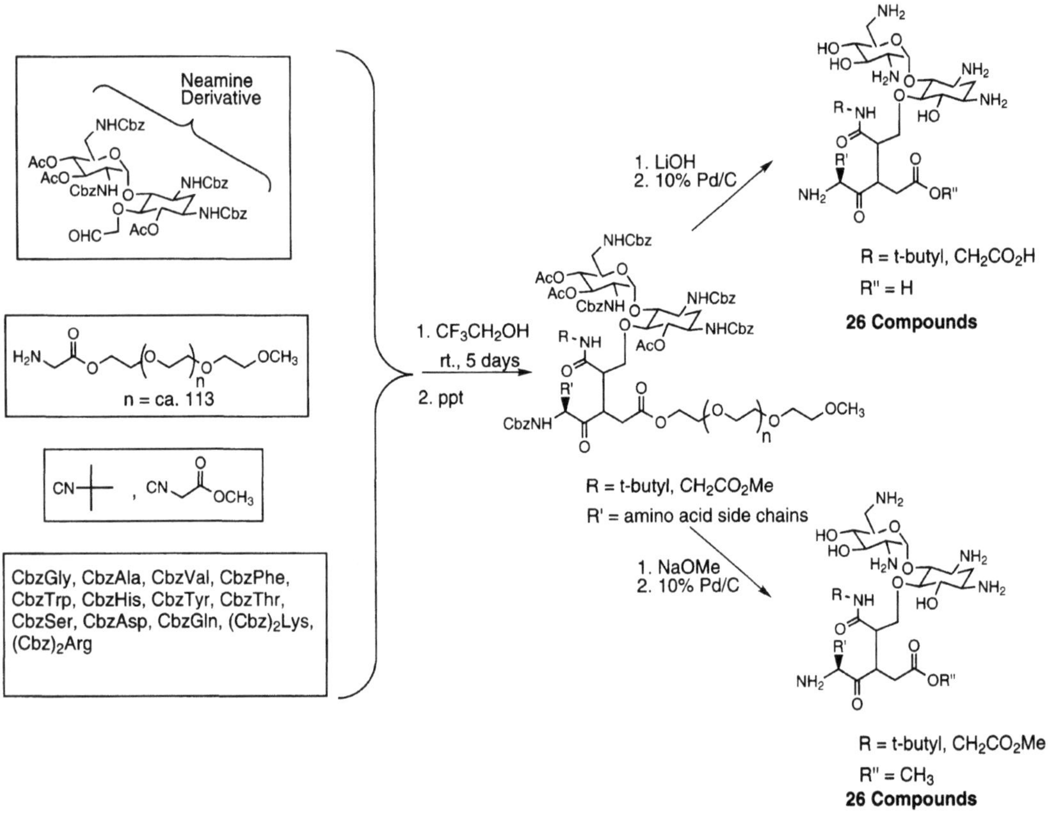

Figure 12. A library of 54 neamine conjugates was constructed using a solution phase multicomponent condensation strategy. A protected neamine aldehyde derivative was condensed with two isocyanates and thirteen protected amino acids to produce the PEG linked products. Cleavage from the PEG support and protecting group removal gave the neamine conjugates as either the free acid or methyl ester.

densation reaction [60] with either C-1 acetaldehyde C-glycosides or C-1 acetic acid C-glycosides as one of the building block components, they were able to produce conjugates which contained a C-glycoside moiety (see Figure 11). The library they described was composed of 96 discrete sialyl Lewis^x glycomimetics constructed from five Rink polymer bound amino acids, eight alkyl diacids of various chain lengths, two isocyanates and fucose acetaldehyde C-glycoside. The products were obtained in approximately 71% yield and ≥95% purity as determined by LC/MS.

Solution phase approaches

The application of soluble polymers for the synthesis of oligosaccharides has been described by several groups [61]. The interest in using soluble polymers over insoluble polymers is in their ability to provide a solution-type environment for chemistry to occur, eliminating the need for chemical reengineering. In their efforts to identify neomycin B mimetics which bind to the Rev responsive element of HIV mRNA (RRE), Wong and coworkers took advantage of soluble polymer synthesis methods and the Ugi-type multiple component condensation (MCC) strategy to prepare a small glycoconjugate library from a neamine derivated core structure [62]. The library was constructed using a C-5 acetaldehyde derivatized neamine, tert-butyl isocyanate and isocyano acetic acid methyl ester, glycine-polyethylene glycol and 13 N-protected amino acids as the MCC reaction components (see Figure 12). The diversity of the resulting 52 compound library was provided by variations in the peptide unit attached to a common neamine unit. The products of the MCC were isolated by ether precipitation followed by base treatment to cleave both the protecting groups and to remove the product from the PEG soluble polymer. Four products from this library were found to inhibit the interaction of the full-length RRE with HIV-1 at >80% at a concentration of 200 μM.

Conclusions

The ability to explore carbohydrate molecular diversity has taken great strides in recent years. Viable solid phase and solution phase methods now exist for rapidly accessing some of the broad universe of carbohydrate-based molecular structures. It is now the job of chemists to exploit these methods for the discovery of novel medicinal, agricultural and industrial agents.

References

1. Gordon, E.M., Barrett, R.W., Dower, W.J., Fodor, S.P.A. and Gallop, M.A., J. Med. Chem., 37 (1994) 1385.
2. Gallop, M.A., Barrett, R.W., Dower, W.J., Fodor, S.P.A. and Gordon, E.M., J. Med. Chem., 37 (1994) 1233.
3. Nefzi, A., Ostresh, J.M. and Houghten, R.A., Chem. Rev., 97 (1997) 449.
4. Varki, A., Glycobiology, 3 (1993) 97.
5. Rademacher, T.W., Parekh, R.B. and Dwek, R.A., Annu. Rev. Biochem., 57 (1988) 785.
6. Petitou, M., In Kovac, P. (Ed.), Synthetic Oligosaccharides: Indispensable Probes for the Life Sciences, ACS Symposium Series 560, American Chemical Society, Washington, DC, 1994, p. 19.
7. Musser, J.H., Anderson, M.B. and Fugedi, P., Pharmaceutical News, 3 (1996) 11.
8. Yan, L., Taylor, C.M., Goodnow Jr., R. and Kahne, D., J. Am. Chem. Soc., 116 (1994) 6953.
9. Danishefsky, S.J., McClure, K.F., Randolph, J.T. and Ruggeri, R.B., Science, 260 (1993) 1307.
10. Randolph, J.T., McClure, K.F. and Danishefsky, S.J., J. Am. Chem. Soc., 117 (1995) 5712.
11. Hunt, J.A. and Roush, W.R., J. Am. Chem. Soc., 118 (1996) 9998.
12. Rademann, J. and Schmidt, R.R., J. Org. Chem., 62 (1997) 3650.
13. Nicolaou, K.C., Winssinger, N., Pastor, J. and DeRoose, F., J. Am. Chem. Soc., 119 (1997) 449.
14. Ito, Y. and Ogawa, T., J. Am. Chem. Soc., 119 (1997) 5562.
15. Rodebaugh, R., Joshi, S., Fraser-Reid, B. and Geysen, H.M., J. Org. Chem., 62 (1997) 5660.
16. Frechet, J.M. and Schuerch, C., J. Am. Chem. Soc., 93 (1971) 492.
17. Frechet, J.M. and Schuerch, C., Carbohydr. Res., 22 (1972) 399.
18. Frechet, J.M. and Schuerch, C., J. Am. Chem. Soc., 94 (1972) 604.
19. Halcomb, R.L., Huang, H. and Wong, C.-H., J. Am. Chem. Soc., 116 (1994) 11315.
20. Wong, C.-H., Halcomb, R.L., Ichikawa, Y. and Kajimoto, T., Angew. Chem., Int. Ed. Engl., 34 (1995) 521.
21. Schuster, M., Wang, P., Paulson, J.C. and Wong, C.-H., J. Am. Chem. Soc., 116 (1994) 1135.
22. Liang, R., Yan, L., Loebach, J., Ge, M., Uozumi, Y., Sekanina, K., Horan, N., Gildersleeve, J., Thompson, C., Smith, A., Biswas, K., Still, W.C. and Kahne, D., Science, 274 (1996) 1520.
23. Ohlmeyer, M.J., Swanson, R.N., Dillard, L.W., Reader, J.C., Asouline, G., Kobayashi, R., Wigler, M. and Still, W.C., Proc. Natl. Acad. Sci. USA, 90 (1993) 10922.
24. Muller, C., Kitas, E. and Wessel, H.P., J. Chem. Soc., Chem. Commun., (1995) 2425.
25. Nicolaou, K.C., Florke, H., Egan, M.G., Barth, T. and Estevez, V.A., Tetrahedron Lett., 36 (1995) 1775.
26. Suhara, Y., Hildreth, J.E.K. and Ichikawa, Y., Tetrahedron Lett., 37 (1996) 1575.
27. Kanie, O., Barresi, F., Ding, Y., Labbe, J., Otter, A., Forsberg, L.S., Ernst, B. and Hindsgaul, O., Angew. Chem., Int. Ed. Engl., 34 (1995) 2720.
28. Ding, Y., Kanie, O., Labbe, J., Palcic, M.M., Ernst, B. and Hindsgaul, O., In Alavi, A. and Axford, J.S. (Eds.) Glycoimmunology, Plenum, New York, NY, 1995, p. 261.
29. Boons, G.-J., Heskamp, B. and Hout, F., Angew. Chem., Int. Ed. Engl., 35 (1996) 2845.
30. Sutherlin, D.P. and Armstrong, R.W., J. Org. Chem., 62 (1997) 5267.
31. Sofia, M.J., Mol. Div., 3 (1997) 75.
32. Sofia, M.J., Drug Discov. Today, 1 (1996) 27.
33. Meldal, M., Curr. Opin. Struct. Biol., 4 (1994) 710.
34. Andrews, D.M. and Seale, P.W., Int. J. Pept. Protein Res., 42 (1993) 165.
35. Vetter, D. and Gallop, M.A., Bioconj. Chem., 6 (1995) 316.
36. Vetter, D., Tumelty, D., Singh, S.K. and Gallop, M.A., Angew. Chem., Int. Ed. Engl., 34 (1995) 60.
37. Vetter, D., Tate, E.M. and Gallop, M.A., Bioconj. Chem., 6 (1995) 319.
38. Chan, T.Y., Chen, A., Allanson, N., Chen, R., Liu, D. and Sofia, M.J., Tetrahedron Lett., 37 (1996) 8097.
39. Allanson, N., Chan, T.Y., Chen, A., Chen, R., Liu, D., Baizman, E., Mintz, C., Axelrod, H. and Sofia, M.J., 211th ACS National Meeting, Abstract No. 199, New Orleans, LA, U.S.A., March 24–28, 1996.
40. Jansson, A.M., Meldal, M. and Bock, K., Tetrahedron Lett., 31 (1990) 6991.
41. Meldal, M. and Bock, K., Tetrahedron Lett., 31 (1990) 6987.
42. Bielfeldt, T., Peters, S., Meldal, M., Bock, K. and Paulsen, H., Angew. Chem., Int. Ed. Engl., 31 (1992) 857.
43. Jansson, A.M., Meldal, M. and Bock, K., J. Chem. Soc., Perkin Trans. 1, (1992) 1699.
44. Reimer, K.B., Meldal, M., Kusumoto, S., Fukase, K. and Bock, K., J. Chem. Soc., Perkin Trans. 1, (1993) 925.
45. Christensen, M.K., Meldal, M. and Bock, K., J. Chem. Soc., Perkin Trans. 1, (1993) 1453.
46. Christiansen-Brams, I., Meldal, M. and Bock, K., J. Chem Soc., Perkin Trans. 1, (1993) 1461.
47. Jensen, K.J., Meldal, M. and Bock, K., J. Chem. Soc., Perkin Trans. 1, (1993) 2119.
48. Christensen, M.K., Meldal, M., Bock, K., Cordes, H., Mouritsen, S. and Elsner, H., J. Chem. Soc., Perkin Trans. 1, (1994) 1299.
49. Meldal, M., Auzanneau, F.-I., Hindsgaul, O. and Palcic, M.M., J. Chem. Soc., Chem. Commun., (1994) 1849.
50. Paulsen, H., Bielfeldt, T., Peters, S., Meldal, M. and Bock, K., Liebigs Ann. Chem., (1994) 369.
51. Seitz, O. and Kunz, H., Angew. Chem., Int. Ed. Engl., 34 (1995) 803.
52. Polt, R., Szabo, L., Treiberg, J., Li, Y. and Hruby, V.J., J. Am.

Chem. Soc., 114 (1992) 10249.
53. Szabo, L., Li, Y. and Polt, R., Tetrahedron Lett., 32 (1991) 585.
54. Broddefalk, J., Bergquist, K.-E. and Kihlberg, J., Tetrahedron Lett., 37 (1996) 3011.
55. Peters, S., Bielfeldt, T., Meldal, M., Bock, K. and Paulsen, H., J. Chem. Soc., Perkin Trans. 1, (1992) 1163.
56. Meldal, M., Bisgaard Holm, C., Bojesen, G., Jakobsen, M. H. and Holm, A., Int. J. Pept. Res., 41 (1993) 250.
57. Rio-Anneheim, S., Paulsen, H., Meldal, M. and Bock, K., J. Chem. Soc., Perkin Trans. 1, (1995) 1071.
58. Frische, K., Meldal, M., Werdelin, O., Mouritsen, S., Jensen, T., Galli-Stampino, L. and Bock, K., J. Pept. Sci., 2 (1996) 212.
59. Sutherlin, D.P., Stark, T.M., Hughes, R. and Armstrong, R.W., J. Org. Chem., 61 (1996) 8350.
60. Armstrong, R.W., Combs, A.P., Tempest, P.A., Brown, S.D. and Keating, T.A., Acc. Chem. Res., 29 (1996) 123.
61. Gravert, D.J. and Janda, K.D., Chem. Rev., 97 (1997) 489.
62. Park, W.K.C., Auer, M., Jaksche, H. and Wong, C.-H., J. Am. Chem. Soc., 118 (1996) 10150.

M.R. Pavia and W.H. Moos (eds.), Annual Reports in Combinatorial Chemistry and Molecular Diversity, Vol. 2, 53–58.
KLUWER/ESCOM

Patent strategies in molecular diversity

Grant D. Green
Fish & Richardson, 2200 Sand Hill Road, Menlo Park, CA 94025, U.S.A.
E-mail: green@fr.com

Key words: anticipation, collaboration, combinatorial library, interference, license, obviousness, partnership, patentability, prior art, trade secret, written description

Summary

The patentability of combinatorial libraries is examined, requirement by requirement. In general, combinatorial libraries can be patented, as long as the library meets the statutory criteria of utility, novelty, and non-obviousness, and the application meets the standards of enablement, best mode, and written description. Licensing and alternatives to patenting are also considered, along with potential problems unique to combinatorial chemistry agreements.

Introduction

Combinatorial chemistry, or diversity-based drug discovery, is one of the fastest developing fields of research. The ability to make and screen vast numbers of compounds by new methods has resulted in a new segment of the research industry. With this new technology comes new questions of patentability. Is a library of compounds patentable? Who is the inventor? How can it be licensed?

Patentability

In the United States, three statutes set forth the elements required in order for an invention to be patentable. 35 U.S.C. §101 requires that the invention be 'described in writing', 'patentable subject matter', and that it be 'useful'. 35 U.S.C. §102 requires the invention to be novel, or new, in the sense of 'never before made, sold, or described'. 35 U.S.C. §103 requires that the invention, in addition to being new, also be non-obvious in view of the state of the art. Most countries that grant patents have similar requirements, although they may be stated in different terms.

Patentable subject matter

'Patentable subject matter' includes 'everything under the Sun that is made by Man' [1]. Chemical compounds and compositions have long been considered patentable. However, not all compounds or compositions are patentable subject matter in a broad, generic sense.[1] Many compounds occur in Nature, and are thus not patentable *per se*. To overcome this problem, the patent practitioner must add limitations to the claim, and seek coverage for a form of the compound that does not occur naturally. For example, a naturally occurring protein such as human interleukin-1 ('hIL-1') cannot be claimed *per se* because it occurs in the body, and was not invented by Man. However, hIL-1 is not found in a pure state, but rather at low concentration in the body. Thus, *purified* hIL-1 is a composition not found in Nature, and is patentable subject matter.

Libraries and mixtures of compounds that are chemically synthesized are clearly 'made by Man'. However, libraries derived from natural sources may not be patentable subject matter. For example, the unpurified mixture of compounds in a broth or culture supernatant resulting from a mixed bacterial culture or cell culture must be distinguished from such mixtures as they occur in Nature. This is sometimes accomplished by claiming the supernatant 'free of cells', or otherwise conditioning the state of the composition with a feature that requires some human intervention.

Written description

To be patented, the invention must be 'described in writing' [2], a task which may be problematic for a diverse mixture of compounds. The 'written description' requirement may be satisfied by (a) listing all the compounds that are (or may be) part of the mixture, or (b) describing the compounds in terms of one or more generic formulas. Where the compounds are all structurally related (e.g., they all derive from a common set of reactions, or share a common nucleus), they may be described using simple generic formulas such as:

A mixture of compounds of the formula:

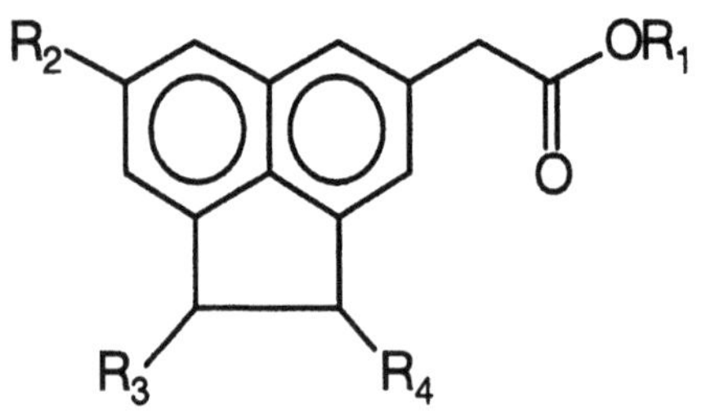

wherein R_1 is H or lower alkyl, R_2 is OH or NO_2, and R_3 and R_4 are each independently COOH, COOR, or CN, where R is H or lower alkyl; wherein said mixture has at least N compounds of formula 1.

In some cases, the members of the library can be described with several generic formulas, e.g., 'A compound library of the invention comprises a mixture of at least two compounds selected from the group consisting of compounds of formula 1, compounds of formula 2, and compounds of formula 3, ...'.

Combinatorial libraries may also be described and claimed in terms of the process of production, known in the patent field as a 'product by process'. Product by process coverage is generally not preferred because there is still some uncertainty as to the scope of the resulting claims. There is legal precedent to the effect that the process limitations in the claims only characterize the final product, and thus that any product that otherwise meets the description of the claim infringes that claim [3]. Other cases take the position that the process limitations are actual limitations, and that for a product to infringe, it must be made by the process recited in the claim [4].

Utility

To be patentable in the United States, an invention must be 'useful'. To be 'useful' within the requirements of the statute, an invention must be capable of some benefit to society, and capable of achieving at least one of its objects [5]. In chemical practice, this means that a new compound must have a stated utility (e.g., as lubricant, filler, or chemotherapeutic), and must be capable of fulfilling that utility. The stated utility or utilities should be chosen carefully, because they are considered when the Examiner determines against which yardstick the claims are to be measured. Thus, for example, claims to a compound capable of inhibiting the binding of a cytokine to its receptor may be evidenced by appropriate *in vitro* experimental data, whereas a claim to treating cancer in a mammal using the same compound may require *in vivo* data.

There are at least two utilities that may be articulated for most combinatorial libraries. First, the combinatorial library is useful *per se* as a tool for research, e.g., to determine what form of molecules bind to a given target molecule. In this aspect, analysis of library compounds that bind and do not bind may impart important information regarding the conformation of the target.

Second, the library, if well designed, should contain one or more active compounds. This aspect may be demonstrated by including data from successful use of the library to identify a lead compound.

Novelty

All modern patent statutes require that the claimed invention be novel, *i.e.*, not previously patented or described in a single reference. A patent claim is not novel (is 'anticipated') if the claim limitations would include subject matter already in the public domain. For example, a claim reading

3. A composition comprising acetyl salicylic acid.

is anticipated by the public use, sale, and literature description of aspirin, as aspirin tablets contain ('comprise') acetyl salicylic acid.

Combinatorial libraries may experience novelty problems due to the number of compounds that make up the library. One must be careful to claim a combinatorial library as a mixture of compounds that includes one or more novel compounds, or a mixture of old compounds that are not likely to have been described in association previously. For example, a claim such as

4. A composition, comprising: a compound of formula 3, and at least three other compounds.

may be novel if the compound of formula 3 is novel. However, if the compound is known, this claim might be anticipated by the prior description of the compound prior to purification (e.g., the compound, the unreacted reactants still present in the pot, and the solvent). This can be avoided by (a) increasing the number of compounds required in the mixture, and/or (b) specifying that the other compounds have a particular structure not likely to have been reported in association with the compound of formula 3. If one has patented several combinatorial libraries with members in common, it may be necessary to specify a large number of compounds, so that the claim describes a set of compounds that is not common to any of the previously claimed libraries.

Non-obviousness

In addition to utility and novelty, an invention must be non-obvious[2] to be patentable. Under U.S. case law, the claimed invention is examined against the prior art existing at the time the invention was made. If the invention is novel over the prior art, the art is examined again to see if the invention is an insignificant modification of the prior art, or if several references taken together suggest the invention.

At the beginning of the combinatorial chemistry field, nonobviousness was not problematic: the very idea that one should deliberately make mixtures of pharmaceutical candidates seemed contrary to decades of pharmaceutical research. Today, however, combinatorial chemistry is well established, and there are numerous references available to 'suggest' the preparation of combinatorial libraries from any given genus of compounds. Some combinatorial libraries may in fact be obvious now. However, many are still patentable, as the superficial idea that one can prepare a useful library out of any given group of compounds may be rebutted, e.g., by demonstrating the non-obvious modifications to the synthesis needed in order to obtain a diverse mixture of compounds in a reasonable range of concentrations.

Alternative protection

The fundamental problem with patenting combinatorial libraries is that the library is seldom the product marketed. More often, the library is a means to obtain lead compounds, or to perform basic research. In such cases, infringement may never be detected: it may not be apparent that a new pharmaceutical compound was developed using combinatorial chemistry techniques. Thus, in many cases it may be more advantageous to maintain the library as a trade secret.[3]

Trade secrets have a number of advantages over patents in this context. First, a trade secret is not disclosed, in contrast to the patent, which is typically published before any patent rights are actually granted. Thus, with a trade secret one's competitors are not alerted to the avenue of research one is pursuing, and are not taught how to follow the same path. Second, trade secrets only expire if disclosed. In contrast, a patent provides protection for a period ending twenty years after the filing date of the patent application.

However, patents are still useful for combinatorial chemistry where, for example, one's market is developing and selling combinatorial technology, and there are competitors who would otherwise market the same technology. Patents also serve as a useful measure of the value of a company, sometimes the primary value of a startup.

Licensing

Diversity-based drug discovery and combinatorial libraries present special problems in licensing. A license, in this context, may include the sale or 'lease' of a physical collection of compounds, the patents and other intellectual property rights to discoveries made using the library, training and knowhow transfer, and grants of exclusivity.[4] Licenses in this field can be categorized as technology transfer agreements, compound generation collaborations, or a combination of both.

Technology transfer agreements may be fairly straightforward, although they may require careful consideration of patent rights, software copyrights, and knowhow. The difficulty that arises is often deciding how to calculate compensation. The technology transferred may be used only at the early stages of lead compound discovery: it may not be apparent from the final commercial product whether or not the licensed technology was used. The Licensor may be unable to determine whether it is being fairly compensated, without subjecting the Licensee to an audit. On the other hand, by the time a commercial product is launched, the Licensee may feel that the technology played an insignificant role, in hindsight. For this reason, it is often preferable to structure such agreements as a 'paid-up' license, with one or more payments that are negotiated, rather than calculated as a royalty based on product sales. As 'infringement' may be difficult to determine, a technology transfer license is sometimes closer to a knowhow or trade secret license than to a traditional patent license.[5]

Compound generation arrangements, however, are host to a number of problems peculiar to diversity-based drug discovery, a few of which are outlined below.

The 'many partners' problem

A problem that often arises for the diversity-based discovery company that wishes to outlicense its technology is the 'many partners' problem. The company may want to maximize the value of its libraries by licensing use of the libraries to other companies, or collaborating with other companies for drug discovery. Ideally, the company would like to license the same libraries (it may have only a few) to as many third parties as possible. However, the third parties are likely to require exclusive rights to any lead compounds identified, which leads to the problem.

For example, suppose that Company A licenses its libraries to Company B to screen for leads against one or more targets. Company B promptly determines that several compounds in several of the libraries are active in its assays, and desires an exclusive license to those compounds, and all compounds that are structurally related. A has probably already obligated itself to grant such a license, and is happy to do so until Company C approaches A with another proposed screening arrangement. At this point, A has a number of libraries that contain compounds owned by B. B has expressly obligated A not to provide licensed compounds to any competitor of B. Thus, A must either resynthesize those libraries without including any of the licensed compounds (which may or may not be feasible or possible), or obtain permission from B, or forego providing any of the affected libraries to C. All of these options are likely to be unattractive to A. This particular complication may be avoided by (a) providing that A will own all libraries, and all compounds contained within the libraries, and will retain 'research rights' for all licensed compounds, (b) designing libraries without overlapping compounds (compounds common to two or more libraries), or that permit exclusion of particular compounds, or (c) not granting exclusive rights for any library compound.

The solutions all have problems for either A or B, but can usually be mitigated during negotiation of the original agreement. For example, B may balk at the prospect of not having exclusive rights to a compound within a library. However, if B plans to further elaborate any lead compound discovered in a library, the need for exclusive rights diminishes. If B is still not satisfied, A might offer exclusive rights for the method of using any 'licensed library compounds' for the specific target: e.g., if B is seeking an IgE receptor antagonist for treatment of asthma, A may grant an exclusive license to the method of using identified library compounds for the treatment of asthma, and an exclusive license for composition coverage to compounds based on library compounds (but not found within the library).

Resynthesizing the libraries is sometimes practical, but deliberately omitting a portion of the library is less than satisfactory because it possibly prevents one from discovering lead compounds that may be structurally related but not licensed.[6]

The 'converging partners' problem

Another problem sometimes occurs where diversity-based company A has granted exclusive rights to license compounds in its libraries to several other companies, and two or more of these partners identify and claim the same compound. This can happen where partners have defined their fields of exclusivity by different mechanisms (e.g., partner B claims 'IL-10R antagonists', while partner C claims 'dihydropyridine derivatives', or 'immune modulators'), or where the selected targets respond to the same compound (e.g., where new, poorly characterized receptors turn out to be related or identical). When this happens, one has partners converging on the same compound (or similar compounds).

'Converging partners' can be a problem even when different partners are given non-overlapping libraries. For example, it is not unusual for a collaboration to provide that the licensor will make 'follow-up' libraries, based on the data collected from the first library. If the targets selected by two different partners will bind the same or similar compounds, successive libraries prepared for each partner may converge on the same compound or group of compounds. Obviously, partners B and C cannot simultaneously have completely exclusive rights. One or both partners may wish to have patent applications filed, resulting in interference,[7] or at least disclosure of information one party may wish to keep confidential. Partner A is caught in the middle, unable to even disclose the existence of the problem without breaching the confidentiality of one or more partners.

This is fundamentally a multiple partner problem, and thus should be addressed during negotiation of each agreement, and a consistent solution adopted and applied to all future agreements. It is difficult to avoid the problem by simply selecting partners or projects that are not likely to converge, as one often does not have sufficient information to reach an accurate conclusion. Waiting until sufficient data *is* available retards research, and thus may obviate one of the primary advantages of diversity-based discovery. One can attempt to minimize the possibility of converging partners by carefully defining partners' fields of exclusivity in mutually exclusive terms, but the possibility will still exist.

A solution is likely to require either (a) mediation or arbitration of conflicting rights, (b) automatic 'sharing' of rights between partners B and C, possibly with some compensatory reduction in royalty rates reflecting the 'co-exclusive' (instead of fully exclusive) position, or (c) awarding all rights to the first partner to determine the compound's activity. In any event, the agreement should provide that it is not a breach of confidentiality for A to disclose the situation to B and C if it arises.

The Swiss cheese effect

Granting exclusivity over library compounds to a licensee may have adverse effects on the utility of the library, depending on the design and construction of the library. A diverse library may have a large variety of compounds, with one or few compounds representing numerous chemical characteristics. For example, a library might contain only a few compounds having an extended aromatic ring system with a phenolic hydroxy substitution. If these compounds demonstrate activity in the Licensee's assays, the Licensee may well want exclusive rights to those compounds, and in fact may be entitled to such rights under their agreement. Granting the licensee exclusive rights to those compounds leaves a 'hole' in the library, a portion of the library devoid of useable compounds. If the licensee patents the compounds, or is otherwise granted exclusivity for them, the hole will generally not consist only of the individual library compounds, but will also include a generic 'space' around the compounds – typically a much larger hole. With repeated use of the library, holes may accumulate to the point where the library fails to represent the degree of diversity originally intended. The resulting 'Swiss cheese' library may fail to bind targets due to the unrepresented areas of diversity.

It may be difficult to convince the Licensee to forego patenting active lead compounds discovered in the library. However, if the Licensor retains at least the right to make and use the compounds for research purposes (including the right to sublicense and/or conduct research for third parties), the library utility may be substantially preserved. While the individual library compounds may be rendered unavailable for other third parties to develop, at least the drug discovery pathways leading through those areas are preserved.

Conclusions

Combinatorial chemistry and combinatorial libraries are patentable in general, although subject to a few unusual wrinkles in the process. As happened with recombinant DNA technology several years ago, combinatorial chemistry presents new questions to the patent granting agencies, with answers yet to be codified. Care should be exercised during patent application drafting to insure that the likely rejections are anticipated and disarmed. One should carefully consider whether to file a patent application at all, or whether one's purposes are better served by trade secret protection. Finally, one should approach combinatorial library licenses with caution, keeping in mind that combinatorial libraries have characteristics distinct from the individual compound or target.

Notes

[1] The broadest possible generic coverage for a compound is found in claims such as the following: 'A compound of formula 1,

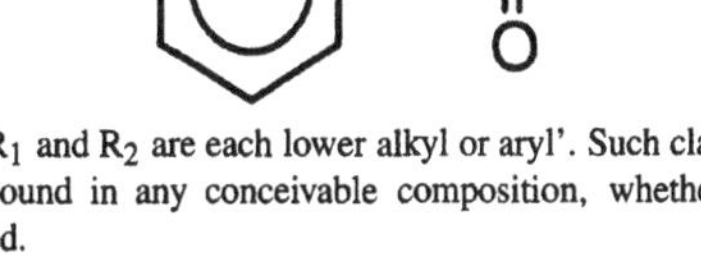

wherein R_1 and R_2 are each lower alkyl or aryl'. Such claims cover the compound in any conceivable composition, whether pure or formulated.

[2] Outside the United States, this is usually expressed as a requirement for an 'inventive step'.

[3] A 'trade secret' is business information that confers a competitive advantage, that is maintained in secrecy by the owner.

[4] 'Exclusivity' may include the Licensor's agreement not to collaborate with third parties with respect to the same targets, not to distribute selected compounds or libraries, and not to use information obtained during the agreement, amongst other possibilities.

[5] In a 'traditional' patent license, the licensee pays a royalty (or some form of compensation) for each product it sells that would otherwise infringe a valid claim of the licensor's patent. For products sold on the open market, it is not difficult for the licensor to see that royalty-bearing products are being sold, and determine whether or not it is being adequately compensated. Where, however, the licensed subject matter is a trade secret or knowhow, or is a manufacturing process, it is often not obvious whether or not a marketed product was made using infringing technology. In such cases, the licensor may prefer compensation in the form of a fixed amount, not dependent on the number of products sold.

[6] See 'The Swiss cheese effect'.

[7] A patent interference is a procedure conducted before the United States Patent and Trademark Office to determine which of two or more parties is entitled to a US patent. When two or more patent applications are filed by different applicants, and claim the same subject matter, the interference is intended to determine which applicant was the first to invent the subject matter. The time and expense involved in conducting an interference can approach that required for litigation.

References

1. Diamond v. Chakrabarty, 100 S.Ct. 2204, 206 USPQ 193 (Sup. Ct. 1980).
2. 35 U.S.C. §101.
3. See, e.g., *Scripps Clinic & Research Foundation v. Genentech, Inc.*, 927 F.2d 1200, 18 USPQ2d 1016 (Fed. Cir. 1991).
4. See, e.g., *Atlantic Thermoplastics Co., Inc. V. Faytex Corp.*, 970 F.2d 834, 23 USPQ2d 1481 (Fed. Cir. 1992).
5. *Carl Zeiss Stiftung v. Renishaw PLC*, 945 F.2d 1173, 20 USPQ2d 1094 (Fed. Cir. 1991).

M.R. Pavia and W.H. Moos (eds.), Annual Reports in Combinatorial Chemistry and Molecular Diversity, Vol. 2, 59–70.
KLUWER/ESCOM

Patent tactics in molecular diversity: Things that get around, come around

(Or: There are always exceptions to the rule)

Karl Bozicevic
Bozicevic & Reed LLP Intellectual Property Law, 285 Hamilton Avenue, Suite 200, Palo Alto, CA 94301, U.S.A.
E-mail: bozicevic@brpatent.com

Key words: 35 U.S.C. §271(e)(1), exemption, FDA approval, infringement, patent law

Summary

Certain activities which would constitute infringement of a U.S. patent can come within a statutory exemption when the activity is carried out to obtain FDA approval. Exemptions to infringement provided by 35 U.S.C. §271(e)(1) are discussed along with recent cases interpreting this statute. The evolution of the statute in providing an exemption to activities necessary to obtain FDA marketing approval is described to provide background, legislative intent and show a comparative of the law before and after 1984 when 271(e)(1) was enacted. By analyzing recent cases, criteria for qualifying an activity as exempt from infringement are described – as are activities that do not qualify for the 271(e)(1) exemption. Overall there appears to be a trend to more broadly allowing for activities to qualify as exempt from infringement. This trend may affect the value of patents and determine which technologies a company will seek patent protection on.

Introduction

Companies involved in combinational chemistry obtain patients for a variety of reasons including (1) their affect on raising capital; (2) cross-licensing technology with competitors; and (3) acquiring a degree of market exclusivity. Regardless of the reason for obtaining a patent, patentees ultimately want the ability to enforce their patent against infringers. When attempting enforcement most patentees recognized that the validity of their patent can be challenged. However, few recognize that if their patent is found valid and infringed that an exemption can be claimed under 35 U.S.C. §271(e)(1). Even fewer patentees recognize the full scope of that exemption which are examined here via different proposed scenarios.

Combinatory chemists can produce a range of different inventions including methods of synthesis, libraries, and assay methods. These inventions are often used to find a compound which is useful as a pharmaceutical drug. In certain circumstances (described here) a patented invention can be used by others when the use is related to obtaining FDA approval of a drug. Such use is exempted from patent infringement, thereby having the potential of significantly reducing the value of the patent. The following is (1) an analysis of this exemption to patent infringement along with (2) an explanation of actions wich can be taken by companies involved in combinatorial chemistry to avail themselves of the exemption.

Factual scenarios

XYZ is a combinatorial chemistry company which has isolated a peptide which activates a receptor specific for an important biological response. The isolated peptide is patentable but not commercially useful as a therapeutic. The peptide and variants thereof, are produced to provide an assay which will hopefully result in isolating and identifying receptor agonists and antagonists which are of considerable commercial interest. The assay is patented and used by XYZ.

After patenting the assay XYZ discovers that its competitor ABC is using the assay for the same purpose, i.e., to find therapeutic agonists and antagonists.

Patentee XYZ sues ABC for infringing the assay patent and ABC claims their use of the assay is exempt from infringement action under 35 U.S.C. §271(e)(1). What are the legal criteria for coming in within a 271(e)(1) exemption? If the exemption does not apply in the above scenario might it apply if XYZ had patented a process for making therapeutic peptides and ABC was using the patented process? Alternatively, what if XYZ invented a new reactor or transgenic animal that produced a therapeutic and ABC used such to produce quantities of the therapeutic sufficient to carry out extensive clinical trials to get FDA approval?

There are no definitive answers to these questions. However, the case law does provide some guidance on when a 271(e)(1) exemption applies. The exemption does not apply when the infringer is actually selling product with FDA approval. However, use of the invention prior to FDA approval can seriously impair the competitive advantage of the patents. To understand the law let's first review its background.

Background of 35 U.S.C. §271(e)(1)

Lobbyists for generic pharmaceutical companies were at least partially responsible for 35 U.S.C. §271(e)(1) being enacted. The lobbyist sought to overrule the Federal Circuit's decision in *Roche Products, Inc. v. Bolar Pharmaceutical Co.*, 733 F.2nd 858, 221 USPQ (BNA) 937, cert. denied, 469 U.S.856 (1984). In the *Bolar* case, Bolar, a generic manufacturer, was infringing a Roche patent on a pharmaceutical compound, flurazepam HCl. The patent was about to expire and Bolar's infringement was carried out to accumulate data for submission to the FDA so that Bolar could market the drug immediately upon expiration of the patent. Bolar wanted to market a generic version of Roche's drug when the patent expired. Bolar argued that they should be allowed to carry out such activity prior to expiration of Roche's patent (under an experimental use exception). Bolar pointed out that if they and other generic manufacturers were prevented from using the patented compound to obtain data for FDA approval the result would be an extension of Roche's patent monopoly since Bolar would be required to spend time after the patent expired to carry out testing to obtain data for submission to the FDA to obtain the requisite FDA marketing approval. The FDA approval process could add years to the market exclusivity.

However, the experimental use exception is very narrow. Prior to enactment of 35 U.S.C. §271(e)(1), the use of a patented drug compound by a generic competitor (in order develop information necessary to obtain FDA marketing approval after expiration of the patent) constituted infringement if such use occurred prior to expiration of the patent. Thus, prior to the passage of 35 U.S.C. §271(e)(1), the patent term of a patented pharmaceutical compound was *de facto* extended, and the extension was often for substantial periods of time beyond the term of the patent, since acquiring information necessary to obtain approval for generic competing products could not be commenced until after the patent expired. Further, the information in the FDA file showing the efficacy and safety of the therapeutic could not be used to obtain a subsequent approval of a generic. After the *Bolar* case, 35 U.S.C. §271(e)(1) addressed both issues by (1) allowing approval by the filing of an Abbreviated New Drug Application (ANDA) which allow the ANDA files to cross-reference data in the master NDA file and (2) exempting from infringement certain activities which were infringement prior to 271(e)(1). Most would agree that the statute exempts the testing of a generic equivalent of a patented pharmaceutical to generate data for submission to the FDA. However, the statute is written in broad terms and as such may provide the exemption to a much broader scope of activities.

When interpreting 35 U.S.C. §271(e)(1) Supreme Court has explained that the exemption from infringement applies to activities reasonably related to the generation of information necessary to obtain FDA regulatory approval for *all FDA regulated products*, not just pharmaceuticals. See 496 U.S. 661, 15 USPQ2nd supra at 1128. See also Chisum, Patents, §16.03[1] for a further discussion of the policies behind the enactment of 35 U.S.C. §271(e)(1). Thus, in a narrow sense the policy behind 35 U.S.C. §271(e)(1) is to promote the marketing of generic versions of a patented FDA regulated product as soon as the patent on the initial name brand drug expires. More generally, the policy behind §271(e)(1) was to create competition among manufacturers of any regulated drug or medical device so that consumers would have more drug options available – hopefully at a lower cost. This policy is achieved by allowing others to gather data needed to obtain FDA marketing approval for competitive products prior to expiration of the patent. Although the name brand pharmaceutical manufacturers lost some market exclusivity via this policy they were appeased somewhat by a portion of the legislation which allowed for the extension of a patent term in certain situations.

The background of 271(e)(1) shows that it was originally intended to provide an exemption in a very limited situation, i.e. to allow a generic drug maker to carry out work to obtain FDA approval on a patented drug. However, the language of the law is much broader. For reasons described below, the statute could be interpreted as providing an exemption to infringement of all patents if the activity was to obtain FDA approval. However, this interpretation might be seen as overly broad because it could provide an exemption to infringement to patents which do not cover products but only methods which would be used only in obtaining FDA approval. Applying the exemption to such patents would render the patent useless – the exemption applies before approval and is not needed after approval. This suggests a narrower interpretation may have been intended. For example, the exemption is applied only if it is needed before and after approval.

The relevant portions of 35 U.S.C. §271

There are subsections (a)–(i) in which 35 U.S.C. § 271 defines patent infringement and states under subsection (a): 'whoever without authority makes, uses, offers to sell, or sells any patented invention, within the United States or imports into the United States any patented invention during the term of the patent therefore, infringes the patent'. Although 35 U.S.C.§271(a) provides broad protection to patent owners, there are exceptions which render certain otherwise infringing acts non-infringing. Specifically, 35 U.S.C. §271(e) provides for one such exception and reads in relevant part as follows:

> It shall not be an act of infringement to make, use, offer to sell, or sell within the United States or import into the United States *a patented invention ... solely for uses reasonably related to the development and submission of information* under a Federal law which regulates the manufacture, use or sale of drugs or veterinary biological products. 35 U.S.C. §271(e)(1).

The terms 'solely' and 'reasonably related' are focused on by courts deciding cases involving the interpretation of §271(e)(1). The cases show a definite trend toward relying on the 'reasonably related' phrase and not the 'solely' wording, meaning that courts are giving a broader interpretation to what activities are covered by the §271(e)(1) exemption.

Is a 35 U.S.C. §271(e)(1) exemption needed?

Section 271 of title 35 is not easy to read and understand. As pointed out by the U.S. Supreme Court: 'No interpretation we have been able to image can transform §271(e)(1) into an elegant piece of statutory draftsmanship'. Notwithstanding such interpreting §271(e)(1) cases do provide sufficient information to set out some criteria for applying the exemption. Some initial issues to resolve in interpretation are whether (1) the activity is of the general type exempted and (2) if collateral activities such as fund raising activities (which are not directly for FDA approval) taint other activities (which are for obtaining approval) to cause a loss of the §271(e)(1) exemption.

When interpreting §271(e)(1) the Supreme Court has held that the phrase 'a Federal law' refers to the entire Food, Drug and Cosmetics Act. 496 U.S. 661 (1990). Thus, the Federal Law referenced under §271(e)(1) is the F, D &C Act, and the statute should be read, as indicated by the Supreme Court, as exempting uses related to the development and submission of information to the FDA. Because so many things must be done to obtain FDA approval of a drug or device this interpretation makes the exemption quite broad.

If an act does not constitute infringement then no exemption to infringement is needed under 271(e)(1). Specifically, it is well settled that only those acts which initially qualify as infringing acts under 35 U.S.C. §271(a) are potentially non-infringing acts under 35 U.S.C. §271(e)(1). See *Telectronics Pacing Systems Inc. v. Ventritex, Inc.*, 982 F.2nd 1520, 25 USPQ2nd (BNA) 1196 (Fed.Cir. 1992). Other acts, such as the dissemination of data for collateral non-FDA related purposes that was generated by the potentially infringing use, do not qualify as infringing acts under 35 U.S.C. §271(a) and therefore do not need the protection of §271(e)(1). Id. See also *Intermedics Inc. v. Ventritex, Inc.*, 775 F.Supp 1269 (N.D.Cal. 1991), *aff'd* 991 F.2nd 808 (Fed.Cir. 1993) and *NeoRX Corp v. Immunomedics, Inc.*, 877 F.Supp 202, 31 USPQ2nd (BNA) 1423 (D.N.J. 1994). This issue is addressed because the patentees in *NeoRX* argued that acts which are non-infringing under §271(a) (such as using data generated to raise money) caused a loss of the exemption for acts that were infringing under §271(a) because the acts in total were not 'solely' related to

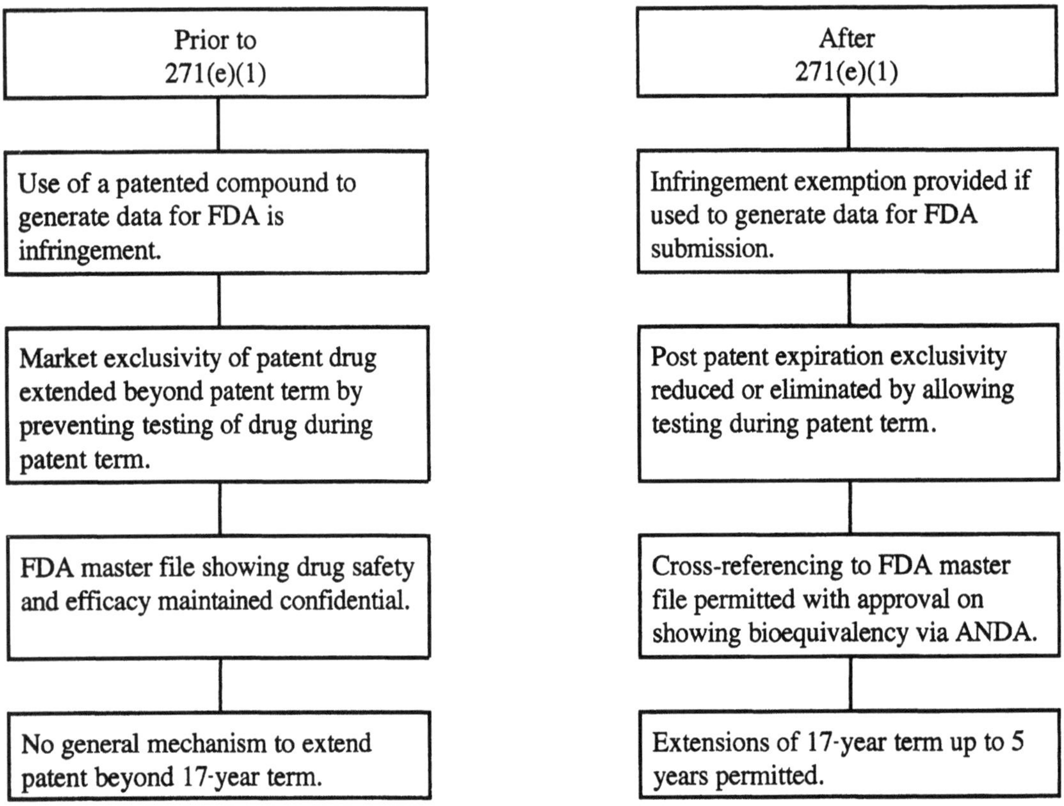

Figure 1. Before and after the enactment of 'The Drug Price Competition – Patent Term Restoration Act of 1984'.

obtaining FDA approval. However, these arguments were not ultimately persuasive in *NeoRX* and have not been accepted by other courts.

The cases discussed immediately above may be of particular relevance to the business development efforts of ABC. These cases support the position that ABC could use data generated from a patented device or process and the use of such data (such as for a business development presentation) would not be found to be an act of infringement – no making, using or selling of the claimed invention so no infringing act under §271(a). Thus, no §271(e)(1) exemption is required.

Thus, in looking at a particular potentially infringing activity under a §271(e)(1), one must first determine whether that activity would even qualify as a 'making, using, selling or offer to sell' of the patented invention, since only those activities are infringing under §271(a) need protection under 35 U.S.C. § 271(e)(1). Dissemination of data to obtain investment funds or other collateral purposes, are not a making, using, selling or offering to sell of the patented invention in question. Such activities are not infringing activities under §271(a) and do not become such merely because the data was generated under the protection of 35 U.S.C. §271(e)(1). Further, such activities do not taint other activities which do require the §271(e)(1) exemption thereby causing a loss of the exemption.

When is the 271(e)(1) exemption needed?

The statute does refer to timing, i.e., does indicate the exemption should be applied before FDA approval is obtained. No exemption is provided after filing a request for approval with the FDA, and no exemption exits after approval is obtained. As described in the background section above, the statute is structured this way for a reason. Specifically, the exemption is provided before FDA approval so that the patent term is not effectively extended by requiring that one wait for the patent to expire before carrying out the lengthy

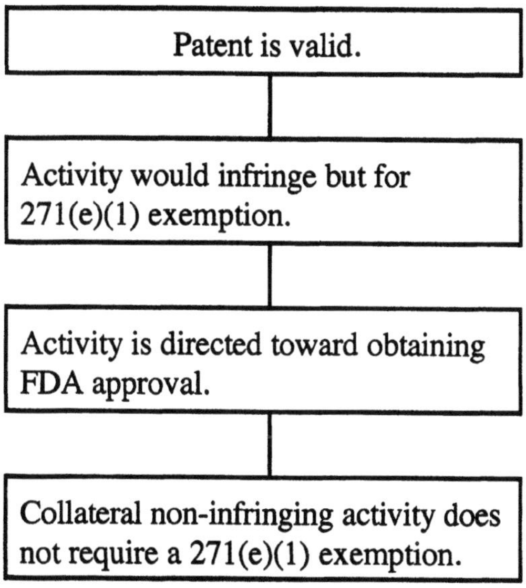

Figure 2. The need for a 271(e)(1) exemption.

activities needed to obtain approval. This suggests that congress's intent was to avoid extending the patent terms vis-a-vis the FDA approval process. It also suggests that congress did not intend to render any class of patent worthless. To fulfill these intentions, it is this author's opinion that the exemption should be applied only to patents which require the exemption before *and* after FDA approval, to avoid infringement. For example, the exemption should apply to a patent on a drug per se, as such would be needed (to avoid infringement) before and after FDA approval. The exemption would also apply to a patent on a method for making a drug, when that method was to be used to manufacture the drug, after FDA approval was obtained. However, the exemption should not apply to an assay for finding a drug, in that such an assay would not be used after the drug was found, and as such; would not be used after FDA approval was obtained. Many companies involved in combinatorial chemistry obtain a significant portion of their patents on assay invention where the assay is used to find pharmaceutically active drugs. If others could use the technology covered by these assay patents by availing themselves of a 271(e)(1) exemption, the patents would be of little value.

What can be a 'Patented Invention' under 271(e)(1)?

The 271(e)(1) exemption would clearly apply to the use of a patented therapeutic to produce information for submission to the FDA to demonstrate bioequivalence with a therapeutic for which FDA marketing approval has already been obtained by a third party. Thus, the 'patented invention' can be a patented drug with FDA approval. However, at present, no cases have denied the exemption to a patented device or process to produce information for submission to the FDA for an as yet unapproved but previously identified FDA regulatable therapeutic – and the *NeoRX* case appears to support the position that such use would qualify for 35 U.S.C. §271(e)(1) protection. Although it is less clear it appears the exemption would also apply to the use of a patented assay to find or process to prepare as yet previously unidentified FDA regulatable therapeutics. Assay patents are often obtained by companies involved in combinatorial chemistry. Current decisions would appear to favor providing the exemption to assay patents. Such a position seems likely because (1) there appears to be no case law which holds that 271(e)(1) would not apply under such facts; and (2) the cases suggest that the courts will give a liberal interpretation to §271(e)(1) in an attempt to enable new drugs to be marketed more cheaply and quickly.

Whether a particular type of activity qualifies for protection under 35 U.S.C. §271(e)(1) depends on the nature of the invention and the type of use. Generally, as the invention used becomes more and more divergent from the FDA regulatable product for which information is being generated and/or the use is to generate information on an as yet unapproved, or further as yet unidentified, FDA regulatable product, it becomes increasingly uncertain as to whether the 35 U.S.C. §271(e)(1) exemption will apply.

Patentees could argue that the patented inventions were being used only to find drugs and that after finding a drug an exemption might apply but not before. However, such arguments are (1) unsupported by any case law; (2) contrary to the overall policy of the legislation as interpreted by the U.S. Supreme Court as being 'to enable new drugs to be marketed more cheaply and quickly'; and (3) against the current trend of Court decisions which have liberally applied the exemption.

After 271(e)(1) was enacted it could be said that the conventional thinking was that it only applied to patented drugs, i.e., that the 'patented invention' referred to in the statute was a patented therapeutic compound approved for use by the FDA. As discussed further below this interpretation is too narrow in that it would not fully serve the statutory purpose of making new drugs available more cheaply and quickly. Thus in interpreting 35 U.S.C. §271(e)(1), it is necessary to determine the full breadth of the term 'patented inventions' under the statute. Interpreting the term 'patented invention' arose in the context of determining whether the patented invention could be a medical device. Courts have held that a 'patented invention' may be a patented medical device that required FDA approval. However, the broader question is whether the phrase 'patented invention' includes any and all patented inventions that might be used to generate the relevant information (i.e., assays used to find therapeutics, machines used to test the compound, software used to prepare the data for submission to the FDA, etc.), or does the phrase 'patented invention' encompass only those inventions which themselves are FDA regulated products (e.g. pharmaceuticals or medical devices)?

While not specifically addressing the broader question the Supreme Court explained in *Eli Lilly v. Medtronic* that the 'phrase "patented invention" in §271(e)(1) is defined to include all inventions, not drug-related inventions alone'. 496 U.S. 15 USPQ2nd supra at 1125, (citing to 35 U.S.C. §100(a) which states that 'When used in this title, unless the context otherwise indicates...[t]he term "invention" means invention or discovery'.). This holding was made in the context of deciding whether the exemption could be applied to a medical device. Thus the holding is not directly on point for applying the exemption to a patented process or for using an assay to find a therapeutic. However, (1) there are no cases indicating that the phrase 'patented invention' does not include inventions such as a process or assay; and (2) the language of the holding as well as the policy reasoning behind the holding indicates the Court wanted the phrase to encompass *all* inventions. Although the cases discussed below do not specifically address this issue they support a conclusion that the patented invention could be any invention by applying the exemption to inventions, other than FDA approved drugs and medical devices.

Courts have found that a 'patented invention' under 35 U.S.C. §271(e)(1) may be: implantable cardiac defibrillators (*Eli Lilly, supra, Telectronics, supra* and *Intermedics, supra*); a plasma sterilization device (*Abtox, Inc. v. Exitron, Inc.*, 888 F.Supp 6, 35 USPQ 2nd (BNA) 1508 (D.Mass. 1995)); a purified preparation of Factor VIII:C (*Scripps Clinic & Research Found. v. Genentech, Inc.*, 666 F.Supp. 1379, 3 USPQ2nd (BNA) 1481 (N.D.C. 1987)); a female condom (*Chartex International PLC v. M.D. Personal Products Corporation*, 1993 U.S. App. LEXIS 20560 (Fed. Cir 1993) (unpublished)); and a transdermal nicotine patch (*Elan Transdermal Inc. v. Cygnus Therapeutic Systems*, 24 U.S.P.Q.2nd 1926 (Fed. Cir. 1992) (unpublished)).

NeoRX Corp. v. Immunomedics, Inc., 877 F.Supp 202, 31 USPQ2nd (BNA) 1423 (D.N.J. 1994), (discussed in greater detail below), further expands the interpretation of a 'patented invention' to include a process for labeling antibodies with radioactive isotopes, which process the defendant employed to produce therapeutic antibodies which it then tested in order to obtain information to submit to the FDA in order to obtain marketing approval for a therapeutic. It is important to note that, contrary to the other cases discussed in this analysis where the patented invention had previously been approved for marketing by the FDA and the accused infringer was intending to market a 'generic' equivalent, such was not the case in *NeoRX*. In *NeoRX*, the patented invention was not an FDA regulatable product in its own right, but only a process which the defendant was using to make, test and obtain quantities of an FDA regulatable product. Thus, *NeoRX* holds that the §271(e)(1) exemption applies even when the patented invention is a process which is used to obtain a product which has not previously been submitted for FDA approval. Such a holding comes with the Supreme Court holding in *Eli Lilly* which held that a patented invention was any invention used to generate information for FDA submission.

The phrase 'patented invention' under 35 U.S.C. §271(e)(1) might ultimately be held to include only those inventions which are FDA regulatable products in their own right, e.g. a patented process for making a drug is not covered under 35 U.S.C. §271(e)(1) when the process itself does not require FDA approval. In making such an interpretation, it would be necessary to argue that the Supreme Court's statement that patented inventions means *all* inventions should not be taken literally. More specifically, it would be necessary to argue that such a literal interpretation would lead to results where seemingly unrelated products and processes could be used and exempted from infringement merely because they were somehow related to the submission of information to the FDA.

Instead, under this interpretation, it would be argued that the Supreme Court meant that only those patented inventions which are themselves actually subject to FDA regulatory approval would be eligible for the safe harbor of 35 U.S.C. §271(e)(1), i.e. all inventions means all FDA regulatable inventions and not inventions used to find, test, or make FDA regulatable drugs or devices. Assays for finding drugs are not in and of themselves regulatable by the FDA. Thus, patentees could argue they are not patented inventions under 35 U.S.C. §271(e)(1) and therefore their use would never qualify under the safe harbor of 35 U.S.C. §271(e)(1). However, the *Eli Lilly* Court's dicta concerning the policy behind 35 U.S.C. §271(e)(1) emphasized that the exemption applies to all inventions regulated under the F, D & C act.

In *Eli Lilly,* the Supreme Court clearly stated that the phrase 'patented invention' for purposes of 35 U.S.C. §271(e) includes all inventions, without any qualification or limitation. Thus, the *Eli Lilly* decision supports an interpretation that all inventions, as long as they are patented, qualify as patented inventions under 35 U.S.C. §271(e)(1), and that their use is therefore non-infringing if the other requirements of the statute are met. Such an interpretation would be completely consistent with respect to the policy behind 35 U.S.C. §271(e)(1). More specifically, such an interpretation would allow others to pursue any and all work needed toward accumulating data for FDA approval so that a competing product could be commercialized quickly after the patent expired. The use of the patented invention would have to be somehow related to the submission of information to the FDA to obtain regulatory approval for an FDA regulated product, making any use of the patented invention of a limited nature. If the §271(e)(1) exemption did not apply when the patented invention was not an FDA regulated product itself (but was nonetheless necessary to generate the kinds of information required to obtain FDA regulatory approval for an FDA regulated product), such would be contrary to the policy behind 35 U.S.C. §271(e)(1) because the patent term could, in some circumstances, be extended. In *Eli Lilly* the Court made its holding based on general policy which was 'to enable new drugs to be marketed more cheaply and quickly'. See *Eli Lilly,* 15 USPQ2nd at 1129.

The *NeoRX* decision supports an interpretation of 35 U.S.C. §271(e)(1) applying the exemption to all patented inventions. As mentioned above, in *NeoRX* the patented invention was a process used to manufacture the FDA regulated product for which information was to be submitted. Under *NeoRX* it did not matter that the patented invention (i.e., the process) was not an FDA regulatable product itself. Furthermore, in *NeoRX* there was no lead product which the accused infringer was using to generate this information. Therefore, in view of *Eli Lilly* and *NeoRX*, the sounder interpretation would appear to be that all patented inventions are to be interpreted as patented inventions under 35 U.S.C. §271(e)(1) exempting their use from infringement, i.e. the 35 U.S.C. §271(e)(1) exemption applies even to patented process and products which themselves do not require FDA approval to market.

The phrase 'patented inventions' as used in 35 U.S.C. §271(e)(1) clearly includes FDA regulatable products for which FDA regulatory approval has already been obtained by a third party. While the risk of a contrary conclusion increases as one moves further away from this core class of patented inventions, it is also likely that patented inventions includes FDA regulatable products for which no previous FDA regulatory approval has been sought or obtained. In view of the *NeoRX* decision, processes used to prepare FDA regulatable products will also qualify as patented inventions. Finally, while even further divergent from the FDA regulatable product, it seems safe to conclude that compositions of matter which are used in such processes also qualify as patented inventions.

Therefore, a court would most likely rule that processes and assays are patented inventions, the use of which is exempted from infringement under 35 U.S.C. §271(e)(1) if the use is 'reasonably related' to obtaining FDA approval. As such, the sole remaining issue is whether their use would meet the remaining requirement of the statute, i.e., be an activity which is 'reasonably related' to obtaining FDA approval.

In Figure 3 the probability of the exemption being applied decreases from top to bottom.

Interpreting 'Solely' and 'Reasonably Related'

35 U.S.C. §271(e)(1) makes otherwise infringing acts non-infringing if such acts are conducted 'solely for uses reasonably related to the development and submission of information' to the FDA. Therefore, if a use of a patented invention is reasonably related to the development and submission of information to the FDA in order to obtain marketing approval for an FDA regulated good, the user of the patented invention will not be liable as an infringer due to the exemption.

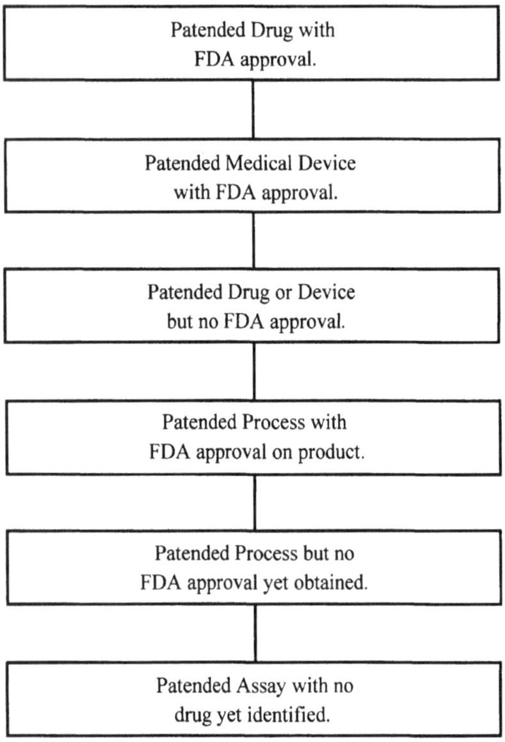

Figure 3. Does the patent cover a 'patented invention' under 35 U.S.C. §271(e)(1).

In one of the first cases interpreting the meaning of 'solely for uses reasonably related to the to the development and submission of information,' the Northern District of California held that 'the use of patented invention is protected so long as that use is solely for purposes reasonably related to meeting the reporting requirements of federal drug laws'. *Scripps Clinic & Research Found. v. Genentech, Inc.*, 3 USPQ2nd at 1493. Under *Scripps* the test is whether the accused infringer can demonstrate that it made and used the patented invention solely for the purpose of meeting FDA reporting requirements. Thus, the *Scripps* test holds that one must look at not only the actual use of the patented invention, but also the subjective intent of the user, i.e. one must ask whether there were other non FDA related collateral reasons that the patented invention was made and used. If there were other purposes behind the activities in question, then activities which would infringe but for an exemption would not qualify for the 35 U.S.C. §271(e) safe harbor exemption.

Cases decided after Scripps do not favor an interpretation which focuses on the purposes of the activities. The *Telectronics Pacing Systems* court held that the dissemination of data for non-FDA related purposes which is obtained from an otherwise §271(e)(1) qualifying use of a patented invention does not render the exemption inapplicable. 25 USPQ2nd at 1200. In coming to this conclusion, which is distinguishable from the *Scripps* decision, the *Telectronics* court cited with approval the *Intermedics, Inc. v. Ventritex Co., Inc.* decision. *Intermedics* explained that in determining whether the use of patented invention is solely for uses reasonably related to the submission of information to the FDA, one should ask whether it would 'have been reasonable, objectively, for a party in defendant's situation to believe that there was a decent prospect that the "use" in question would contribute (relatively directly) to the generation of kinds of information that was likely to be relevant in the processes by which the FDA would decide whether to approve the product?' 775 F.Supp 1269, at 1280. In supporting its position, the *Intermedics* court concluded that Congress intended the test to be objective rather than subjective, and that one should thereby focus on the actual uses of the patented invention and not purposes behind the uses and apply the above test. Id. at 1278. Other contemporary unpublished Federal Circuit decisions have held that one does not look at the purposes behind the uses, but objectively at the uses themselves. See *Chartex* and *Elan Transdermal Ltd, supra.* Subsequent lower court cases have also followed this interpretation. See *NeoRX* and *Abtox, supra.*

The Courts have focused on the phrase 'reasonably related' within 271(e)(1) in that this phrase appears to demonstrate congressional intent. Specifically, the Courts have held that:

> In this regard, the phrase 'reasonably related' communicates congresses' 'intention that the courts give parties some latitude in making judgements about the nature and extent of the otherwise infringing activities they would engage in as they sought to develop information to satisfy the FDA'. *Intermedics*, 775 F.Supp. at 1280.

The Courts have also indicated that if a party were to lose the §271(e)(1) exemption every time a business purpose was detectable in its otherwise infringing activities, the exemption would virtually never be applicable and thus would fail to achieve congresses' objective.

In *NeoRX* there are a number specific factual situations described wherein the court held that the conduct was 'reasonably related' to obtaining FDA approval and thereby afforded the protection of §271(e)(1). For example the court held that the manufacturing of commercial scale batches of product was within the exemption in that scaling up to a commercial production process would be part of obtaining FDA approval for commercial sales. Further, the court held that making product for use at trade shows still fell within the exemptions provided that most of the product made was made in order to generate data for use for submission to the FDA. However, when a party produced product and sent the product to foreign regulatory agencies for approval outside of the United States the court held that such activity was beyond the protection of §271(e)(1).

In evaluating whether a particular use of a patented invention is a non-infringing use under §271(e)(1), one should look at the use objectively, disregarding any subjective purposes behind the use in question. A determination should be made as to whether the accused infringer could reasonably view the use as a use which would contribute relatively directly to the generation of the kinds of information required by the FDA for regulatory approval of an FDA regulated product.

Applications of §271(e)(1) exemptions

A number of courts have evaluated a variety of different uses to determine whether they qualify as protected uses under §271(e)(1). In *Telectronics*, the Federal Circuit held that the demonstration of a potentially infringing defibrillator at a medical conference in order to obtain clinical investigators, even though some non-physicians were able to view the demonstration, qualified as a use reasonably related to the development of submission of information and therefore was a non-infringing activity under 35 U.S.C. §271(e). Id. at 1523.

In *Intermedics,* the Court held that the following activities were non-infringing under 35 U.S.C. §271(e)(1):

(1) manufacturing defibrillators to be used to generate data for the FDA, Id. at 1282;
(2) sales of the defibrillators to U.S. hospitals where the sold devices were used solely in clinical trials, Id. at 1282;
(3) sales of the defibrillators to international distributors, where every such defibrillator was resold to an FDA approved clinical investigator, Id. at 1283;
(4) sales of the defibrillators to German investigators conducing clinical trials in Germany, where such information is potentially to be submitted to the FDA, Id. at 1284;
(5) demonstrating defibrillators at trade shows and medical conferences in order to identify potential clinical investigators even though many of the people at the trade show could not be clinical investigators. Id. at 1287.

The *NeoRX* court held that the following specific activities were non-infringing under 35 U.S.C. §271(e)(1):

(1) the production of commercial scale quantities of a therapeutic antibody using a patented process, since the FDA requires a demonstration of capability of manufacturing commercial scale batches and it is uncertain how much data the FDA will require for approval, Id. at 206;
(2) the shipment of infringing product to foreign clinical investigator where data was to intended to be submitted to FDA, Id. at 208;
(3) shipment of product to foreign commercial distributor where distributor was to conduct clinical tests which would be submitted to the FDA, Id. at 211.

The *NeoRX* Court also held that the following activities would not qualify as non-infringing under the 35 U.S.C. §271(e)(1) exemption:

(1) the shipment of infringing product to a foreign clinical investigator where there was no evidence that the data generated from the investigator's use was to be submitted to the FDA, Id. at 209;
(2) the shipment of infringing product to foreign regulatory agencies in order to obtain foreign marketing approval, Id. at 207.

The *NeoRX* case was a District Court decision. There is no Appellate Court decision on point. Thus, the law remains unsettled as to whether a qualifying use of a patented invention would include the use of a patented assay or process for using such to identify as yet previously unknown FDA regulatable products. However, in arguing that the use of a patented assay and a process for using the assay to obtain information on a previously unknown FDA regulatable product should fall within the safe harbor of 35 U.S.C. §271(e)(1), one could rely on an extension of the *NeoRX* case. Specifically, it could be argued that in using the patented invention, such as an assay or process of using the assay, to identify as yet unknown FDA regulatable products such as therapeutics, one is still just using the patented inventions to generate information for submission to the FDA. Furthermore, the case law supports the position that exempting such a use from infringement under 35 U.S.C. §271(e)(1) would be consistent with the policy of 35 U.S.C. §271(e)(1), since the as yet unidentified invention could be marketed immediately upon expiration of the patent if such use were permitted. Such a holding would be compatible with the Legislative history of §271(e)(1) and the *Eli Lilly* Supreme Court holding – such a holding arguably promotes the development of more drugs and make those drugs available in a shorter time frame.

However, the owners of assay and process patents may argue that the exemption does not apply in this situation, as there is no FDA regulatable product in existence. Therefore, it would be argued that the use of the patented invention could not be to generate information for submission to the FDA, since there is no product on which to submit the information. The current trend of decisions on 271(e)(1) does not support such a position. Regardless of the trend of the decisions, a rational approach to deciding whether to apply the exemption would be to determine if applying the exemption would essentially eliminate the value of the patent. If the exemption would not be needed after FDA approval is obtained, it should not be given before FDA approval is obtained. Otherwise, applying the exemption eliminates any use for the patent.

How much information about a particular FDA regulatable product must be in existence for the FDA regulatable product to be known is unsettled. On the one hand, if a specific potential therapeutic has been prepared and reported in the literature, as well as methods of how the therapeutic was obtained and the specific assay or process used to produce the therapeutic, then that therapeutic is known and the use of the patented process and assay inventions will fall within the 35 U.S.C. §271(e)(1) exemption, assuming the other requirements of the statute are met. Because there are no cases dealing with assays for therapeutics there is less certainty that the exemption will apply where one only has identified a class of compounds to be tested. In this situation, the FDA regulatable product would be considered unknown, since the therapeutic has not been previously produced in isolateable form and one is practicing the invention without sure knowledge that the potentially FDA regulatable product can even be produced. However, there do not appear to be cases suggesting that the exemption should not be applied under such facts. Further, the trend of the cases is to liberally apply the 271(e)(1) exemption. Finally, the following specific uses are illustrative of what in all likelihood will and will not qualify for protection under 35 U.S.C. §271(e)(1).

Qualifying uses

(1) Using patented assays and processes to generate sufficient quantities of potential therapeutics to provide safety and efficacy data; and
(2) Using patented assays and processes to generate sufficient quantities of potential therapeutics to demonstrate scale up potential.

Non-qualifying uses

(1) Making, using or selling therapeutics that are produced using the claimed invention during its term for a profit; and
(2) Testing the therapeutics to generate data that has no reasonable potential to be submitted to the FDA, e.g. data that is generated solely to obtain foreign regulatory approval.

In addition to the above protected and non-protected uses, collateral uses of the information generated by the above protected uses would also be permitted. Such collateral uses would include:
(1) dissemination of the information in the form of journal articles;
(2) dissemination of the information to potential investors; and
(3) dissemination of the information to business partners.

Subjective criteria

In trying to determine if a 271(e)(1) exemption will apply it will be necessary to closely analyze (A) the statute; (B) case law interpreting the statute; and (C) how the particular facts relate to each. However, in all probability such an analysis will not provide a definitive answer and subjective criteria will effect the judge's decision. These criteria may include (1) the relative size of the parties; (2) the advocacy skills of their representatives; (3) the need for new drugs in the area; (4) the current cost of drugs in the area; (5) the complexity of the technology; and (6) the time available to the judge.

In general judges do not like patent cases because they are often legally and technically complex and can take up large amounts of time. The biggest part of the time consuming, document intensive aspects of a patent case relates to the issue of patent validity. If the exemption under 271(e)(1) is liberally applied validity and infringement issues can become moot. Specifically, a patent might be assumed valid and infringed and the case would still have no merit and could be readily dismissed if the otherwise infringing activity were exempt under 271(e)(1). Such might become a significant factor in liberally applying the exemption in complex patent cases within an overcrowded court system.

Conclusions

Companies involved in combinatorial chemistry may wish to use the patented technology of their competitors while claiming a 217(e)(1) exemption – this exemption can provide a broad and powerful shield. Combinatorial chemists in a company attempting to use the 271(e)(1) exemption as a shield should be careful to direct their actions to activities seen as being reasonably related to the generation of information needed to obtain FDA approval. Indications in company memos and notebooks that an experiment or project is being carried out for the ultimate purpose of obtaining FDA approval could later be found to be useful evidence that the 271(e)(1) shield should be applied.

In making decisions of what inventions should be protected by a patent, companies involved in combinatorial chemistry should consider the possibility that their competitors will attempt to avail themselves of the 271(e)(1) exemption. Although it is not presently clear how broad the exemption will be applied, court decisions are trending toward a broad interpretation of the exemptions. Combinatorial chemists should keep the following points in mind when considering what invention to patent.

(1) Assays and processes are likely to be viewed as patented inventions under 35 U.S.C. §271(e)(1); i.e., a 'patented invention' is not only a patented therapeutic or medical device with FDA approval.
(2) If the exemption is not needed after FDA approval is obtained, the exemption should not be applied before FDA approval or such would completely destroy the value of the patent.
(3) Use of these patented inventions will be protected as non-infringing under §271(e)(1) as long as such uses are reasonably viewed objectively as relatively directly related to the generation of information needed by the FDA to approve an already identified FDA regulatable product, such as a known therapeutic.
(4) The applicability of the exemption is less clear if the patented inventions are used to identify as yet unknown FDA regulated products, (i.e., find a new therapeutic) even though there are no cases supporting this narrow interpretation.
(5) The trend of the courts and the subjective criteria support liberally applying 271(e(1) exemptions.

In sum, it seems well settled that using a patented invention to generate bioequivalency data for submission to the FDA on a previously FDA approved FDA regulatable product will qualify for protection under 35 U.S.C. §271(e)(1). As one diverges from this core protected class of uses, the probability that 35 U.S.C. §271(e)(1) will not apply increases. Members of the bar, in general, have a narrow view of how broadly the exemption should be applied. Most attorneys are of the opinion that the exemption is limited to bioequivalency testing of a patented drug. Such a limited interpretation does not comport with recent decisions or the overall trend of the Courts on interpreting 271(e)(1). Due largely to a failure to appreciate decisions such as *NeoRX* and *Eli Lilly* it will be argued that the likelihood of fitting within the 35 U.S.C. §271(e)(1) exemption decreases where: (1) there is no FDA approved product in existence at the time the activity in question takes place; (2) the information needed for approval is arguably not needed as no such product exists; and (3) the patent covers an invention such as a process which itself does not need FDA approval. Although these factors should be considered, even when all three factors exist Courts have indicated a willingness to liberally apply the exemption in order to provide the general policy behind §271(e)(1) i.e., allow for the faster development of and sale of more drugs to the public at lower costs.

Appendix

The most relevant case giving a narrow interpretation to §217(e)(1) exemptions is *Scripps Clinical Research Found. v. Genentech, Inc.*, 666 F.Supp. 1379, 3 USPQ2nd (BNA) 1481 (N.D.C. 1987) and that case – although not explicitly overruled – has not been followed.

(1) The Supreme Court has held that the general policy behind §271(e)(1) is 'to enable new drugs to be marketed more cheaply and quickly'. Allowing

others to use a patented assay or process to develop drugs for FDA approval would serve that general policy.

(2) The Supreme Court has held that the exemption applies to *all* inventions subject to federal approval regulations. Although this holding was in the context of applying the exemption to medical devices (in addition to drugs) the tenor of the opinion suggests that the Court would apply the exemption to processes for making drugs and lower courts have since applied the exemption to processes.

(3) Acts such as using data to encourage investments are not infringing acts under §271(a) – and such acts do not result in a loss of the exemptions for other acts which would be infringing under §271(a) but for the exemption.

(4) Cases tend to emphasize the domination of the phrase 'reasonably related' as opposed to the word 'solely' in order to find that uses of patented inventions are entitled to a §271(e)(1) exemption.

(5) Those wishing to come within a 271(e)(1) exemption should be careful to direct all activities covered by an assay or process patent toward obtaining FDA approval of a therapeutic – i.e. have as its ultimate goal – obtaining an NDA on one or more therapeutics.

M.R. Pavia and W.H. Moos (eds.), Annual Reports in Combinatorial Chemistry and Molecular Diversity, Vol. 2, 71–92.
KLUWER/ESCOM

Advances in diversity profiling and combinatorial series design

Dimitris K. Agrafiotis*, James C. Myslik & F. Raymond Salemme
3-Dimensional Pharmaceuticals, Inc., 665 Stockton Drive, Exton, PA 19341, U.S.A.

Key words: analog design, combinatorial chemistry, experimental design, high throughput screening, QSAR, series design, similarity

Summary

Rapid advances in synthetic and screening technology have recently enabled the simultaneous synthesis and biological evaluation of large chemical libraries containing hundreds to tens of thousands of compounds, using molecular diversity as a means to design and prioritize experiments. This paper reviews some of the most important computational work in the field of diversity profiling and combinatorial library design, with particular emphasis on methodology and applications. It is divided into four sections that address issues related to molecular representation, dimensionality reduction, compound selection, and visualization.

Introduction

There is only atoms and space. The rest is opinion.
Democritus

Diversity is a much sought after commodity these days. Governments, universities and corporations alike espouse the benefits of biological, cultural, intellectual and commercial diversity. The problem is that no two individuals or organizations can seem to agree upon a common definition for this term. What one person perceives a being diverse, another sees as being homogeneous. Maybe that's the point. Diversity, after all, is based upon human perception and is, therefore, inherently subjective. It is a quality rather than a quantity. However, with the advent of combinatorial chemistry and high-throughput screening as tools for massively parallel drug discovery, computational chemists have found themselves in need of quantitative measures that can be used to distinguish between good and mediocre experiments.

Historically, drug discovery has been based on a serial and systematic modification of chemical structure guided by the 'similar property principle', i.e. the assumption that structurally similar compounds tend to exhibit similar physicochemical and biological properties. New therapeutic agents are typically generated by identifying a lead compound, and creating variants of that compound in a systematic and directed fashion. The first phase of this process, known as *lead generation*, is carried out by random screening of large compound collections, such as natural product libraries, corporate banks, etc. The second, known as *lead optimization*, represents the rate-limiting step in drug discovery, and involves the elaboration of sufficient SAR around a lead compound and the refinement of its pharmacological profile. Prior to the arrival of combinatorial chemistry, this process involved a simple prioritization of synthetic targets based on preexisting structure-activity data, synthetic feasibility, experience, and intuition.

However, revolutionary advances in synthetic and screening technology have recently enabled the simultaneous synthesis and biological evaluation of large chemical libraries containing hundreds to tens of thousands of compounds. These tools haven't changed the fundamental way in which drugs are discovered, but they have changed the way in which chemists contemplate experiments. With the expansion of our knowledge base of solid and solution-phase chemistry and the continuous improvement of the underlying robotic hardware, combinatorial chemistry has moved beyond its traditional role as a source of compounds

* To whom correspondence should be addressed.

for mass screening, and is now routinely employed in lead optimization and SAR refinement. This has led to the conceptual division of combinatorial libraries into 1) *exploratory* or *universal* libraries which are target-independent and are designed to span a wide range of physicochemical and structural characteristics, and 2) *focused* or *directed* libraries which are biased against a specific target, structural class, or known pharmacophore.

Most work to date has focused on lead generation, using molecular diversity as a means to design and prioritize experiments. The need for careful experimental design became apparent as the initial euphoria that accompanied the advent of this exciting new technology gave way to the realization of its practical limitations. Indeed, it is clear that at least for now the number and types of molecules that can be synthesized using parallel synthesis still represents a small fraction of all the compounds of potential therapeutic interest, and that the quality of the biological response degrades rapidly with throughput. Thus, the synthesis and testing of a chemical library should ideally be planned in a way that renders maximum information about the underlying biological target. To achieve this goal, it is no longer sufficient to examine the properties of individual compounds, but it is also important to assess the *collective* quality and information content of our libraries. In this respect, molecular diversity can be viewed as one design strategy for maximizing the hit-rate of high-throughput screening experiments. It represents a generalization of the concept of molecular similarity from individuals to collections, and its effective use can reduce the redundancy and cost of experiments and substantially increase the odds of discovering new drugs. It is closely related to molecular similarity, structure-activity correlation and statistical series design, which are thoroughly reviewed in numerous texts [1,2]. The scale and complexity of the problem is daunting for all but the most gifted chemists, and makes it ideally suited to computation.

This paper reviews recent advances on computational aspects of molecular diversity, with particular emphasis on methodology and applications. There are probably as many different definitions of diversity as the number of researchers who are active in the field, and it is our belief that theoretically the subject borders on religion. However, many of the concepts described herein are now routinely employed in combinatorial library design, and the choice of methods has definitive and measurable practical implications. The article is divided into three main sections that address issues related to molecular representation, compound selection, and visualization. The reader is also referred to various reviews [3–5,70].

Molecular encoding

Rabbi Raditz of Poland was a very short rabbi with a long beard, who was thought to have inspired many pogroms with his sense of humor. One of his disciples asked,

'Who did God like better – Moses or Abraham?'

'Abraham,' the Zaddik said.

'But Moses led the Israelites to the Promised Land,' said the disciple.

'All right, so Moses,' the Zaddik answered.

'I understand rabbi, it was a stupid question.'

'Not only that, but you are stupid, your wife's a meeskeit, and if you don't get off my foot you are excommunicated.'

Woody Allen, Getting Even.

Conceptually, the problem of quantifying molecular diversity involves two parts: the first is the definition of chemical distance, and the second is the selection of a representative set of compounds from a (typically) much larger collection. This section reviews a number of methods that have been proposed to encode molecular structures in a way that is suitable for numerical processing. It is organized in three parts, which discuss descriptors derived from the topology, three-dimensional structure, and physicochemical and electronic properties of the molecules.

Two-dimensional descriptors

Molecular connectivity indices

When asked to describe a molecule, a chemist will instinctively reach for a pen and some paper and draw a picture. The picture shows how constituent atoms are interconnected to form the molecule (i.e. the topology of the molecule). Mathematically, this is equivalent to a molecular graph. Molecular connectivity or topological indices are numerical values calculated from certain invariants (characteristics) of a molecular graph [13,14] which encode features such as number of atoms, branching, ring structures, heteroatom content, and bond order. They are attractive for quantifying molecular diversity because they are inexpensive to compute, and have been validated through years of use in the field of structure-activity correlation. In particular, the widespread availability of Kier and Hall's Molconn-X program [15], which calculates indices based on connectivity, shape, sub-graph counts, topological equivalence, electrotopological state and information content, has done much to promote the use of molecular connectivity indices.

Topological indices may be classified into four groups based on their logical derivation:

1) Those derived from the adjacency matrix: the total adjacency index, the Zagreb group indices, the Randic connectivity index, the Platt index, the compatibility code, and the largest eigenvalue index;
2) Those based on the topological distance matrix, including the Wiener index, the polarity number, the distance sum, the Altenburg polynomial, the mean square distance, the Hosoya index, and the distance polynomial;
3) Centric indices, including the generalized graph center;
4) Information-theoretic indices, including the Shannon index, the chromatic information index, the orbital information index, the topological information superindex, the electropy index, and the Merrifield and Simmons indices [13,14].

Binary descriptors

Binary descriptors include substructure keys and hashed fingerprints. Substructure keys encode molecular structures as bitstrings, each binary digit of which indicates the presence or absence of a selected structural feature or pattern. Typical target features might include the number of occurrences of a particular element (e.g. the presence of at least 1, 2 or 3 nitrogen atoms), electronic configurations or atom types (e.g. sp^2 nitrogen or aromatic carbon), common functional groups such as alcohols, amines etc., and ring systems. Features that are rare enough not to merit an individual bit, yet extremely important when they do occur, are assigned a common bit which is set if any one of the patterns is present in the target molecule (a 'disjunction'). Substructure keys were originally developed for rapid searching of large databases, but have also proven effective in similarity applications. Generating substructure keys is time-consuming, requiring a substructure search for each target pattern in every molecule in a database. However, once generated, they allow a database to be searched by means of Boolean operations upon the keys; a process performed very rapidly by digital computers.

Database designers tailor substructure keys to minimize molecule retrieval time. As such, the choice of encoded structural features tends to be application-specific. For example, keys employed in drug databases encode functional groups of particular interest to medicinal chemists, while those used in databases of organometallic compounds contain features related to metal-carbon bonding. Despite this specificity, substructure keys contain sufficient information about the molecular structures to permit meaningful similarity comparisons.

Like structural keys, hashed fingerprints are bitstrings derived directly from the connection table and were developed primarily for database searching. They differ from structural keys in that they do not depend on pre-selected structural fragments to perform the bit assignment. Instead, every pattern in the molecule up to a predefined path length is systematically enumerated. The resulting set of patterns serves as input to a hashing algorithm that turns 'on' a small number of bits at pseudo-random positions in the bitstring. Because the number of possible patterns far exceeds the length of the fingerprint, many patterns are mapped onto a single bit. In practice, this does not pose a problem for database searching. Every bit that is set in a fingerprint of the target pattern will also be set in that of a molecule which contains the pattern, making database screening deterministic and fast. For similarity comparisons, this is not the case. While two different molecules may have exactly the same fingerprint, the probability of this occurring is extremely small for all but the simplest cases. Experience has shown that the similarity between two fingerprints is a good indicator of the similarity between the two structures. Hashed fingerprints have the additional characteristic that as structures become more complex the density of encoded information increases. A number of studies have shown that fingerprints and substructure keys are equally effective for the purpose of diversity analysis (see section Descriptor Validation below).

While most of the descriptors require an enumerated structure (i.e. the full connection table of a molecule), for combinatorial libraries, this is not strictly necessary. Downs and Barnard [16] have recently presented an elegant method to compute molecular fingerprints based on reaction precursors, using techniques developed for Markush structure handling in chemical patents.

A number of similarity metrics have been proposed for binary descriptors [6]. The most frequently used ones are the normalized Hamming distance:

$$H = \frac{|\mathrm{XOR}(x, y)|}{N} \tag{1}$$

where x and y are two binary sets (encoded molecules), XOR is the bitwise exclusive OR operation (a bit in the result is set if the corresponding bits in the two operands are different), and N is the number of bits in each set. The result, H, is a measure of the number of bits that are dissimilar in x and y; the Tanimoto or Jaccard coefficient:

$$T = \frac{|\mathrm{AND}(x, y)|}{|\mathrm{IOR}(x, y)|} \tag{2}$$

where AND is the bitwise AND operation (a bit in the result is set if both of the corresponding bits in the two operands are set) and IOR is the bitwise inclusive OR operation (a bit in the result is set if the either of corresponding bits in the two operands are set). The result, T, is a measure of the number of substructures shared by two molecules relative to the ones they *could* have in common; and the Dice coefficient:

$$T = \frac{2|\mathrm{AND}(x, y)|}{|x| + |y|} \tag{3}$$

Another popular metric is the Euclidean distance which, in the case of binary sets, can be recast in the form:

$$E = \sqrt{N - |\mathrm{XOR}(x, \mathrm{NOT}(y))|} \tag{4}$$

where $\mathrm{NOT}(x)$ denotes the binary complement of x, and the expression $|\mathrm{XOR}(x, \mathrm{NOT}(y))|$ represents the number of bits that are identical in x and y (either 1's or 0's). The Euclidean distance is a good measure of similarity when the binary sets are relatively rich, and is mostly used in situations in which similarity is measured in a relative sense. Of all the indices defined above, Tanimoto is perhaps the most commonly used. As we will see below (section Descriptor Validation), these simple binary descriptors, when used with an appropriate clustering methodology, have been surprisingly successful in discriminating active from inactive compounds.

Molecular holograms

Molecular holograms are an extension of hashed fingerprints that are based on fragment counts rather than simple yes/no information alone. As such, they are encoded as relatively short vectors of integers rather than bitstrings. As a consequence, much more information about fragment patterns in the molecule is retained upon hashing. Holograms can also incorporate structural information that ordinary fingerprints can not, such as branching, cyclic structures, hybridization patterns and chirality [74]. They have proven quite useful in QSAR studies [75] and once a QSAR model is in hand, can be used to screen databases of compounds which are likely to be active.

Atom pairs and topological torsions

Atom pairs and topological torsions are two related types of descriptors which represent another attempt to eliminate the subjectivity inherent in substructure keys [17]. Atom pairs are patterns of the form $a_i - d - a_j$, where a_i and a_j are the types of atoms i and j, respectively, and d is the topological distance between the atoms (the number of bonds along the shortest path connecting these atoms). A molecule with N atoms has $N(N-1)/2$ atom pairs, although some of them may not be unique. Topological torsions take the form $a_i - a_j - a_k - a_m$, where i, j, k and m are sequentially bonded atoms, and a_i is again the type of the i-th atom. Originally, atoms were assigned types based on atomic number, number of neighbors and number of π-electrons. More recent studies have utilized physicochemical properties [18] and geometric features [19] as well. Physicochemical atom pairs include binding properties, atomic log$P$ contributions and partial atomic charges. Binding properties categorize atoms into seven classes: anions, cations, neutral hydrogen bond donors, neutral hydrogen bond acceptors, polar atoms, hydrophobic atoms, and other. Geometric atom pairs are similar to the regular atom pairs with the exception that the topological distance is replaced by the through-space distance of the corresponding atoms in some low-energy conformation of the target molecule. Geometric and topological atom pairs are equally effective in similarity searching, but the new generation of descriptors seem to perform worse than the original ones in their overall ability to discriminate biologically active from inactive compounds. As one would expect, which set does better than another varies greatly from probe to probe, and is very difficult to predict *a priori*.

The similarity between two structures described by atom pairs or topological torsions is measured by:

$$s(i, j) = \frac{\sum_{k=1}^{K} \min(f_{ik}, f_{jk})}{0.5\left[\sum_{k=1}^{K}(f_{ik} + \sum_{k=1}^{K} f_{jk}\right]} \tag{5}$$

where f_{ik} is the count of the k-th descriptor in the i-th structure, and K is the union of all unique descriptors in i and j. This index ranges from 0 to 1, with 1 indicating complete identity and 0 indicating that the two structures have nothing in common.

Atom layers

Martin et al. [8] developed the concept of atom layers in an attempt to account for the topological distribution of chemical features around a combinatorial core that are believed to play a critical role in receptor binding. Atom layers are based upon the premise that atoms of a substituent which are close to the point of attachment to the core contribute differently to binding than those that are more distant. They are constructed by summing a given property over all atoms in a substituent at a given number of bonds distant from the attachment point. Properties considered by the Chiron group were atomic radius, acidity, basicity, the ability to serve a hydrogen bond donor or acceptor, and aromatic character. The similarity between two substituents was computed by comparing the corresponding atom layer tables element by element, and dividing the sum of the minimum by the sum of the maximum values in each cell.

2D autocorrelation vectors

Moreau [20] has proposed an autocorrelation function to encode the topology of a molecular graph:

$$A(d) = \sum_{i,j} p_i p_j \qquad (6)$$

where p_i and p_j represent the values of an atomic property at atoms i and j, respectively, and d is the topological distance between the two atoms measured in bonds along the shortest path. The function has the useful property that no matter how large and complex the molecule, it can be encoded in a fixed-length vector of small rank. Typically, only path lengths of 2 to 8 are considered. Atomic properties that have been encoded using this method include volume, electronegativity, hydrogen bonding character and hydrophobicity. Originally, a separate autocorrelation vector was computed for each property, and the resulting set was reduced into a smaller number of variables using principal component analysis.

Topological autocorrelation vectors were also used by Gasteiger [21] as input to a Kohonen network which successfully separated dopamine from benzodiazepine receptor agonists, even when these compounds were embedded in a large and diverse set of chemicals extracted from a commercial supplier catalog. Gasteiger [22] has also extended the concept to three dimensions, by introducing a spatial autocorrelation vector based on properties measured on the molecular surface. These spatial autocorrelation vectors were used to model the activity of 31 steroids against the corticosteroid binding globulin, and the cytosolic Ah receptor activity of 78 polyhalogenated aromatic compounds. These descriptors were also found to be effective in describing the diversity of combinatorial libraries, also through the use of Kohonen networks (see section Self-Organizing Maps below) [23].

B-Cut values

B-Cut values were developed by Pearlman [24] as an extension of Burden's concept of molecular identification numbers [25] to multiple dimensions. Burden represented the hydrogen-suppressed connection table of a molecule as a symmetric NxN matrix in which atomic numbers were placed in the diagonal elements and the off-diagonal elements were assigned a value of 0.1 times the nominal bond order if the corresponding atoms were bonded or 0.001 if not. An additional score of 0.01 was added to the (off-diagonal) terminal nodes. A molecular identification number was then calculated from the two lowest eigenvalues of the matrix. Rusinko and Lipkus used molecular identification numbers for similarity searching of a 60 000-membered subset of the CAS registry. They found this method compared well with results obtained from an established similarity searching procedure.

Pearlman broadened this approach to include properties deemed significant to protein-ligand binding. Three classes of matrices were constructed placing atomic charge, polarizability and hydrogen bonding characteristics in the diagonal elements and a combination of interatomic distances, overlaps, computed bond orders in the off-diagonal elements. A six-dimensional property space was then defined using the lowest and highest eigenvalues (B-Cut values) of a representative matrix from each of these three classes. Optimal combinations of on- and off-diagonal properties were selected by their ability to produce a uniform distribution of molecules in the property space, as determined by a χ^2 criterion. The resulting six-dimensional space is small enough to permit diversity analysis based on partitioning (binning). This is described in greater detail below (section Partitioning Techniques). Pearlman concluded that B-Cut values based solely on the connection table were proven satisfactory for most diversity profiling tasks.

Three-dimensional descriptors

3D structural keys

Geometric structural keys are direct analogues of topological (two-dimensional) substructure keys in three dimensions. They are binary sets tailored to minimize compound retrieval time from three-dimensional molecular databases [26,27]. Typically, the keys encode the Euclidean or angular distance between pairs of selected features such as atom types, centroids of aromatic rings, ring normals, or attachment points of functional groups. The distance between the members of the pair is divided into a fixed number of ranges and a bit is assigned for each range. In forming a key, if a molecule contains a pair of selected features, the distance between the members of the pair is calculated and the bit corresponding to the range into which the distance falls is set.

The success of 3D structural keys, or any three-dimensional descriptors, depends upon their ability to account for the possibility of multiple molecular conformations. Early implementations of three-dimensional molecular database systems stored molecules as single low-energy conformations, determined experimentally through X-ray crystallography or calculated using a fast structure-generation software package [28]. Later implementations employed conformational search procedures, but, in the interest of speed, these searches were relatively crude and did not rule out highly improbable conformations. Other problems associated with 3D structural keys are poor representation of shape and chirality, and the limit to the number of features that can be encoded in a finite length string. Shape and chirality are concepts that are undefined in the single dimension defined by a pair of features. To overcome these limitations, several groups have developed a related class of descriptors defined by sets of three or four selected features, known as 3D pharmacophore keys.

3D pharmacophore keys

Pharmacophore keys, introduced by Sheridan and coworkers [29], are 3D structural keys which incorporate features of functional importance to macromolecular recognition including hydrogen bond donors and acceptors, centers of positive charge, aromatic ring centers and hydrophobic centers. A pharmacophore is defined as a combination of 3 or 4 such loci, forming a triangle or tetrahedron, respectively, and is characterized by the set of distances between the loci. As with structural keys, the distances are divided into ranges, and a bit is assigned to each range.

As is the case for structural keys, pharmacophore keys can be readily extended to account for multiple conformations. Additionally, because pharmacophores are two- and three-dimensional objects, they are able to capture information on molecular shape and chirality. Three-point pharmacophore keys also lend themselves well to visualization *via* three-dimensional scatter plots (see section Visualization without Dimensionality Reduction below). Sheridan's original work has been extended by a number of groups, most notably those at Chemical Design [27], Rhone-Poulenc [30], and Abbott [10]. Davies and Briant [31] have employed pharmacophore keys for similarity/diversity selection using an iterative procedure that takes into account the flexibility of the compounds and the amount of overlap between their respective keys (see section Boolean Logic).

Molecular hashkeys

Defining and measuring the three-dimensional surface properties and similarities of molecules are also the key to MetaXen's approach for predicting biological properties of molecules as well as in determining the diversity of a population of molecules. Since determining pairwise surface similarity measurements for large sets of molecules is computationally prohibitive, Sage and co-workers have extrapolated the similarity measure to a novel representation of molecular surface properties which they called a molecular hashkey [76]. A molecular hashkey is a real-valued vector of fixed dimension, that is used to represent information about the surface properties of a molecule. The molecular hashkey is much smaller than a complete 3D surface representation of the molecule. The term 'hashkey' is borrowed from computer science, and represents a compact numerical representation of an object that is used to solve indexing problems by storing objects using their hashkeys as memory addresses. A molecular hashkey has the property that molecules with similar hashkeys will appear similar based on observation of their surfaces. Molecules with identical surface properties will have identical hashkeys, independent of the underlying atomic scaffolding. Given a molecule M, its N-dimensional hashkey ($H_1, H_2, \ldots, H_N$) is computed by calculating its molecular surface similarity to a set of N basis molecules. The basis molecules are in arbitrary fixed conformations. M is flexibly aligned to each B_i of the basis molecules ($B_{1\ldots N}$) to maximize molecular similarity, and the best match yields the similarity value that becomes H_i. The molecular hashkey has been used successfully in combination with machine-learning techniques for developing predictive models of biological properties of molecules. The hashkey technique has also been used to create diverse subsets of molecules from a starting population by maximizing the molecular hashkey distance between the molecules in the subset.

Physicochemical and electronic descriptors

Physicochemical properties

Physicochemical properties have long been used to develop structure-activity relationships. They quantify a large number of molecular characteristics known to determine the transport and binding of a drug to its target. It is natural, then, that the first attempts to quantify molecular diversity were based upon physicochemical properties. These properties can be calculated using standard molecular modeling and quantum mechanical packages. They include the number of filled orbitals, HOMO and LUMO energies, standard deviation of partial atomic charges and electron densities, dipole moment, ionization potential, heat of formation, total energy, molecular weight, octanol-water partition coefficient (logP), molar refractivity, van der Waals volume and surface area, and many others. Molecular property descriptors have been used for diversity profiling by Willett et al. [6,7], Martin et al. [8], Lewis et al. [9], Brown et al. [10,11], and many others. They have been extensively reviewed by Kubinyi [12].

Electronic fields

Cramer and co-workers [33] have developed descriptors for combinatorially generated molecules based upon steric field methods employed in 3D-QSAR [32]. Their procedure attempts to find a representative conformation for each substituent group pendant upon a particular point of variation of a combinatorial template. The process begins with a low energy conformation generated by a model-building routine, which is then fitted as a rigid body onto the combinatorial template using least-squares minimization. The torsional angles of the rotatable bonds within the substituent are then sequentially adjusted, starting from the bond closest to the template, using a simple set of topological precedence rules. Once aligned, the steric field of the substituent group is computed using a CoMFA-like approach. Two compounds may be compared, for example, by calculating the root-square-differences in steric field values summed over all lattice points in the CoMFA region.

These descriptors have been used to classify 736 commercially available thiols into 231 bioisosteric clusters, consistent with results obtained using molecules encoded as 2D fingerprints and compared with the Tanimoto similarity coefficient. While such an alignment procedure has utility in comparing members of a single combinatorial library, it is unclear how it could be applied to comparison of compounds which belong to different libraries, to heterogeneous compound collections, or to libraries having a variable template.

Affinity fingerprints

Affinity fingerprints represent an entirely different class of molecular descriptor than those previously presented. They are based upon the measured affinity of a molecule for a set of target proteins rather than the structure of a molecule itself. The resulting vector of affinities – the affinity fingerprint – may be used in the same way as structure-derived descriptors to discern similarities between molecules and to quantify molecular diversity. This functional approach was pioneered by Terrapin [34], based upon the company's fluorescent polarization high-throughput screening technology. The significance of the results depends critically upon the selection of an appropriate basis set of proteins. The set of proteins must be able to recognize a wide variety of organic compounds and its members binding specificities must be uncorrelated. A systematic analysis of several hundred candidates resulted in a set of 18 proteins, which are now used routinely at Terrapin for screening new compound collections.

Affinity fingerprints provide an empirical way to assess the molecular diversity of a chemical library. Kauvar [34], for example, suggests that an estimate of the diversity of a given set of compounds can be determined based on the maximum separation and most frequently occurring distance between two affinity vectors in the collection. Although this is a rather qualitative measure, the mathematical nature of the affinity fingerprint makes possible more quantitative measures of diversity (see section Compound Selection). From the standpoint of drug design, affinity fingerprints have one significant limitation: they do not have the ability to predict the affinity profile of an untested class of compounds. It is, however, conceivable that it may be possible to predict the biological profiles of tested compounds against new protein targets through traditional regression techniques.

Validation

While no generally accepted theoretical definition for molecular diversity exists, there is agreement on the criterion for success in choosing descriptors and diversity metrics: a successful choice should be able to discriminate between biologically active and inactive compounds. The most comprehensive study specifically designed to address this issue was reported by Brown and Martin [10] of Abbott Laboratories. What makes their contribution unique is the size of the data set used in their analysis. More than 20 000 structures were analyzed, including three different sets of compounds tested against monoamine oxidase and two other proprietary enzyme targets, and a collection of over 16 000 compounds that were tested over the years in Abbott's high-throughput screens. Seven types of descriptors (MACCS, Unity and Daylight fingerprints, Unity 3D rigid and flexible descriptors, and two pharmacophore descriptors developed at Abbott), and 4 different clustering methodologies (Jarvis-Patrick, Ward, group-average and Guenoche) were evaluated based upon their abilities to map active and inactive compounds to different regions of descriptor space. The results indicated that the two-dimensional descriptors (the fingerprints) were considerably more effective than the three-dimensional ones. Among the clustering techniques, Ward's hierarchical agglomerative algorithm prevailed. In a follow-up study, they found that this ranking held for the ability of these descriptors to predict individual receptor-ligand binding determinants such as hydrophobicity, dispersion, electrostatics, and steric and hydrogen bonding capability. These results were consistent with those reported previously by the Sheffield group [7].

Patterson et al. [35] reported an alternative method for validating descriptors based on the concept of a 'neighborhood radius'. Their approach was to plot differences in the value of a descriptor against those in biological activity for a set of related compounds. If the descriptor is to be useful as a measure of similarity, the resulting plot should exhibit a characteristic trapezoidal distribution revealing a 'neighborhood behavior' for that descriptor. The method was applied to 20 data sets, and 11 descriptors were ranked by performance. They concluded that 2D fingerprints and 3D CoMFA fields far outperformed physicochemical properties such as logP and molar refractivity, while topological descriptors such as connectivity indices, atom pairs and auto-correlation vectors fell in the middle of the spectrum. Interestingly, they also found that, for combinatorially generated molecules, 2D fingerprints based on the whole molecule performed worse than those based on the substituents alone. They attributed this to a 'diluting' effect due to the presence of an identical template in each molecule. However, Patterson's study considered one descriptor at a time, and did not account for the possibility of correlations between two or more descriptors.

Dimensionality reduction

The high-dimensional data representations that are commonplace in molecular diversity/similarity analyses pose a number of problems. Firstly, as the number of variables used to describe data increases, the likelihood that some of the variables are correlated dramatically increases. While certain applications are more sensitive to correlation than others, in general, redundant variables tend to bias the results. Secondly, the amount of the computational effort needed to perform the analysis increases in proportion to the number of dimensions. Finally, visualization of the results in a concise and intuitive manner rapidly becomes impossible.

Fortunately, most multivariate data in $\Re^d$ are almost never d-dimensional. That is, the *underlying structure* of the data is almost always of dimensionality lower than d. To simplify the analysis and representation of the data, it is often desirable to reduce the dimensionality of the space by eliminating dimensions that add very little to the overall picture. We must stress that none of the methods that will be discussed here guarantees to extract the most important features for the application at hand. There is always the possibility that some critical piece of information is left behind, buried under a pile of redundancies. Experience in many different application areas has shown that, in practice, this situation does not arise often.

This discussion will focus on three main techniques to perform the reduction: 1) principal component analysis, 2) factor analysis, and 3) multidimensional scaling. Other approaches such as nonlinear mapping and Kohonen networks are discussed in greater detail in the section Visualization below.

Principal component analysis

Principal component analysis (PCA) [8,40–41] takes as its input a set of vectors described by partially cross-correlated variables and transforms it into one described by a smaller number of orthogonal variables (principal components) without a significant loss in the variance of the data. Principal components correspond to the eigenvectors of the covariance matrix, m_{ij}, a square symmetric matrix that contains the variances of the variables in its diagonal elements and the covariances in its off-diagonal elements:

$$m_{ij} = m_{ji} = \frac{1}{N}\sum_{k=1}^{N}(x_{ki} - \mu_i)(x_{kj} - \mu_j) \quad (7)$$

where μ_i is the mean value of variable i:

$$\mu_l = \frac{1}{N}\sum_{i=1}^{N} x_{ij} \quad (8)$$

and N is the number of observations in the data set. The eigenvalues of this matrix represent the variances of the principal components. PCA achieves a reduction in dimensionality by filtering out the principal components which contribute the least to the variance of the data (i.e. those with the smallest eigenvalues) until the variance reaches some pre-defined threshold, typically 90–95% of the original value. Finally, the original data are transformed using Equation (9):

$$\mathbf{x}' = \mathbf{V}^T\mathbf{x} \quad (9)$$

where $\mathbf{V}^T$ is the transpose of the filtered eigenvector matrix, $\mathbf{x}$ is the input vector in the original coordinate frame, and $\mathbf{x}'$ are the coordinates of that sample in the transformed frame. The components of $\mathbf{x}'$, therefore, are linear combinations of the original, cross-correlated variables.

The main advantage to PCA is that it makes no assumptions about the underlying probability distributions of the original variables. The primary disadvantage is that it is sensitive to outliers, missing data and poor correlations between variables due to poorly distributed variables.

Factor analysis

Factor analysis (FA) is a closely related technique that attempts to extract coherent subsets of variables that are relatively independent from one another [40]. It is often the case in science that the variable we are interested in is not directly observable. However, it is often possible to measure other quantities that reflect the underlying variable of interest. Factor analysis is an attempt to explain the correlations between variables in the form of underlying factors, which are themselves not directly observable, and which are thought to be representative of the underlying process that has created these correlations.

Factors are linear combinations of original variables. They may be associated with two or more of these variables (*common factors*) or with a single variable (*unique factors*). The specific association between the original variables and the derived factors is described in the form of *loadings*, which are derived from the magnitude of the eigenvalues of the covariance matrix. Factor loadings are inherently indeterminate. Rotation attempts to put these factors into a simple position, so that each variable is loaded highly on one factor, and all factor loadings are either large or near zero. A number of different rotation methods are available, including varimax, quartimax, and equimax. The varimax method maximizes the variance of the loadings, and is the most widely used.

On the surface, factor analysis and principal component analysis are very similar. Both rely on an eigenvalue analysis of the covariance matrix, and both use linear combinations of variables to explain a set of observations. However, in PCA the quantities of interest are the observed variables themselves; the combination of these variables is simply a means for simplifying their analysis and interpretation. Conversely, in factor analysis the observed variables are of little intrinsic value; what is of interest is the underlying factors.

Factor analysis has been used by Cummins et al. [42] to reduce a set of 61 molecular properties to 4 factors, which were then used to compare the diversity of 5 chemical databases (see section Partitioning Techniques below). It was also explored by Gibson et al. [41] in a comparative study of 100 different heterocyclic aromatic systems, but they concluded that FA did not reduce the complexity of the analysis, and did not offer any significant advantages over PCA.

Multi-dimensional scaling

Multi-dimensional scaling (MDS) [65,66] is a positional refinement technique that attempts to map a set of points in a high-dimensional space onto one of lesser dimensionality while preserving, as best as possible, the pairwise Euclidean distances, d_{ij}, between the points. Each iteration of the procedure consists of calculating the distances, δ_{ij}, between each pair of points in a lower-dimensional trial configuration and, using a steepest descent algorithm, shifting the positions of those points so as to create a new configuration characterized by a smaller sum-of-squares difference between δ_{ij} and d_{ij}. Two commonly used objective functions are *Kruskal's stress*:

$$S = \sqrt{\frac{\sum_{i<j}(\delta_{ij} - d_{ij})^2}{\sum_{i<j}\delta_{ij}^2}} \qquad (10)$$

and Lingoes' *alienation coefficient*:

$$S = \sqrt{1 - \frac{\sum_{i<j}(\delta_{ij} \cdot d_{ij})^2}{\sum_{i<j} d_{ij}^2}} \qquad (11)$$

The procedure terminates when the change in the objective function between iterations falls below a user-defined threshold.

Using multi-dimensional scaling, the group at Chiron [8] has shown that the 2048-bit Daylight fingerprints associated with 721 commercially available primary amines could be reduced to only five continuous variables that reproduced all 260 000 original pairwise dissimilarities (distances) with a standard deviation of only 10%. Similarly, only seven dimensions were necessary to represent the 642 000 pairwise similarities among a set of 1133 carboxylic acids and acid chlorides to the same precision. Despite its successes, the substantial computational cost of traditional multi-dimensional scaling makes the technique inapplicable to large data sets, such as those encountered in combinatorial library designs.

Compound selection

> *'There is nothing wrong with shooting, just as long as the right people get shot.'*
>
> Clint Eastwood, Magnum Force.

Clustering

Clustering algorithms attempt to organize elements of a set into groups (clusters) based upon selected characteristics. Members of a cluster must be similar to one another (internally homogeneous) and dissimilar to members of other clusters (externally heterogeneous). Because of its long-standing application in determining molecular similarity, clustering was one of the first selection methods to be applied to diversity analysis [6]. In contrast to other selection algorithms, which are based on statistical theory, clustering is an entirely heuristic approach consisting of four principal steps. First, a set of molecular descriptors must be selected and scaled. Second, the distances between pairs of molecules in the collection are calculated and collected in a similarity matrix. Third, members are assigned to clusters by a set of user-defined criteria. Finally, the clustering is validated by visual inspection or statistical means.

Clustering algorithms may be classified as hierarchical or non-hierarchical based on the way in which the clusters are formed. The end-result of hierarchical clustering analysis is a tree, or dendrogram, the structure of which reflects the organization of all the members of the collection. The dendrogram may be created from the top down beginning with a single cluster which is recursively sub-divided into increasingly smaller groups until each member is a cluster unto itself (a 'singleton'). Alternatively, one could begin with singletons and work up the tree by combining clusters until all the members belong to a single group. Non-hierarchical clustering (also known as k-nearest-neighbor clustering) produces a set of clusters based on some user-defined criteria. The most commonly used member of this class, particularly for diversity applications, is the Jarvis-Patrick algorithm. This method begins by determining the k nearest neighbors of each member of the collection. Members are placed in the same cluster if they have a user-defined number of nearest neighbors in common. The major advantage of this method is its speed; its main disadvantage is its tendency to generate either too many singletons or too few very large clusters depending on the stringency of the clustering criteria.

Willett [7] has performed a systematic evaluation of four different clustering methodologies – the Ward and group-average hierarchical agglomerative methods, the minimum diameter polythetic hierarchical divisive method, and the Jarvis-Patrick nearest neighbor algorithm – for purposes of determining molecular similarity. The test set was comprised of 5982 compounds characterized by 13 molecular descriptors. The results were evaluated by means of simulated property prediction experiments. Willett concluded that any of the three hierarchical methods was preferred to the Jarvis-Patrick (non-hierarchical) algorithm. A subsequent study by Brown and Martin has confirmed these findings [10].

Regardless of the particular algorithm used to cluster the collection, a diverse set is typically created by selecting one member from each cluster (most commonly the centroid). The resulting set may be examined for possible colinearities. If it is found to be non-orthogonal, suspect compounds are replaced with other members of the same cluster, and the new solution is re-evaluated. This cycle continues until a quasi-orthogonal set is identified.

Maximin

The maximin algorithm begins with a randomly selected member of a set of compounds and builds a maximally diverse subset, one compound at a time. At each step, the compound added to the subset is that which is farthest from its nearest neighbor in the subset. While this selection method is conceptually straightforward and easy to implement, it is impractical for use with large collections of compounds since the number of operations scales to the square of the size of the set. Maximin was first applied to diversity selection by Lajiness [45], and has been extended by others including Polinsky [46]. Hassan [43] and Agrafiotis [44] used the maximin criterion itself as a diversity metric that was optimized via simulated annealing.

Chapman [47] has recently reported a maximin algorithm which employs a diversity metric that explicitly accounts for multiple molecular conformations. Each molecule of a collection is subjected to an exhaustive search to identify its lowest energy conformations which, in turn, are aligned with the lowest energy conformations of every other molecule in the collection. Through rigid body rotations, the alignment procedure attempts to maximize a similarity coefficient which is a measure of the extent of steric and charge overlap between the two conformations. At each step of the procedure, the molecule that most increases the diversity of the selected subset, as computed using the following equation, is chosen:

$$D(M) = \sum_{m \in M} \left[\left(\sum_{c \in C(m)} \min_{c' \in C}(d(c, c')) \right) - T \Delta S(m) \right] \quad (12)$$

where M is the set of all compounds, $C(m)$ is the set of all conformers of compound m, and C the set of all conformers of all compounds. $T \Delta S(m)$ is an entropic term proportional to the number of rotatable bonds that penalizes highly flexible compounds.

Chapman applied this approach to two test collections, one of which consisted of naturally occurring amino acids and the other of 1371 commercially available carboxylic acids. In the case of the amino acids, the results were intuitive and suggested that the measure of similarity is indeed a reasonable one. In the case of the carboxylic acids, the selection compared favorably to random controls, and identified reagents that were quite diverse in terms of shape, size and functionality. While the method is intellectually robust and intuitive, it remains to be seen whether the markedly increased computational cost of taking multiple conformations into consideration is offset by any advantage over existing techniques.

Stepwise elimination

Taylor [48] has developed a selection method which sequentially eliminates members from the whole set rather than building up the diverse subset from a single compound. Starting with the symmetric $N \times N$ similarity matrix, the largest off-diagonal element (i.e. the most similar pair of compounds) is identified, and one of the pair of compounds associated with it is eliminated. This process continues until a single compound is left in the set. This algorithm then sorts the compounds, placing the most diverse molecules at the top of the list.

Cluster sampling

Cluster sampling is another nearest neighbor selection algorithm developed by Taylor [48] which, despite its name, does not explicitly partition a set of compounds into clusters. Using a minimum similarity threshold of 0.8, the method begins by generating a list of nearest neighbors for each compound in the set, which are then merged to form the nearest neighbor table (NNT) for the entire set. During each iteration, the procedure selects the compound which occurs most often in the NNT, which corresponds to the compound situated at the center of the most densely populated region (cluster) of property space. All the nearest neighbors of this molecule are then flagged as unavailable for subsequent cycles of selection. The procedure terminates when all the compounds in the set are either selected or flagged as unavailable. Both cluster sampling and stepwise elimination are intuitive and robust procedures, but scale to the square of N, which makes them impractical for large data sets.

Experimental design

In an attempt to provide a rational criterion for selecting a maximally diverse subset of amines and carboxylic acids for use as side chains and capping groups in N-substituted glycine peptoid combinatorial libraries, Martin et al. [8] developed a selection procedure based upon an established method of statistical experimental design known as D-optimal design. D-optimal design strives to identify a subset of compounds which is both diverse (the inter-compound distance in property space is maximized) and orthogonal (the covariances are minimized). The method begins with the empty set or a set of pre-selected compounds. At each step, a compound is chosen which maximizes the determinant of the 'information matrix', $X^T X$, which is equivalent to the volume subtended by the subset in covariance space. The rows of the design matrix, X, index the compounds of the subset, and its columns index either individual properties or higher-order combinations of properties such as their squares, cubes or cross products. The procedure terminates when a pre-determined number of compounds have been selected. This approach is nearly identical to one developed previously by Marsili and Saller [49] for the purpose of selecting multivariate synthetic analogues.

Recent work by Hassan et al. [43] has indicated that this method tends to increase the rank of the selected subset at the expense of spread. They used a Monte-Carlo method to maximize a diversity objective function based on the D-optimal criterion. The resulting sets of compounds were biased toward the periphery of the property space. This bias was especially evident when the number of compounds selected far exceeded the dimensionality of the space. It must be noted that the design matrix used in this study did not include higher-order combinations of properties, which may partially account for the redundancy of the results.

Partitioning techniques

Most of the algorithms discussed to this point scale to the square of the number of molecules in the set considered, making their use impractical for large data sets. In attempts to reduce this computational complexity, a number of groups have investigated simple partitioning (binning) techniques. Each axis of a molecular property space is divided into equal segments, creating a honeycomb of multi-dimensional cells. Compounds are assigned to the cells based on their properties. Many diversity-related tasks are greatly simplified using this approach. For example, a diverse subset of compounds may be created by selecting a pre-determined number of molecules from each cell. Comparison of two compound collections reduces to comparing the number of molecules assigned to corresponding cells in each collection. However, as with any partitioning method, the challenge is to minimize the number of discrete cells without having them become so large as to be useless. Pearlman [24] and Cummins et al. [42] have addressed this problem by reducing the dimensionality of the property space.

Pearlman constructed a 6-dimensional space using eigenvectors corresponding to the largest and smallest eigenvalues of an optimized set of three matrices which encoded atomic charge, polarizability and hydrogen bonding characteristics (B-Cut values). The principal shortcoming of this method is that it lacks a straightforward physical interpretation. Why choose the extreme eigenvalues? Burden, upon whose work this approach is based [25], contests that the smallest eigenvalue contains contributions from every atom, and therefore reflects the topology of the entire molecule. In addition, there is ample evidence that an adequate representation of molecular diversity requires substantially more than six dimensions [8]. Cummins et al. used factor analysis to reduce a 61-dimensional property space to four factors which accounted for 90% of the variance of five chemical databases (CMC, MDDR, ACD, SPECS and the Wellcome Registry) containing in excess of 300 000 compounds. The diversity of each data set was computed as the fraction of the total volume occupied by that set, using a Riemann-style approach. A trimming procedure was also employed to eliminate outliers and focus the analysis on the more densely populated areas of the feature space. The resulting density functions for two of these factors are shown in Figure 1.

Stochastic techniques

All of the algorithms discussed thus far select a diverse subset of molecules from a larger collection, but none guarantee that the subset will be the most diverse possible of a given size. Agrafiotis [44,50] and Hassan et al. [43] have independently proposed using simulated annealing to select an optimal subset of compounds based on an objective function which measures the diversity of any conceivable set of compounds. Simulated annealing is a global, multivariate optimization technique based on the Metropolis Monte-Carlo search algorithm. The method, as it is applied to molecular diversity, starts with a randomly selected subset of molecules. At each step, it makes a small change in the composition of the subset (usually on the order of 1–10% of the constituent members) and evaluates the diversity of the new subset. If the new subset is more diverse than the old, it is retained and the cycle begins

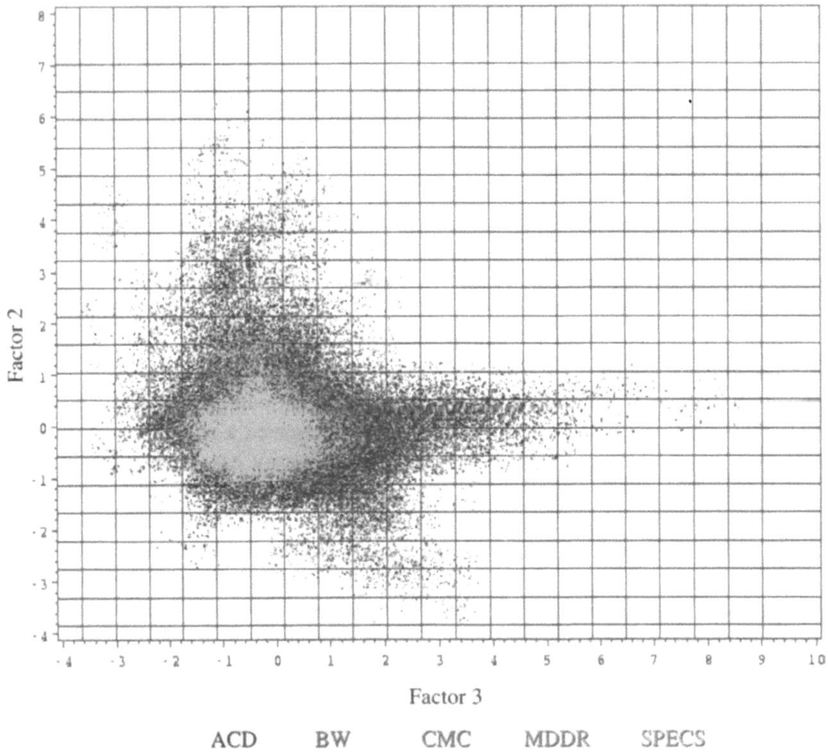

Figure 1. A comparison of the property distributions of five chemical databases using factor analysis.

anew. If it is less diverse, it is not rejected out-of-hand, but is retained with a probability that is inversely proportional to difference in diversity between the two subsets.

Hassan employed the maximin, power-sum and product functions as diversity metrics, while Agrafiotis' original implementation used maximin and a volumetric diversity measure of his own device. Although their methods were similar, the motivations of the two groups appear to be rather different. Hassan et al. wanted to compare the performance of different diversity metrics. Agrafiotis, on the other hand, was interested in developing a generalized selection method based on any number of disparate criteria (e.g. similarity and predicted activity as well as cost and availability of starting materials or reaction block design). This requirement was essential since the algorithms were to be used as part of an iterative drug discovery system known as DirectedDiversity®, in which the selection criteria would typically differ from iteration to iteration [51,52]. He later applied this algorithm to study other diversity metrics with surprising results [53,54] (see the sections on Information Theory and Cosine Coefficient below), and has also reported evolutionary and genetic variants of this general sampling approach.

Optimizable K-dissimilarity selection

K-Dissimilarity selection (OptiSim) is a generalization [71] of maximin which balances diversity against representativeness in the selection set. It starts by selecting a compound at random, then examines *K* other compounds chosen at random from the data set, excluding from consideration any which are too similar to the initial selection. From among these *K* compounds, the one which is most dissimilar to the initial selection is added to the selection set. At each iteration thereafter, a fresh sample of *K* candidates is compared to the compounds which have already been selected, and the most dissimilar among them is added to the selection set; all compounds are considered once before any candidate is considered twice. The published application employs the maximin dissimilarity criterion (the minimum pairwise dissimilarity of the candidate to any compound already in the set) to identify the 'best' candidate from each subset, though other measures [2] can also be used.

The algorithm scales with $K*M^2$, where M is the number of compounds being selected. The smaller the sample size K, the more representativeness is favored in the selection set. As K approaches N, the total number of compounds in the data set, OptiSim reduces to maximin. At modest values of K (2–10), the balance between representativeness and diversity is very similar to that seen for selection based on hierarchical clustering [72]. Multiple passes with different random seeds are even more efficient (order $P*K*M^2$, for P passes of M selections each), and correspond to selecting multiple representatives from each cluster. OptiSim is commercially available from Tripos, Inc. as part of ChemEnlighten [73].

Vector analysis

A couple of groups have reported diversity methods based on an analysis of the spatial relationships of intramolecular functionalities. Boyd [56] reported a method, called HookSpace, that measures diversity based on the spatial distribution of distances between user-defined functional groups. In particular, each pair of functional groups in a given compound was aligned on the xy plane so that one of the groups was placed along the x axis with the head atom at the origin, and the other was positioned on the xy plane, with the head-to-tail vector pointing in the positive z direction. Once the alignment was complete, the position of the head atom of the second group on the xy plane was recorded. This process was repeated for every pair of functional groups in each structure, and for every structure in the database. The xy plane was then partitioned into a finite number of cells, and each cell recorded either the total number of functional groups, or the number of different functional groups at that position. This permitted diversity measurements by computing the percentage of non-vacant cells on the xy plane, similar to the method described by Pearlman. The authors used this approach to compare the structural diversity of the Available Chemicals Directory (ACD), the Cambridge Structural Database (CSD), and a benzodiazepam combinatorial library, using a theoretical reference space. They concluded that the ACD covered 85% of that space, whereas the CSD and the benzodiazepam library covered only 34 and 13% of the space, respectively. It is quite likely, however, that this difference reflects the different origins of the three-dimensional structures of these compounds (computed versus experimental), rather than the intrinsic functional and geometric diversity of the two databases.

In a related approach, Bartlett [57] presented a system that compared the diversity of different combinatorial templates using the angles between the bond vectors connecting the core to the substituent. The method followed the spirit of the Caveat approach, and the results were presented in a visual form.

Minimum spanning trees

At a recent conference, Ruppert and co-workers at Arris presented a novel method of computing molecular diversity based on spanning trees [77]. Their method, which they named IcePick, measures the difference between small molecules by comparing steric and polar features on their (flexible) three-dimensional surfaces. IcePick compares one molecule to another by 'turning it inside out' to form a pocket that perfectly fits around it (i.e. an ideal protein) and then flexibly docking the second molecule into the pocket. The fit is scored by comparing the protein accessible surface and the vector directions available for both hydrogen bond donation and acceptance; this is then averaged over the flexible conformational space of both molecules. From this measure of pairwise dissimilarity, IcePick computes the intrinsic diversity of a set or library of molecules using a 'spread' metric based on minimum spanning trees. The most diverse set is the one maximizing the weight of the minimum spanning tree. Since flexible three-dimensional docking can be computationally expensive, a statistical fingerprinting technique was developed to speed the diversity computation for large libraries.

Information theory

Lin [58] has proposed a diversity metric based upon the premise that maximizing the diversity of a subset of molecules is equivalent to maximizing its information content. The crux of the approach is the postulate that every collection of molecules is composed of a finite number of distinct species, or classes, of molecules, and that the ability to distinguish among these species can be described as a function of their mutual dissimilarity. The more distinguishable the species, the greater their information content of the collection. The diversity of the collection may then be quantified using Shannon's entropy formalism:

$$I = S_{\max} - S \tag{13}$$

where

$$S = -\sum_{i=1}^{N}\sum_{j=1}^{N} p_{ij} \ln p_{ij} \tag{14}$$

and N is the number of molecules in the collection, p_{ij} is the probability of finding the i-th molecule in the j-th species (given some function of their dissimilarity), and $S_{\max}$ is the maximum entropy of the system.

While the use of information theory seems like a natural choice, the actual implementation suffers from a number of disadvantages. In a recent article [53], we reported that a strict application of this approach produced extremely unbalanced designs, and clustered points at maximum separation along the diagonal of the feature space. We believe that this is due to the use of the wrong type of 'information', and to the implicit assumption that ideal designs should be equiprobable (i.e. that the pairwise intermolecular dissimilarities should be as uniform as possible). In a private communication, Lin suggested that our results could be an artifact of the similarity measure used in our study, but a detailed response has yet to appear in print. As of this writing, the debate is still open.

Cosine coefficient

Willett et al. [59] have noted that if the diversity of a set of compounds, $D(A)$, is defined as the complement of the mean pairwise intermolecular similarity:

$$D(A) = 1 - \frac{\sum_{i=1}^{N}\sum_{j=1}^{N} \delta(i, j)}{N^2} \tag{15}$$

where $\sigma(i, j)$ is the similarity between compounds i and j, and N is the number of compounds in A, the expression can be reduced from one that scales with the square of the number of compounds in the set to one that scales linearly by using the cosine coefficient to evaluate the pairwise similarities. The cosine coefficient of similarity is defined as the cosine of the angle between two molecular property vectors:

$$\sigma(i, j) = \cos(i, j) = \frac{\sum_{k=1}^{K} m(i, k)m(j, k)}{\sqrt{\sum_{k=1}^{K} m(i, k)^2 \sum_{k=1}^{K} m(j, k)^2}} \tag{16}$$

where $m(i, k)$ is the k-th component of the i-th atomic property vector, $m(i)$, and K is the dimensionality of the space. Substituting Equation (16) for $\sigma(i, j)$, Equation (15) reduces to:

$$D(A) = 1 - \frac{\mathbf{a}_c \cdot \mathbf{a}_c}{N^2} \tag{17}$$

where

$$\mathbf{a}_c = \sum_{i=1}^{N} w(i)\mathbf{m}(i) \tag{18}$$

and the weights, $w(i)$, are given by:

$$w(i) = \frac{1}{\sqrt{\sum_{k=1}^{K} m(i, k)}} \tag{19}$$

Unfortunately, this dramatic improvement in performance comes at a significant price. Using the stochastic approach outlined above, Agrafiotis showed that the method has a tendency to over-sample the principal axes of the property space [54]. He postulated that this behavior is an artifact of the simple summation function used for the dissimilarity metric and the fact that the cosine coefficient only measures the angle between two property vectors and ignores their lengths, which are necessary to measure spread.

Visualization

The ancients built Valdrada on the shores of a lake, with houses all verandas one above the other, and high streets whose railed parapets look out over the water. Thus the traveler, arriving, sees two cities: one erect above the lake, and the other reflected, upside down. Nothing exists or happens in the one Valdrada that the other Valdrada does not repeat, because the city was so constructed that its every point would be reflected in its mirror, and the Valdrada down in the water contains not only all the flutings and juttings of the facades that rise above the lake, but also the rooms' interiors and ceilings and floors, the perspective of the halls, the mirrors of the wardrobes ...

Valdrada's inhabitants know that each of their actions is, at once, that action and its mirror image, which possesses the special dignity of images, and this awareness prevents them from succumbing for a single moment to chance and forgetfulness.

Italo Calvino, Invisible Cities.

One of the most difficult challenges in data analysis is to be able to represent whatever complexities might be intrinsic to the data in a simple and intuitive form. In fact, one might argue that if the results of an analysis are unable to be conveyed to a target audience in a straightforward manner, the analysis, no matter how thorough, has failed. As will become evident, traditional methods of data visualization are inadequate to represent the extremely large, high-dimensional data sets common to molecular diversity/similarity analyses. In order to minimize the complexity and shear number of individual plots needed to visualize this sort of data, one must attempt to reduce the dimensionality of the representation. The three techniques described below, self-organizing maps, multidimensional scaling and non-linear mapping, use different approaches to achieve dimensionality reduction, while preserving the topology of the original space. That is, points near each other in the high-dimensional space are also near each other in the low-dimensional space.

Visualization without dimensionality reduction

Multivariate data may be visualized up to three dimensions at a time through use of traditional techniques such as histograms and two- and three-dimensional scatter plots. Histograms provide a convenient means of analyzing one variable at a time, but do not reveal any relationships between variables. Scatter plots are more effective in this respect, but they become difficult to interpret if the density of points is high. In addition, if the number of variables exceeds all but a handful, the number of plots needed to represent the data rapidly becomes overwhelming. One particularly useful type of three-dimensional scatter plot is a pharmacophore plot, which is used to visualize the three-point pharmacophore keys (see previously) present in a molecule or set of molecules. Each pharmacophore is positioned in the plot according to its three inter-point distances, is labeled by a symbol which reflects its type, and is color-coded according to whether it contains 1, 2 or 3 identical loci. An example of such a plot generated using Chemical Design's Chem-X suite is shown in Figure 2.

Another interesting approach are the flower plots of Martin et al. [8] which represent aestetically pleasing variants of the traditional star diagrams used in multivariate analysis. Flower plots display all the properties associated with a single compound in one plot. They are circular bar graphs in which each 'petal' of the flower represents a molecular property or descriptor. The value of the property is reflected in the size of the petal and its sign determines whether the petal points outward (positive values) or inward (negative values). In addition, the central sphere is color-coded according to some additional property such a biological activity or similarity to a reference compound. Figure 3 shows the flower plot of tyramine, described by means of five chemical functionality descriptors, five shape descriptors, five atom-layer receptor recognition descriptors and the computed logP.

Because there is a one-to-one correspondence between the number of compounds and the number of plots, flower plots are impractical for visualizing large data sets. Flower plots are particularly useful, however, in assessing the distribution of properties across a small collection of compounds, for example, a set of reagents used in generating a combinatorial library.

Visualization with dimensionality reduction

Self-organizing maps

Self-organizing maps (SOM's) or Kohonen networks [64] belong to a class of neural networks known as competitive learning or self-organizing networks. They were originally developed to model the ability of the brain to store complex information as a reduced set of salient facts without loss of information about their interrelationships. High-dimensional data are mapped onto a two-dimensional rectangular or hexagonal lattice of neurons in such a way as to preserve the topology of the original space.

A Kohonen network is trained in the following manner: Each neuron, i, has an associated vector of weights $m_i = [\mu_{i1}, \mu_{i2}, \ldots, \mu_{iN}]$, where N is the dimensionality of the original space. A randomly chosen training sample, $x = [\xi_1, \xi_2, \ldots, \xi_N]$, is presented to the network, and the weighting vectors of all the neurons are adapted to the input sample through use of a *neighborhood function* or *smoothing kernel*. A widely used kernel is given in Equation (20):

$$h_{ci}(t) = \alpha(t) \exp\left(-\frac{\|\mathbf{r}_c - \mathbf{r}_i\|^2}{2\sigma(t)^2}\right) \quad (20)$$

where c is the neuron whose Euclidean distance to the input sample is the smallest, $\mathbf{r}_c$ and $\mathbf{r}_i$ are the respective locations of the c-th and i-th neurons on the lattice ($\mathbf{r}_c$, $\mathbf{r}_i \in \Re^2$), $\alpha(t)$ is the *learning rate*, and $\sigma(t)$ is the width of the function. This process is repeated until each training sample has been presented to the network, a phase referred to as a training epoch. Typically, many training epochs are necessary to complete the training. Upon completion, each neuron is sensitized to a particular region of the original space. Samples which fall within the same region, whether they were or were not included in the original training set, are mapped onto the same neuron.

Self-organizing maps were first used to visualize collections of molecules by Gasteiger et al. at the Uni-

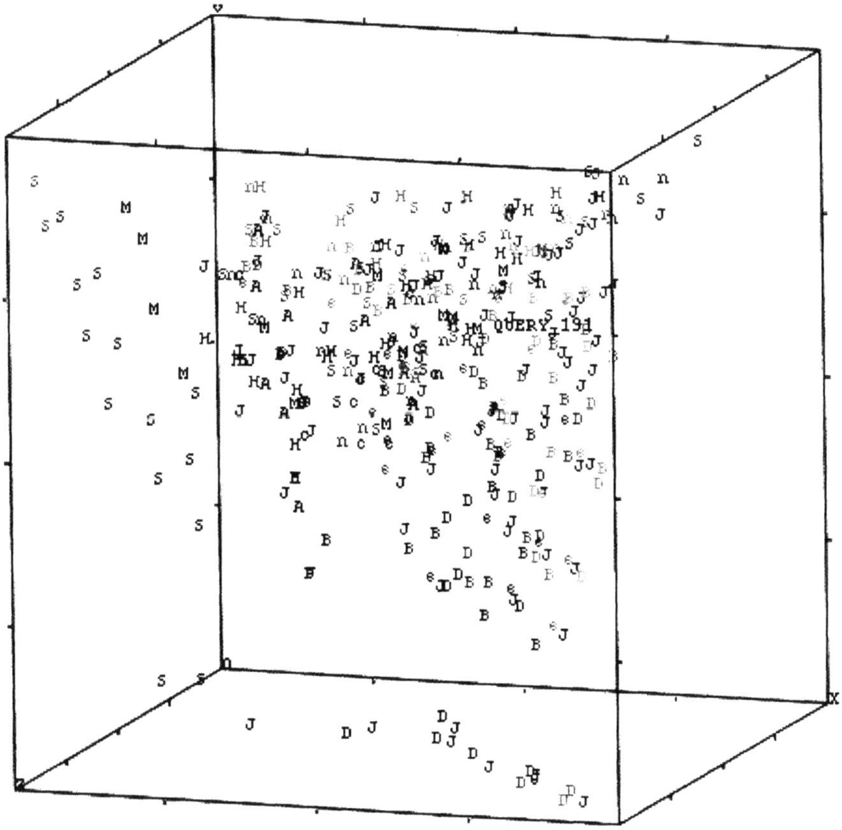

Figure 2. Pharmacophore plot generated by Chemical Design's ChemX software suite.

versity of Erlangen [23]. They validated the utility of the method by showing that it was able to spatially separate two combinatorial libraries designed by Rebek as potential trypsin inhibitors. The libraries were created by combining the tetra-substituted xanthene and cubane scaffolds shown in Figure 4 with a common set of 19 L-amino acids. Thus, while their substituents were structurally identical, the xanthene library arranged its substituents around a common plane, while those of the cubane library were arranged at the vertices of a tetrahedron. In addition, they showed that a third, adamantane-based library, which also oriented its substituents tetrahedrally, mapped to the same region as the cubane library.

Each molecule was described by a 12-dimensional spatial autocorrelation vector that encoded the electrostatic potential at its surface according to Equation (21):

$$A(d_l, d_u) = \frac{1}{N} \sum_{i,j} p_i p_j \qquad (21)$$

where p_i and p_j are the values of the electrostatic potential at two randomly chosen points, i and j, on the molecular surface and N is the number of distance bins on the interval $[d_l, d_u]$ (12 in this case). Two 50×50 neuron Kohonen networks were trained: the first with the cubane and xanthene libraries, and the second with all three libraries. As is shown in Figure 5a, the cubane library (black cells) maps to a region distinct from that of the xanthene library (light gray cells). The overlap between the libraries amounts to only 3% of the total number of neurons, and is confined to the periphery of the cubane cluster. Figure 5b shows the map created by training the network with all three libraries. As expected due to their similar geometry, members of the cubane (black cells) and adamantane (dark gray cells) libraries mapped onto the same region of the network which was, again,

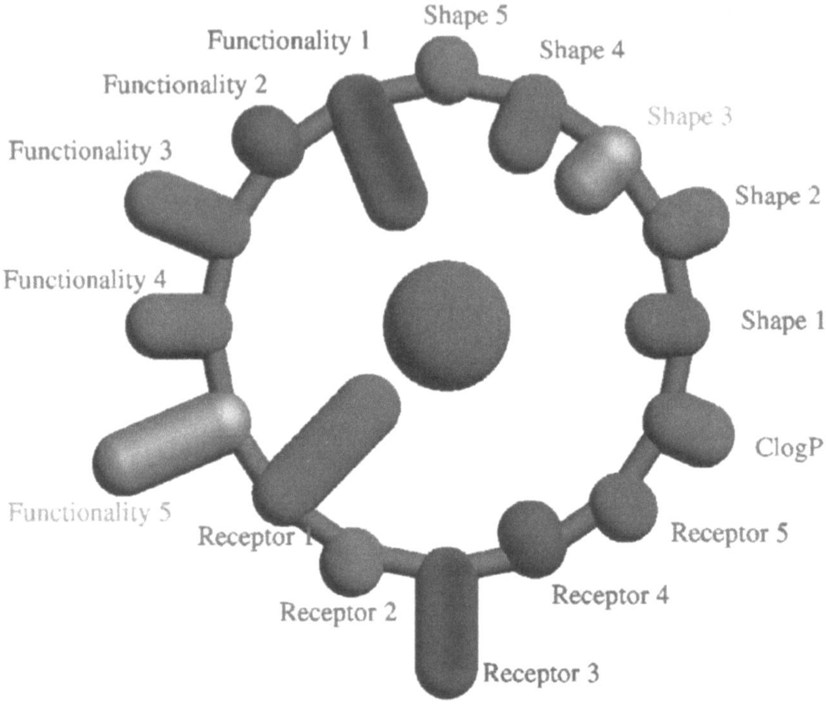

Figure 3. Flower plot of tyramine. There is one petal for each of the five chemical functionality descriptors, five shape descriptors, five receptor recognition descriptors and the computed logP.

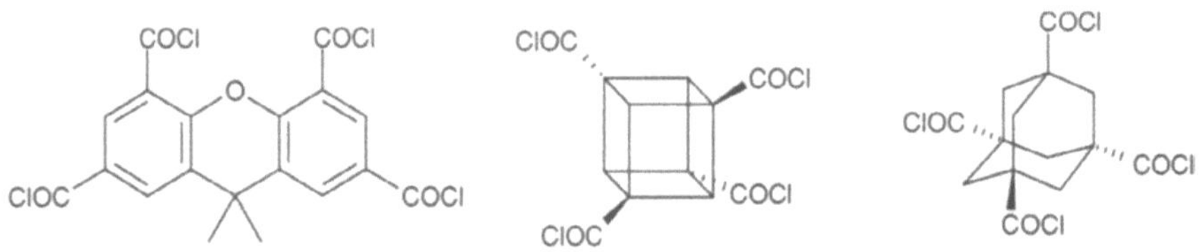

Figure 4. Combinatorial scaffolds used by Sadowski, Wagener, and Gasteiger: (left) xanthene, (middle) cubane, (right) adamantane.

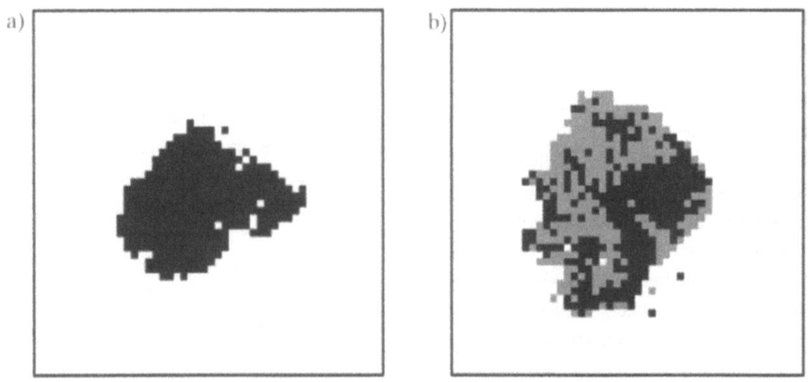

Figure 5. Self-organizing maps of (a) the xanthene (light gray) and cubane (black) libraries, and (b) the xanthene (light gray), cubane (black), and adamantane (dark gray) libraries.

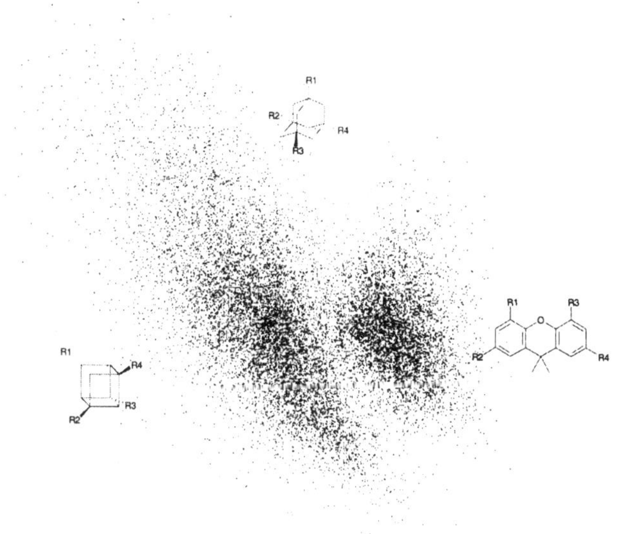

Figure 6. Sammon map of the xanthene, cubane, and adamantane libraries used by Gasteiger et al.

distinct from that of the xanthene library (light gray cells).

Given the simplicity of their output, SOM's can be very effective for visualizing and comparing chemical libraries, particularly when they are coupled with advanced, interactive graphical tools.

Non-linear maps

Non-linear mapping is a multivariate statistical technique that is closely related to multi-dimensional scaling [67]. Just like MDS, the objective is to approximate local geometric relationships on a two- or three-dimensional plot. The difference between MDS and non-linear mapping is in the minimization procedure. Sammon's algorithm is the most commonly used, but it too does not scale gracefully with the size of the data set. Self-organized non-linear mapping is a variant of Sammon's original algorithm that was developed by Agrafiotis [44,50] and is based on a self-organization principle reminiscent of Kohonen's SOM training algorithm. This method belongs to the family of non-metric algorithms, and is therefore applicable to a wide variety of input data. This is particularly useful when the (dis)similarity measure is not a true metric, i.e. it does not obey the distance postulates and, in particular, the triangle inequality (such as the Tanimoto coefficient). Although an 'exact' projection is only possible when the distance matrix is positive definite, meaningful projections can be obtained even when this criterion is not satisfied.

Non-linear maps were introduced by Agrafiotis to visualize protein sequence relationships in two dimensions [68], and were later employed as a means of visualizing and comparing large compound collections, represented by a set of molecular descriptors [44,50]. The advantage of non-linear maps over Kohonen networks is that they provide much greater detail about the individual compounds and their interrelationships. To provide a direct comparison between self-organized and non-linear maps, we applied the Sammon algorithm on the xanthene, cubane and adamantane libraries that were used by Gasteiger et al. in [23] (Figure 6). The projection was carried out using the same 12-dimensional auto-correlation descriptors and the Euclidean metric as a pairwise measure of dissimilarity. The resulting map is shown in Figure 6.

The map is sufficiently faithful, as manifested by a Sammon and Kruskal stress value of only 10 and 8%, respectively. It is evident that the Sammon map is not only capable of reproducing the sharp separation between the planar and tetrahedral systems that was observed in the self-organized maps (Figure 5), but also revealed a more subtle distinction between the cubane and adamantane libraries that was not captured by the Kohonen network.

While the first application of this technique involved continuous molecular descriptors, the programs were later extended to include other molecular representations and molecular similarity metrics such as substructure keys, hashed fingerprints, Tanimoto coefficients etc. [69].

Conclusions

'You will, I am sure, agree with me that if page 534 finds us only in the second chapter, the length of the first one must have been really intolerable.'

Sherlock Holmes, The Valley of Fear.

The study of molecular diversity is an evolving field, fueled by rapid advances in experimental discovery and an urgent need for rigorous statistical experimental design. However, despite its conceptual simplicity, a rigorous mathematical definition remains elusive. Most approaches developed to date are rooted in the fields of molecular similarity and QSAR, and some have shown promise in increasing the hit rates of combinatorial chemistry experiments. On the other hand, there are many examples that suggest that diversity is serendipity in disguise. Validation is critical, but it can only come from comparison with appropriate control experiments which are hard to design and too expensive to execute. Although interest in molecular diversity will continue to grow, we believe that it will eventually become inextricably linked with structure-activity correlation and statistical series design, perhaps giving rise to a new unifying field devoted to the planning and analysis of massively parallel experiments.

References

1. Johnson, M. A. and Maggiora, G. M., Concepts and Applications of Molecular Similarity, Wiley, New York, NY, 1990.
2. Hansch, C. and Leo, A., Exploring QSAR. Fundamentals and Applications in Chemistry and Biology, American Chemical Society, Washington, D.C., 1995.
3. Martin, E. J., Spellmeyer, D. C., Critchlow, R. E. Jr. and Blaney, J. M., *Does combinatorial chemistry obviate computer-aided drug design?*, In Lipkowitz, K. B. and Boyd, D. B. (Eds.) Reviews in Computational Chemistry, VCH, Weinheim, 1997, pp. 75–100.
4. Blaney, J. M. and Martin, E. J., *Computational approaches for combinatorial library design and molecular diversity analysis*, Curr. Biol., in press.
5. Martin, Y. C., Brown, R. D. and Bures, M. G., *Quantifying Diversity*, In Kerwin, J. F. and Gordon, E. M. (Eds.) Combinatorial Chemistry and Molecular Diversity, Wiley, New York, NY, in press.
6. Willett, P., Similarity and Clustering in Chemical Information Systems, Research Studies Press, Letchworth, 1987.
7. Downs, G. M. and Willett, P., *Similarity searching and clustering of chemical-structure databases using molecular property data*, J. Chem. Inf. Comput. Sci., 34 (1994) 1094–1102.
8. Martin, E. J., Blaney, J. M., Siani, M. A., Spellmeyer, D. C., Wong, A. K. and Moos, W. H., *Measuring diversity: experimental design of combinatorial libraries for drug discovery*, J. Med. Chem., 38 (1995) 1431–1436.
9. Lewis, R., McLay, I. M. and Mason, J. S., *Chem. Des. Autom. News*, 10(4) (1995) 37–38.
10. Brown, R. D. and Martin, Y. C., *Use of structure-activity data to compare structure-based clustering methods and descriptors for use in compound selection*, J. Chem. Inf. Comput. Sci., 36 (1996) 572–584.
11. Brown, R. D. and Martin, Y. C., *The information content of 2D and 3D structural descriptors relevant to ligand-receptor binding*, J. Chem. Inf. Comput. Sci., 37 (1997) 1–9.
12. Kubinyi, J., In Manhold, R., Krogsgaard-Larsen, P. and Timmermann, H. (Eds.) Methods and Principles in Medicinal Chemistry, Vol. 1, VCH, Weinheim, 1993, pp. 21–36.
13. Kier, L. B. and Hall, L. H., Molecular Connectivity in Structure-Activity Analysis, Wiley, New York, NY, 1986.
14. Charton, M. and Motoc, I. (Eds.), Steric Effects in Drug Design, Springer-Verlag, Heidelberg, 1983.
15. *Molconn-X*, Haney Associates, Mercer Island, WA.
16. Downs, G. M. and Barnard, J. M., *Techniques for generating descriptive fingerprints in combinatorial libraries,* J. Chem. Inf. Comput. Sci., 37 (1997) 59–61.
17. Carhart, R. E., Smith, D. H. and Venkataraghavan, R., *Atom pairs as molecular features in structure-activity studies: definition and application*, J. Chem. Inf. Comput. Sci., 25 (1985) 64–73.
18. Kearsley, S. K., Sallmack, S., Fluder, E. M., Andose, J. D., Mosley, R. T. and Sheridan, R. P., *Chemical similarity using physicochemical property descriptors*, J. Chem. Inf. Comput. Sci., 36 (1996) 118–127.
19. Sheridan, R. P., Miller, M. D., Underwood, D. J. and Kearsley, S. K., *Chemical similarity using geometric atom pair descriptors*, J. Chem. Inf. Comput. Sci., 36 (1996) 128–136.
20. Moreau, G. and Broto, P., *The autocorrelation of a topological structure: a new molecular descriptor*, Nouv. J. Chim., 4 (1980) 359–360.
21. Bauknecht, H., Zell, A., Bayer, H., Levi, P., Wagener, M., Sadowski, J. and Gasteiger, J., *Locating biologically active compounds in medium-sized heterogeneous datasets by topological autocorrelation vectors: dopamine and benzodiazepine agonists*, J. Chem. Inf. Comput. Sci., 36 (1996) 1205–1213.
22. Wagener, M., Sadowski, J. and Gasteiger, J., *Autocorrelation of molecular surface properties for modeling corticosteroid binding globulin and cytosolic Ah receptor activity by neural networks*, J. Am. Chem. Soc., 117 (1995) 7769–7775.
23. Sadowski, J., Wagener, M. and Gasteiger, J., Angew. Chem. Int. Ed. Engl., 34 (1996) 23–24.
24. Pearlman, R. S., *Novel software tools for addressing chemical diversity*, Network Science, 1996, June, http://www.awod.com/netsci/issues/.
25. Burden, F. R., *Molecular identification number for substructure searches*, J. Chem. Inf. Comput. Sci., 29 (1989) 225–227.
26. Unity Chemical Information Software, Tripos Associates, St. Louis, MO.
27. Murrall, N. W. and Davies, E. K., J. Chem. Inf. Comput. Sci.,

30 (1990) 312–316.
28. Sadowski, J., Gasteiger, J. and Klebe, G., *Comparison of automatic three-dimensional model builders using 639 X-ray structures,* J. Chem. Inf. Comput. Sci., 34 (1994) 1000–1008.
29. Sheridan, R. P., Nilikantan, R., Rusinko, A., Bauman, N., Haraki, K. and Venkataraghavan, R., J. Chem. Inf. Comput. Sci., 29 (1989) 255–260.
30. Pickett, S., Mason, J. S. and McLay, I. M., *Diversity profiling and design using 3D pharmacophores: pharmacophore-derived queries (PDQ),* J. Chem. Inf. Comput. Sci., 36 (1996) 1214–1223.
31. Davies, E. K. and Briant, C., *Combinatorial chemistry library design using pharmacophore diversity,* Network Science, 1995, http://www.awod.com/netsci/issues/.
32. Kubinyi, H., 3D-QSAR in Drug Design: Theory, Methods and Applications, ESCOM, Leiden, 1993.
33. Cramer, R. D., Clark, R. D., Patterson, D. E. and Ferguson, A. M., *Bioisosterism as a molecular diversity descriptor: steric fields of single topomeric conformers,* J. Med. Chem., 39 (1996) 3060–3069.
34. Kauvar, L. M., Higgins, D. L., Villar, H. O., Sportsman, J. R., Engqvist-Goldstein, A., Bukar, R., Bauer, K. E., Dilley, H. and Rocke, D. M., *Predicting ligand binding to proteins by affinity fingerprinting,* Chem. Biol., 2 (1995) 107–118.
35. Patterson, D. E., Cramer, R. D., Ferguson, A. M., Clark, R. D. and Weinberger, L. E., *Neighborhood behavior: a useful concept for validation of molecular diversity descriptors,* J. Med. Chem., 39 (1996) 3049–3059.
36. Bellman, R. E., Adaptive Control Processes, Princeton University Press, Princeton, 1961.
37. Scott, D. W., Multivariate Density Estimation: Theory, Practice and Visualization, Wiley, New York, NY, 1992.
38. Wegman, E., *Hyperdimensional data analysis using parallel coordinates,* J. Ann. Statist., 41 (1970) 457–471.
39. Teig, S., Cambridge Healthtech Institute's Conference on Chemoinformatics, May 12–13, 1997, Arlington, VA.
40. Cooley, W. and Lohnes, P., Multivariate Data Analysis, Wiley, New York, NY, 1971.
41. Gibson, S., McGuire, R. and Rees, D. C., *Principal components describing biological activities and molecular diversity of heterocyclic aromatic ring fragments,* J. Med. Chem., 39 (1996) 4065–4072.
42. Cummins, D. J., Andrews, C. W., Bentley, J. A. and Cory, M., *Molecular diversity in chemical databases: comparison of medicinal chemistry knowledge bases and databases of commercially available compounds,* J. Chem. Inf. Comput. Sci., 36 (1996) 750–763.
43. Hassan, M., Bielawski, J. P., Hempel, J. C. and Waldman, M., *Optimization and visualization of molecular diversity of combinatorial libraries,* Mol. Div., 2 (1996) 64–74.
44. Agrafiotis, D. K., *Stochastic algorithms for maximizing molecular diversity,* J. Chem. Inf. Comput. Sci., 37 (1997) 841.
45. Lajiness, M. S., In Silipo, C. and Vittoria, A. (Eds.) QSAR: Rational Approaches to the Design of Bioactive Compounds, Elsevier, Amsterdam, 1991, pp. 201–204.
46. Polinsky, A., Feinstein, R. D., Shi, S. and Kuki, A., *LiBrain: software for automated design of exploratory and targeted combinatorial libraries,* In Chaiken, I. M. and Janda, K. D. (Eds.) Molecular Diversity and Combinatorial Chemistry, American Chemical Society, Washington DC, 1996, pp. 219–232.
47. Chapman, D., *The measurement of molecular diversity: a 3-dimensional approach,* J. Comput.-Aided Mol. Design, 10 (1996) 501–512.
48. Taylor, R., *Simulation analysis of experimental design strategies for screening random compounds as potential new drugs and agrochemicals,* J. Chem. Inf. Comput. Sci., 35 (1995) 59–67.
49. Marsili, M. and Saller, H., *ANALOGS: a computer program for the design of multivariate sets of synthetic analogs,* J. Chem. Inf. Comput. Sci., 33 (1993) 266–269.
50. Agrafiotis, D. K., 3-rd Electronic Computational Chemistry Conference, http://hackberry.chem.niu.edu/ECCC3/paper48, 1996.
51. Agrafiotis, D. K., Bone, R. F., Salemme, F. R. and Soll, R. M., United States Patent 5,463,564, 1995.
52. Graybill, T. L., Agrafiotis, D. K., Bone, R., Illig, C. R., Jaeger, E. P., Locke, K. T., Lu, T., Salvino, J. M., Soll, R. M., Spurlino, J. C., Subasinghe, N., Tomczuk, B. E. and Salemme, F. R., *Enhancing the drug discovery process by integration of high-throughput chemistry and structure-based drug design,* In Chaiken, I. M. and Janda, K. D. (Eds.) Molecular Diversity and Combinatorial Chemistry: Libraries and Drug Discovery, American Chemical Society, Washington, DC, 1996, pp. 16–27.
53. Agrafiotis, D. K., *On the use of information theory for assessing molecular diversity,* J. Chem. Inf. Comput. Sci., 37 (1997) 576–580.
54. Agrafiotis, D. K. and Lobanov, V. S., *An efficient implementation of distance-based diversity measures based on k-d trees,* J. Chem. Inf. Comput. Sci., in press.
55. Shemetulsksis, N. E., Weininger, D., Blankley, C. J., Yang, J. J. and Humblet, C., *Stigmata: an algorithm to determine structural commonalities in diverse datasets,* J. Chem. Inf. Comput. Sci., 36 (1996) 862–871.
56. Boyd, S. M., Beverly, M., Norskov, L. and Hubbard, R. E., *Characterizing the geometrical diversity of functional groups in chemical* databases, J. Comput.-Aided Mol. Design, 9 (1995) 417–424.
57. Bartlett, P. A., Abstracts of Papers of the American Chemical Society, 1996, 211.
58. Lin, S. K., *Molecular diversity assessment: logarithmic relations of information and species diversity and logarithmic relations of entropy and indistinguishability after rejection of Gibbs paradox of entropy mixing,* Molecules, 1 (1996) 57–67.
59. Turner, D. B., Tyrrell, S. M. and Willett, P., *Rapid quantification of molecular diversity for selective database acquisition,* J. Chem. Inf. Comput. Sci., 37 (1997) 18–22.
60. Chernoff, H., *The use of faces to represent points in k-dimensional space graphically,* J. Am. Statist. Assoc., 68 (1973) 361–368.
61. Andrews, D. F., *Plots of high dimensional data,* Biometrics, 28 (1972) 125–136.
62. Fienberg, S. E., *Graphical methods in statistics,* Am. Statist., 33 (1979) 165–178.
63. Inselberg, A., *The plane with parallel coordinates,* The Visual Computer, 1 (1985) 69–91.
64. Kohonen, T., Self-Organizing Maps, Springer-Verlag, Heidelberg, 1996.
65. Torgerson, W. S., *Multi-dimensional scaling: I. Theory and method,* Psychometrika, 17 (1952) 401–419.
66. Kruskal, J. B., *Non-metric multi-dimensional scaling: a numerical method,* Phychometrika, 29 (1964) 115–129.
67. Sammon, J. W., *A non-linear mapping for data structure analysis,* IEEE Trans. Comp., C-18 (1969) 401–409.
68. Agrafiotis, D. K., *A new method for analyzing protein sequence relationships based on Sammon maps,* Protein Sci., 6 (1997) 287–293.

69. Agrafiotis, D. K. and Lobanov, V. S., unpublished results.
70. Agrafiotis, D. K., *Diversity of chemical libraries*, In Schleyer, P. v. R. (Ed.), Encyclopedia of Computational Chemistry, Wiley, New York, NY, 1998.
71. Clark, R. D., *OptiSim: an extended dissimilarity selection method for finding diverse representative subsets,* J. Chem. Inf. Comput. Sci., 37 (1997) 1181–1188.
72. Holliday, J. D. and Willett, P. J., *Definitions of dissimilarity for dissimilarity-based compound selection*, J. Biomol. Screening, 1 (1996) 145–151.
73. Tripos, Inc., St. Louis, MO.
74. Lowis, D., Tripos Technical Notes, 1 (1977) 2–7.
75. Nayeem, A., Heritage, T. and Hurst, T., IBC 6-th Annual Conference on Rational Drug Design, December 11–12, 1996, Coronado, CA.
76. Sage, C. R. and Ghuloum, A. M., IBC 6-th Annual Conference on Rational Drug Design, December 11–12, 1996, Coronado, CA.
77. Mount, J., Ruppert, J., Welch, W. and Jain, A., IBC 6-th Annual Conference on Rational Drug Design, December 11–12, 1996, Coronado, CA.

M.R. Pavia and W.H. Moos (eds.), Annual Reports in Combinatorial Chemistry and Molecular Diversity, Vol. 2, 93–127.
KLUWER/ESCOM

Comprehensive survey of chemical libraries yielding enzyme inhibitors, receptor agonists and antagonists, and other biologically active agents: 1992 through 1997

Roland E. Dolle
Department of Chemistry, Pharmacopeia, Inc., 101 College Road East, Princeton, NJ 08540, U.S.A. E-mail: roland@pharmacop.com

Key words: agonist, antagonist, antiinfective, combinatorial chemistry, enzyme, inhibitor, kinase, protease, receptor

Summary

This review is a historical accounting of chemical libraries from which biologically active agents have been obtained. The comprehensive tabulation includes citations as early as 1992, when the first descriptions of biologically active libraries were disclosed, and continues through 1997. Four tables are provided listing libraries screened against (1) proteolytic enzymes, (2) non-proteolytic enzymes, (3) G-protein coupled receptors (GPCRs), and (4) other targets not classified in the first three tables (e.g., non-GPCRs, integrins, antiinfectives). A name, generic structure, and size is provided for each library citation, accompanied by the molecular screen and the structure and potency of the most active library member. In total, 86 libraries are presented with 60% of the contributions reported from pharmaceutical and biotechnology companies. Approximately 70% of the libraries have used α-amino acid synthons in their construction and 85% of the libraries include one or more amide bonds.

Introduction

This past decade has witnessed the emanation of combinatorial chemistry as a new technology for the rapid construction of molecular entities [1–3]. Numerous solution- and solid-phase techniques are currently available for constructing chemical libraries for biological screening from the robotic-assisted parallel synthesis of individual compounds or defined mixtures to the more sophisticated techniques of encoding individual members of massive compound libraries created by split-pool synthesis [1]. The application of polymer supports in combinatorial chemistry allows for reactions to be driven to completion with excess reagents and simplifies product isolation. This has provided the impetus for the adaptation of classical organic transformations and the development of new synthetic methodology for resin-based library synthesis [2]. The use of polymer supports has also spawned the continuing development of a broad spectrum of photo-, acid-, base- and fluoride-cleavable linkers [1,2]. Of particular significance is the renewed interest in honing the chemical properties of scavenger resins, which provide a convenient avenue for reaction purification of chemical libraries produced in solution [2e,f].

The principal tenet of combinatorial chemistry is that the expedient assembly and screening of compound libraries will accelerate the drug discovery process [1]. It is this expectation which continues to fuel the development of combinatorics and which affords its broad acceptance within the biotechnology and pharmaceutical industries. In this regard, it is appropriate to comprehensively review those chemical libraries which have successfully yielded biological agents over the past five years [3]. Biologically active libraries reported in the literature vary in structural identity, size and purpose. Their design may be purely random in nature for use as a broad screening tool or more rational in their design, taking full advantage of

Table 1. Chemical libraries targeted for proteases

Serine proteases			
Library 1.1: Dodecapeptide **Size**: ca. 50 000 000 members **Affiliation**: Houghten Pharm. [8a,b] **Year**: 1993	Ac-Aa$_1$-Aa$_2$-Aa$_3$-Ala-Lys-Ile-Tyr-Arg-Pro-Aa$_4$-Aa$_5$-Aa$_6$-OH	→	Ac-Tyr-Tyr-Gly-Ala-Lys-Ile-Tyr-Arg-Pro-Asp-Lys-Met-OH Enzyme: Trypsin Activity: IC_{50} = 10.0 μM
Library 1.2: Cyclic lysine **Size**: 25 230 members **Affiliation**: Houghten Pharm. [9] **Year**: 1994	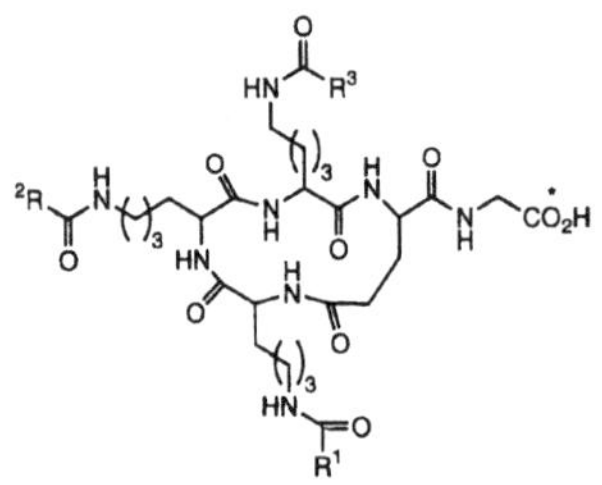	→	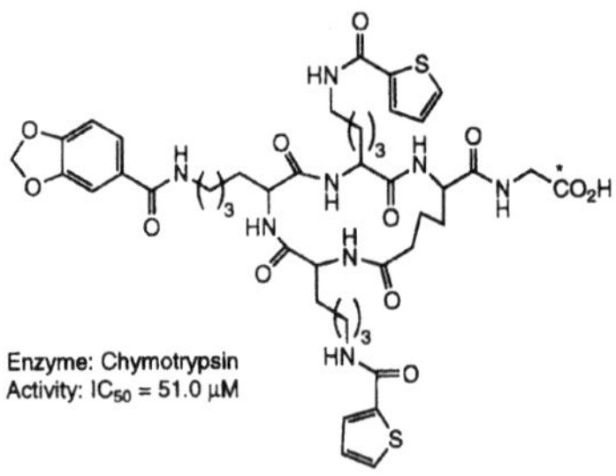 Enzyme: Chymotrypsin Activity: IC_{50} = 51.0 μM
Library 1.3: D-Hexapeptide **Size**: several thousand members **Affiliation**: Houghten, R. A. et al. [8c,d] **Year**: 1995 **Note**: D-Amino acids only.	H-D-Aa$_1$-D-Aa$_2$-D-Aa$_3$-D-Aa$_4$-D-Aa$_5$-D-Aa$_6$-NH$_2$	→	H-D-Arg-D-Tyr-D-Arg-D-Pro-D-Trp-D-Pro-NH$_2$ Enzyme: Trypsin Activity: IC_{50} = 62.0 μM
Library 1.4: Octapeptide **Size**: >10 000 members **Affiliation**: Meldal, M. et al. [20] **Year**: 1995	H-Aa$_1$-Aa$_2$-Aa$_3$-D-Aa$_4$-Aa$_5$-Aa$_6$-Val-Phe-OH	→	H-Ala-Met-Met-D-Cys(*t*-Bu)-Met-Ile-Val-Phe-OH Enzyme: Subtilisin Carlsberg Activity: IC_{50} = 3.0 μM
Library 1.5: Nonapeptide **Size**: 280 054 members **Affiliation**: Selectide Corp. [7] **Year**: 1995	H-D-Phe-Pro-Arg-Pro-Aa$_1$-Aa$_2$-Aa$_3$-Aa$_4$-Aa$_5$-OH	→	H-D-Phe-Pro-Arg-Pro-Phe-Gly-Tyr-Arg-Val-βAla-OH Enzyme: Thrombin Activity: IC_{50} = 40.0 μM
Library 1.6: Kemp's triamide **Size**: not disclosed **Affiliation**: Selectide Corp. [7b,c] **Year**: 1995 **Note**: Amino acid encoding and bead shaving method.	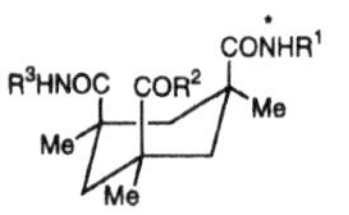	→	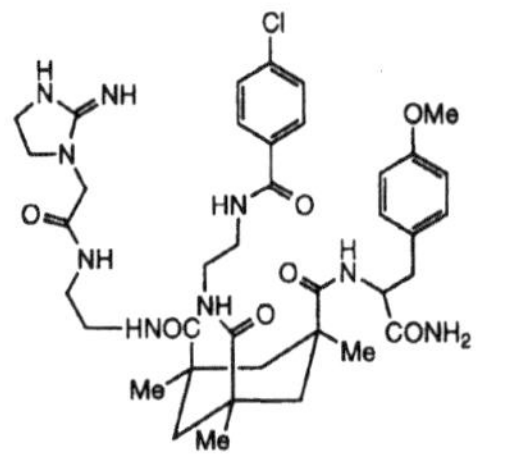 Enzyme: Thrombin Activity: IC_{50} = 4.0 μM
Library 1.7: Xanthene tetra-amide **Size**: ca. 65 000 members **Affiliation**: Rebek, J. et al. [6] **Year**: 1995	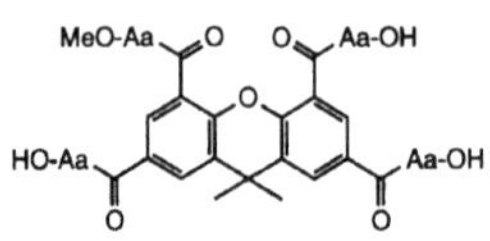	→	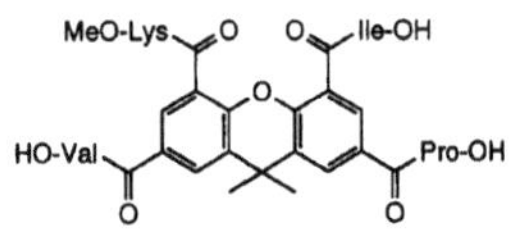 Enzyme: Trypsin Activity: IC_{50} = 9.0 μM

Table 1. (continued)

Serine proteases		
Library 1.8: Peptidyl α-ketoamide Size: 324 members Affiliation: Abeles, R. H. et al. [4] Year: 1995 Note: On-bead assay.	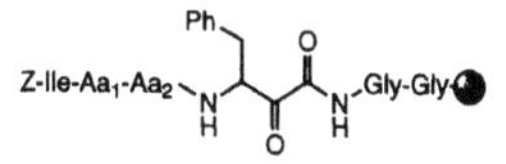	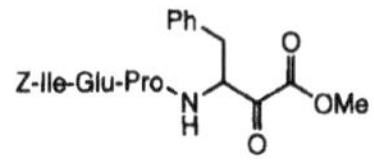 Enzyme: Chymase (human heart) Activity: K_i = 1.0 nM Enzyme: Chymotrypsin Activity: K_i = 10.0 nM
Library 1.9: Peptidyl fluorobenzylpyruvamide Size: 400 members Affiliation: Abeles, R. H. et al. [4] Year: 1995 Note: On-bead assay.	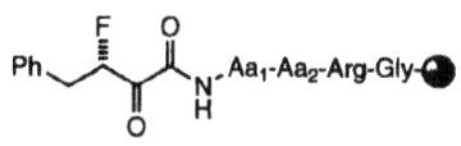	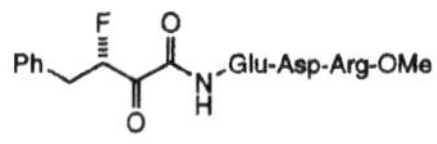 Enzyme: Chymase (human heart) Activity: K_i = 1.0 μM Enzyme: Chymotrypsin Activity: K_i = 100.0 μM
Library 1.10: Cyclonodecapeptide Size: up to 8 000 members Affiliation: Leatherbarrow, R. L. et al. [18] Year: 1996 Note: On-bead assay.	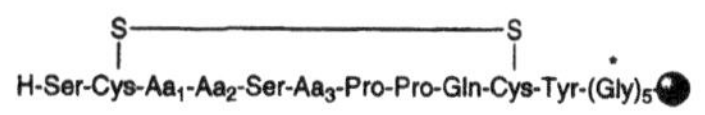	 Enzyme: α-Chymotrypsin (yeast) Activity: K_i = 19.0 nM
Aspartic acid proteases		
Library 1.11: Statine tetrapeptide Size: >200 members Affiliation: Lily [22] Year: 1991	Ac-Aa_1-Aa_2-Statine-Aa_3-NH_2	Ac-Trp-Val-Statine-D-Leu-NH_2 Enzyme: HIV-1 protease Activity: IC_{50} = 200.0 nM
Library 1.12: Aminimide Size: 22 members Affiliation: ArQule, Inc. [25] Year: 1995 Note: Solution-phase synthesis.	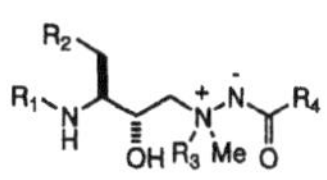	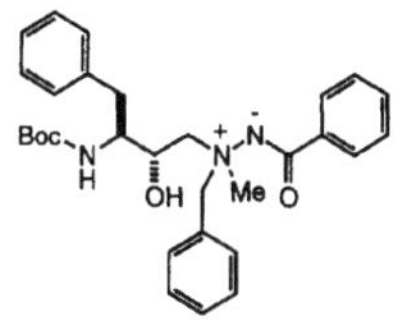 Enzyme: HIV-1 protease Activity: K_i = 137.0 nM
Library 1.13: Diamino-diol, -alcohol Size: *ca.* 300 members Affiliation: Abbott Laboratories [28] Year: 1995	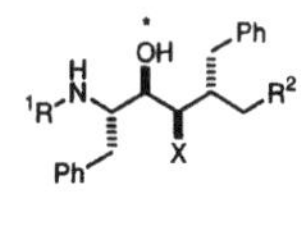	Targeted for HIV-1 protease (no activity data given)
Library 1.14a: Hydroxyethylamine Size: 1 000 members Affiliation: Ellman, J. A. et al. [15, 16] Year: 1997 Note: Structure-based design and computational directed synthons selection.	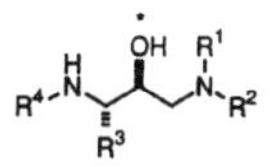	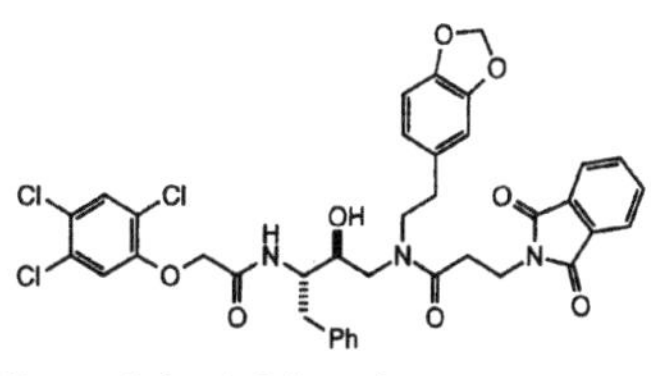 Enzyme: Cathepsin D (human) Activity: K_i = 73.0 nM

Table 1. (continued)

Aspartic acid proteases		
Library 1.14b: Hydroxyethylamine Size: 1 000 members Affiliation: Ellman, J. A. et al. [16] Year: 1997 Note: Computational selection of diverse synthons.	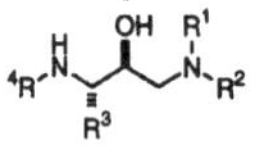	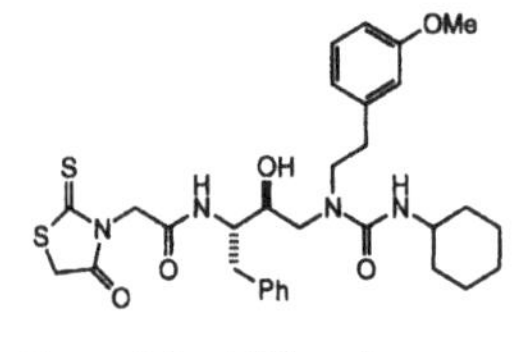 Enzyme: Cathepsin D (human) Activity: K_i = 356.0 nM
Library 1.15: Hydroxyethylamine Size: 39 members Affiliation: Ellman, J. A. et al. [16] Year: 1997 Note: Focused library based on lead structure discovered from library 1.14.	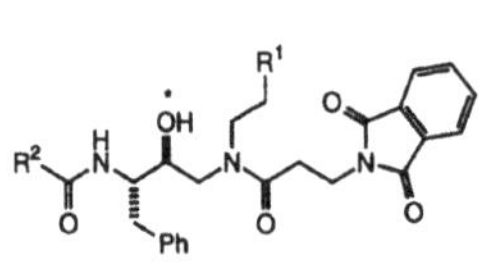	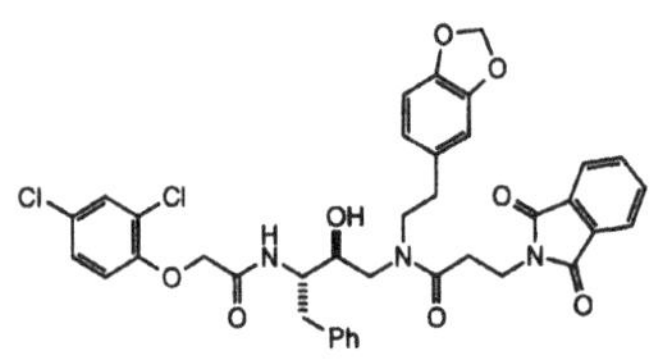 Enzyme: Cathepsin D (human) Activity: K_i = 9.0 nM
Metallo-proteases		
Library 1.16: Tetrapeptide Size: 2 500 000 members Affiliation: Ferry, G. et al. [11] Year: 1995	H-Aa$_1$-Aa$_2$-Aa$_3$-Aa$_4$-OH	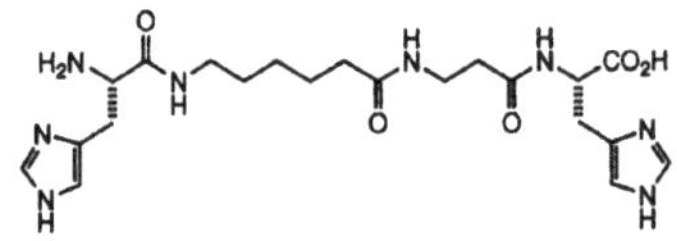 Enzyme: MMP-2 (human lung) Activity: IC_{50} = 40.0 μM Enzyme: MMP-9 (human lung) Activity: IC_{50} = 60.0 μM
Library 1.17: Peptidyl phosphonic acid ester Size: 540 members Affiliation: Affymax [5] Year: 1995	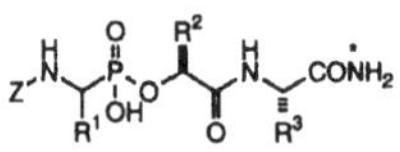	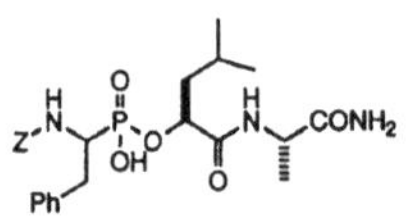 Enzyme: Thermolysin Activity: K_i = 49.0 nM
Library 1.18: Mecaptoacyl proline Size: 240 members Affiliation: Affymax [19, 21] Year: 1995, 1997 Note: Library was prepared twice: originally without encoding [21], then with encoding [19].	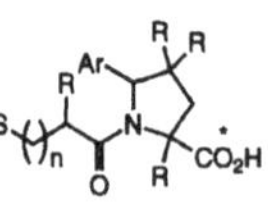	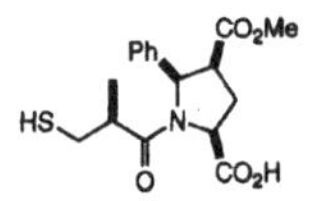 Enzyme: Angiotensin converting enzyme Activity: K_i = 0.61 nM (mixture of diastereomers); K_i = 0.16 nM (pure diastereomer)

Table 1. (continued)

Metallo-proteases

Library	Library structure	Active compound
Library 1.19: Peptidyl phosphinic acid Size: ca. 800 members Affiliation: Dive, V. et al. [14] Year: 1995	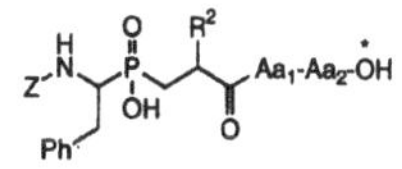	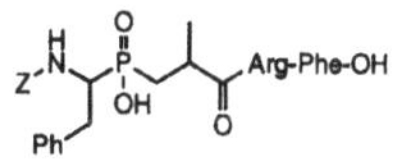 Enzyme: Thimet oligopeptidase (rat brain) Activity: K_i = 0.16 nM Enzyme: Neurolysin (rat brain) Activity: K_i = 530.0 nM
Library 1.20: Peptidyl phosphinic acid Size: ca. 800 members Affiliation: Dive, V. et al. [13] Year: 1996	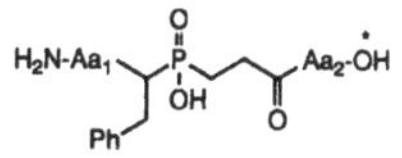	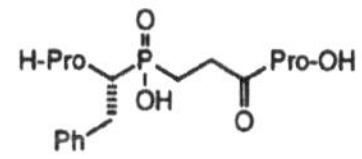 Enzyme: Neurolysin Activity: K_i = 4.0 nM Enzyme: Thimet oligopeptidase Activity: K_i = 8100.0 nM
Library 1.21: Cysteine dipeptide Size: 'several hundred' members Affiliation: Glaxo Wellcome [12] Year: 1996	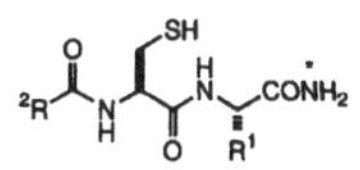	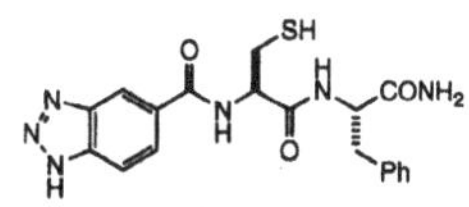 Enzyme: Collagenase (human) Activity: IC_{50} = 8.0 nM Enzyme: Gelatinase (human) Activity: IC_{50} = 24.0 nM
Library 1.22: Peptidyl phosphinic acid Size: 23 members Affiliation: Bristol-Myers Squibb [17] Year: 1996	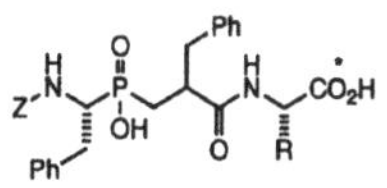	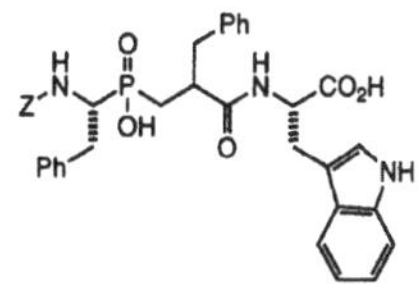 Enzyme: Endothelin converting enzyme (human) Activity: IC_{50} = 1.1 μM Enzyme: Neutral endopeptidase (human) Activity: IC_{50} = 1.1 μM
Library 1.23: N-Carboxyalkyl amino acid Size: ca. 100 members Affiliation: DuPont Merck [24] Year: 1996 Note: The library active was further optimized through structure-based design to yield a selective MMP-3 inhibitor, K_i = 9.0 nM. [24]	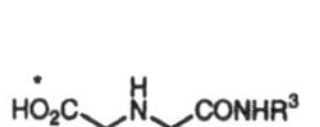	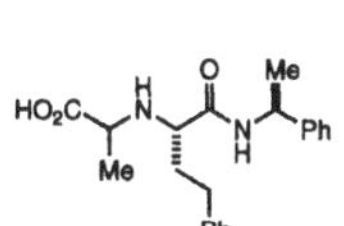 Enzyme: Stromelysin (MMP-3; human) Activity: 33% inhibition at 100 μM
Library 1.24: N-Succinylalkyl dipeptide Size: 324 members Affiliation: Affymax [26] Year: 1997	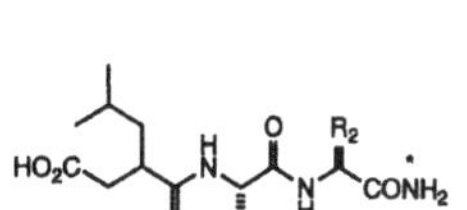	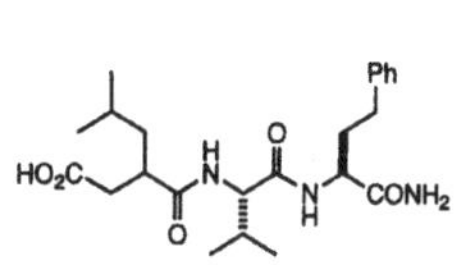 Enzyme: Matrilysin (MMP-7; human) Activity: IC_{50} = 165.0 nM

Table 1. (continued)

Metallo-proteases

Library 1.25: N-Carboxyalkyltripeptide
Size: 20 000 members
Affiliation: Merck [10]
Year: 1997

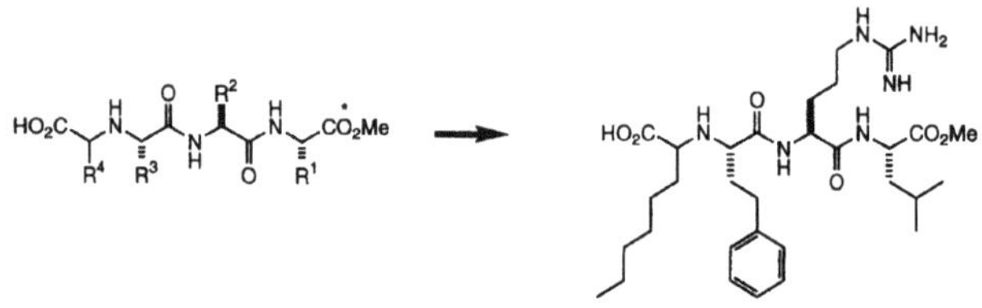

Enzyme: MMP-3
Activity: IC_{50} = 0.4 μM

Cysteine proteases

Library 1.26: N-Substituted glycine
Size: >10 000 members
Affiliation: Chiron Corp. [27]
Year: 1992
Note: Cross-reference: libraries 2.12, 4.8.

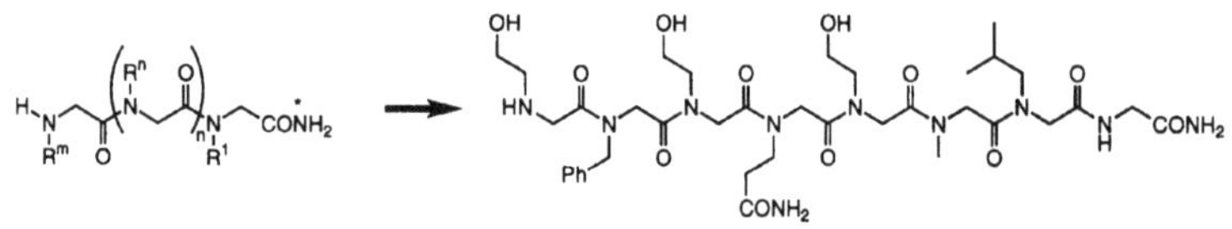

Enzyme: 3C protease (Hepatitis A virus)
Activity: K_i = 2000.0 μM

Library 1.27: Coumarin tetrapeptide
Size: 24 000 members
Affiliation: Merck [23a]
Year: 1997
Note: A substrate library from which was derived Ac-Trp-Glu-His-Asp-CHO, a potent inhibitor of ICE; K_i = 0.056 nM.

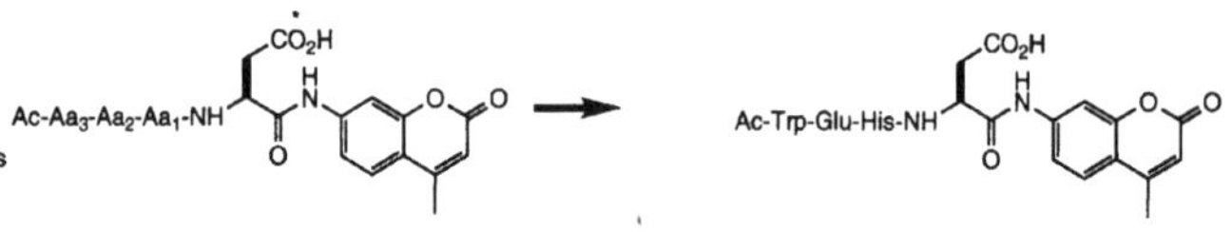

Enzyme: Interleukin-1β converting enzyme (ICE, human)
Activity: k_{cat}/K_m = 33.4 x 10^{-5} $M^{-1}s^{-1}$

• Represents the point of attachment to the solid support.

any known structural or mechanistic nuances of a particular molecular target. Library constructs have been evaluated in mechanism- and cellular-based screens, as well as directly against microorganisms affording potentially novel anti-infective agents. The early successes in this field serve to support the principal tenet and continued application of combinatorial chemistry.

Library descriptions and biological activity

Over the past five years, numerous chemical libraries have been constructed for the purpose of biological screening [1–3]. Due largely to proprietary issues, the biological activity of a library is not always disclosed to the scientific community. However, there are now many instances where biological activity is ascribed to a library, and it is these libraries that are tabulated in Tables 1–4. Library citations are divided into four main categories and are presented in chronological order within each category. Tables 1 and 2 compile those libraries screened against proteolytic [4–28] and non-proteolytic enzymes [27,29–45], respectively. Table 3 lists libraries yielding G-protein coupled receptor (GPCR) agonists and antagonists [30,42,46–68]. Table 4 delineates libraries evaluated against non-GPCRs and other biological targets not classified in Tables 1–3, e.g., the integrins, transportors, and whole organisms [27,60–80]. The tables include only those library constructs prepared by solution- or solid-phase techniques, but not through phage displays or other

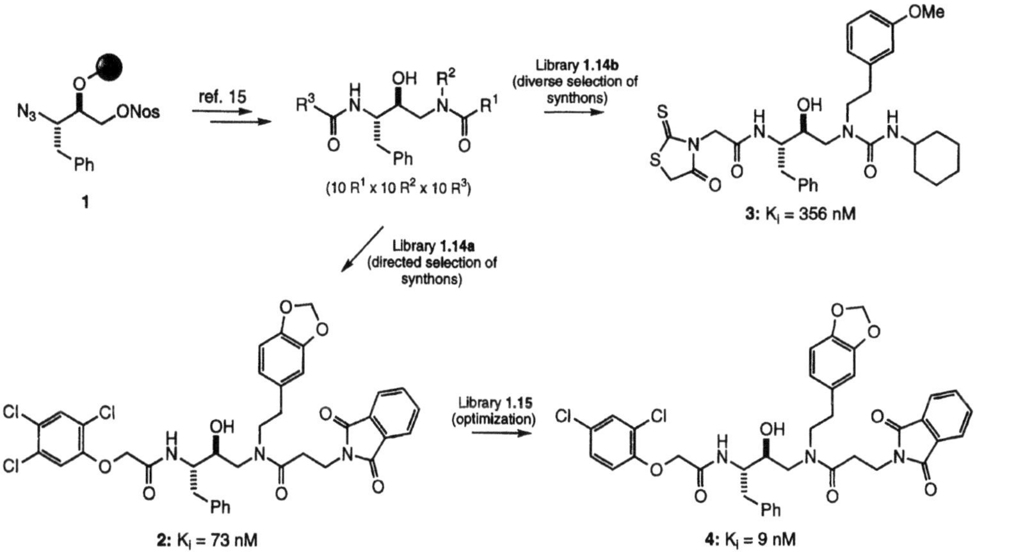

Figure 1. Ellman's and Roe's structure-based designed libraries for the inhibition of human cathepsin D [16].

biological methods. A name and generic structure for each library is provided as well as its size, affiliation, and the year of its disclosure. The affiliation is given either as a company name, for those libraries produced in industry, or a specific author, designating the senior author on libraries produced at academic or other institutions. Each library is accompanied by the name of the molecular target against which it was evaluated, the biological source of the molecular target, and, as best assessed from the literature, the structure and potency of the most active library member. In certain instances, a note field is given, typically citing a cross-reference for libraries which were screened in multiple biological assays. Finally, each library is given a reference code indicating the table in which it is listed and its corresponding entry number within that table. For example, Library 2.12 refers to library entry 12 in Table 2.

Libraries directed toward enzyme targets

Libraries yielding proteolytic enzyme inhibitors are listed in Table 1 and are sub-divided according to mechanistic class: serine proteases (libraries 1.1–1.10), aspartic acid proteases (libraries 1.11–1.15), metallo-proteases (libraries 1.16–1.25) and cysteine proteases (libraries 1.26 and 1.27). Libraries yielding non-proteolytic enzymes are found in Table 2 and are sub-divided further by kinases and phosphatases (libraries 2.1–2.6), viral enzymes (libraries 2.7 and 2.8) and other enzymes (libraries 2.9–2.19). The tables are more comprehensive in scope than in a previous review [3a] and are updated through 1997. The new entries for 1997 include libraries 1.14a,b, 1.15, 1.18, 1.24, 1.25, and 1.27, targeted against the aspartic acid, metallo- and cysteine proteases, and libraries 2.6, 2.18 and 2.19, targeted against protein phosphatase 2, an aldose, and adenosine deaminase.

Ellman published the biological activity of two 1000 member hydroxyethylamine libraries (Libraries 1.14a,b) against the human lysosomal protease, cathepsin D [16]. The chemical synthesis of the transition-state mimetic libraries was carried out as outlined two years earlier [15]. As a starting point for the selection of synthons R^1–R^3 for coupling to the resin-bound transition state isostere intermediate **1** (Figure 1), a list of amines, carboxylic acids, sulfonyl chlorides, and isocyantes with molecular weights <275 Da was generated from commercial sources. The search yielded some 700 amines and 1900 acylating agents which were deemed compatible with the library chemistry. The master list was reduced to two $10(R^1) \times 10(R^2) \times 10(R^3)$ final synthon lists used to construct libraries 1.14a (structure-based directed library) and 1.14b (diverse library). One list of synthons for library 1.14a was obtained using a computational screening process, CombiBuild, a feature of the BUILDER molecular modeling program. In essence, this process begins by placing the hydroxyethylamine transition state isostere into the active site of the enzyme, guided by the crystal structure of a pepstatine-cathepsin D complex. Synthons are then added on to the N- and C-termini and are scored on the basis of their 'fit'. The synthons with the best scores were included in the library. The second list of synthons for library 1.14b was selected solely on the basis of synthon diversity. This was accomplished with the Jarvis-Patrick algorithm using the Daylight connectivity measure of similarity and a binary Tanimoto metric. Screening the structure-based directed library 1.14a yielded a hit rate of 6–7% at 1 μM, while the more diverse library 1.14b yielded at hit rate of only 2–3%. The most active member of each library was confirmed by resynthesis: inhibitor **2** (K_i = 73 nM, library 1.14a) and **3** (K_i = 356 nM, library 1.14b). Inhibitor **2** was then subjected to further optimization through the construction of library 1.15. The strategy for optimization involved expanding on the salient structural elements observed in the library 1.14a. through re-examination of the original clustered library components. Library 1.15, composed of 39 members, furnished inhibitor **4** (K_i = 9 nM) with an 8-fold increase in activity versus inhibitor **2**. Inhibitors **2** and **4** differ from one another only by a single chlorine atom. The hydrophobic nature of the inhibitors may be a reflection of the hydrophobicity of the enzyme's active site.

Affymax described the use of their secondary amine-based tags to encode a library of highly functionalized pyrrolidines, the mercaptoacyl prolines (library 1.18) [19]. The construct was identical to the library generated several years earlier without encoding technology [21]. The key transformation in the synthesis was the 1,3-dipolar cycloaddition reaction of resin-bound azomethine ylides (derived from imines) to electron deficient olefins generating the pyrrolidine ring (Scheme 1). Split synthesis was used to prepare the library wherein four amino acids, four aldehydes, five olefins and three mercaptoacyl chlorides furnished 480 distinct products as the cycloaddition reaction is not stereospecific. Because the libraries were screened against the same target, angiotensin-converting enzyme (ACE), it allowed for the first time a direct comparison of iterative deconvolution and encoding technologies. The most potent compound **5a** was found in both libraries, but substantially more SAR information could be obtained from the encoded library. For example, analogs **6–8** were decoded from the library, demonstrating the effect of the methyl groups on activity and the tolerance of *ortho*-substitution in the phenyl ring. It was concluded that in comparison to the iterative deconvolution protocol, '... *the encoded screening strategy is a much more efficient means for extracting information from such compound collections, producing more data on a larger number of active structures*'.

Affymax also described the synthesis of an encoded 324 member library (1.24) of isobutyl succinate dipeptides [26]. The library was used in a model study to demonstrate the utility of a novel miniaturized high-throughput screening format developed at the company. The format employs a high-density 6500 well array on a standard 96-well microtiter footprint and utilizes ultra-low assay volumes (0.37 μL). In this study, the library compounds were released *in situ* by photochemical cleavage, after distribution of the beads and enzyme into the wells. Following the addition of a fluorogenic substrate, the extent of inhibition was determined using fluorescence emission/UV irradiation. The SAR data obtained in this assay format was consistent with previously established SAR, validating the utility of the high-throughput approach for screening bead-based libraries.

Combinatorial chemists at Merck described the synthesis of a mechanism-based library (1.25) of matrix metallo-protease inhibitors combining both mix and split and indexed combinatorial techniques [10]. The results found in the study were consistent with the known SAR for this class of inhibitor.

In a separate study, a Merck group constructed a protease substrate library 1.27 of P_1 aspartic acid tetrapeptide coumarins [23a]. They used this library to profile the specificity of interleukin-1β converting enzyme, leading to the identification of a novel substrate, Ac-Trp-Glu-His-Asp-coumarin. The discovery is noteworthy because a new definition of Glu as a preferred P_3 residue was established. The corresponding S_3 subsite on the enzyme was previously thought to be specific for hydrophobic residues such as valine. The new specificity information was in turn used to synthesize the corresponding tetrapeptide aldehyde **9**, Ac-Trp-Glu-His-Asp-H. The K_i determined for **9** was 56 pM, and it is the most potent tetrapeptide aldehyde yet to be described for this enzyme. A crystal structure of the inhibitor-ICE complex revealed a charge-charge interaction with the γ-carboxylate of P_3 Glu and Arg^{341} present in the enzyme's S_3 subsite. In addition, the partially negative center of P_4 Trp interacts with the guanidine side chain of Arg^{383}. The unusual high affinity of the peptide aldehyde would most likely not have been predicted on the basis of the known

Table 2. Chemical libraries targeted for non-proteolytic enzymes

Library	Structure		Result
Kinases and phosphatases			
Library 2.1: Pentapeptide Size: 2 500 000 members Affiliation: Samson, I. et al. [40] Year: 1995 Note: On-bead assay.	$H\text{-}Aa_1\text{-}Aa_2\text{-}Aa_3\text{-}Aa_4\text{-}Aa_5\text{-}$●	→	H-Asn-Trp-Met-Met-Phe-OH Enzyme: Glycosomal phosphoglycerate kinase (*Trypanosoma brucei*) Activity: IC_{50} = 80.0 μM
Library 2.2: Dibenzylaniline Size: 60 members Affiliation: ARIAD Pharm. [33] Year: 1995	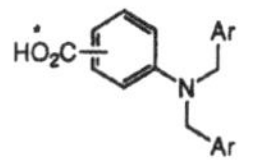	→	Enzyme: tyrosine kinases
Library 2.3: Phenol Size: 600 members Affiliation: Sphinx Pharm. [37] Year: 1995	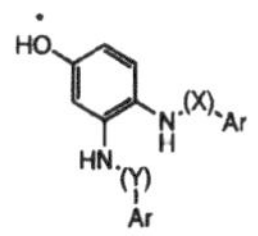	→	Enzyme: tyrosine kinases, serine/threonine kinases
Library 2.4: Peptidyl cinnamic acid Size: 125 members Affiliation: Armstrong, R. W. et al. [38] Year: 1995	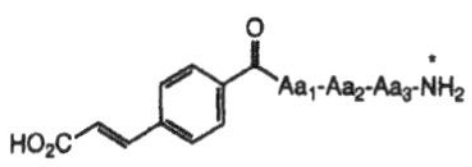	→	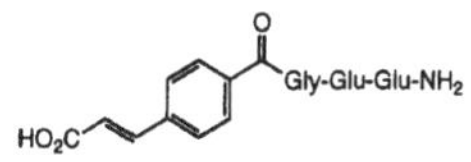 Enzyme: Protein tyrosine kinase PTP1B Activity: K_i = 79.0 nM
Library 2.5: Aminopurine glycinamide Size: 348 members Affiliation: Schultz, P. G. et al. [39] Year: 1996	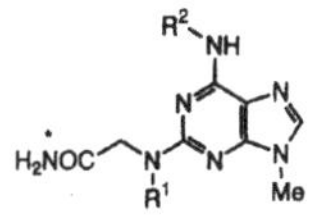	→	Enzyme: Kinases (CDK2)
Library 2.6: Calyculin A-based Size: 18 members Affiliation: Wipf, P. et al. [44] Year: 1997	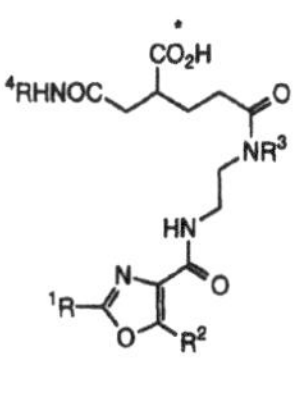	→	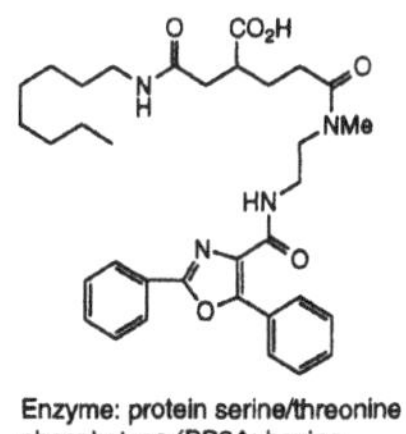 Enzyme: protein serine/threonine phosphatase (PP2A; bovine cardiac muscle) Activity: IC_{50} = ~50.0 μM
Viral enzymes			
Library 2.7: Hexapeptide Size: >2 million members Affiliation: Houghten Pharm. [36] Year: 1995	$H\text{-}Aa_1\text{-}Aa_2\text{-}Aa_3\text{-}Aa_4\text{-}Aa_5\text{-}NH_2$	→	$H\text{-}His\text{-}Cys\text{-}Lys\text{-}Phe\text{-}Tyr\text{-}Tyr\text{-}NH_2$ Enzyme: HIV-1 integrase Activity: IC_{50} = 2.0 μM
Library 2.8: Aminothiazole Size: ca. 400 members Affiliation: Pfizer [42] Year: 1996 Note: Solution-phase synthesis.	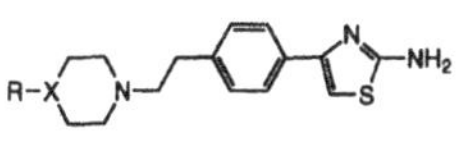	→	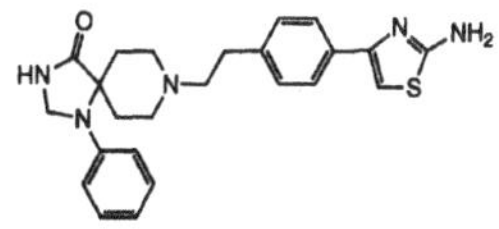 Enzyme: HSV-1 helicase ATPase Activity: IC_{50} = 1.1 μM

Table 2. (continued)

Other enzymes		
Library 2.9: Cyclohexapeptide **Size**: up to 47 000 000 members **Affiliation**: Eichler, J. et al. [31] **Year**: 1995	cyclo(Aa_1-Aa_2-Aa_3-Aa_4-Aa_5-Aa_6)	cyclo(D-Ile-Trp-D-Tyr-Arg-D-Tyr-Asn) Enzyme: α-Glucosidase (bakers yeast) Activity: K_i = 1.7 μM
Library 2.10: Oligo-nucleoside phosphate **Size**: 20 736 members **Affiliation**: Isis Pharm. [30] **Year**: 1995 Note: Cross-reference: library 3.22.	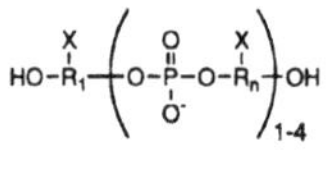	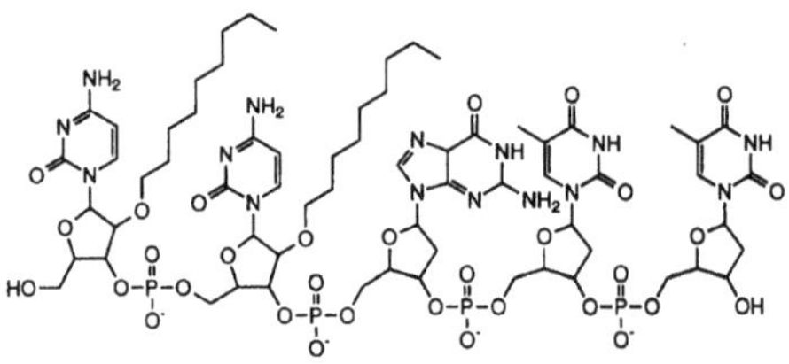 Enzyme: Phopholipase A_2 (PLA_2; human type II) Activity: IC_{50} = 2.0 μM
Library 2.11: Phosphate diester **Size**: 810 members **Affiliation**: Isis Pharm. [34] **Year**: 1996	$R\text{-}OP(O)(O^-)O\text{-}X_1\text{-}OP(O)(O^-)O\text{-}(CH_2)_2\text{-}\dot{O}H$	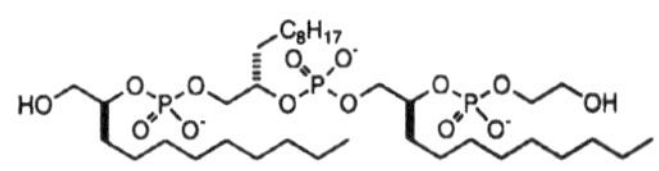 Enzyme: PLA_2 (human type II) Activity: IC_{50} = 5.0 μM
Library 2.12: N-Substituted glycine **Size**: >10 000 members **Affiliation**: Chiron Corp. [27] **Year**: 1992 Note: Cross reference: libraries 1.26, 4.8.	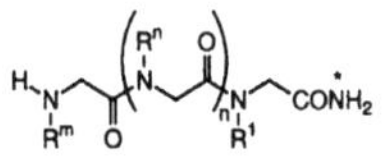	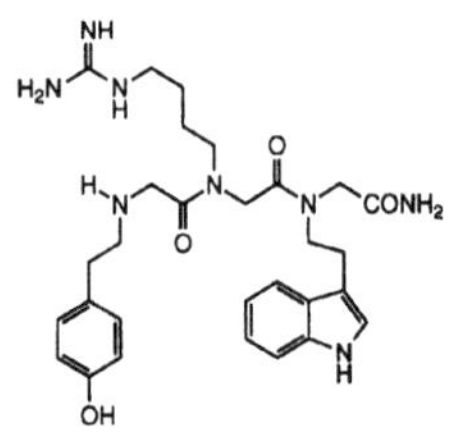 Enzyme: α-Amylase Activity: K_i = 100.0 μM
Library 2.13: Amino acid amide **Size**: 6 727 members **Affiliation**: Pharmacopeia, Inc. [29] **Year**: 1995	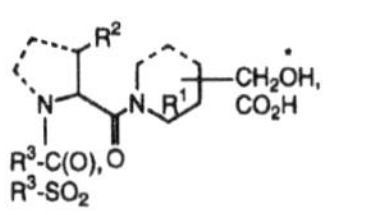	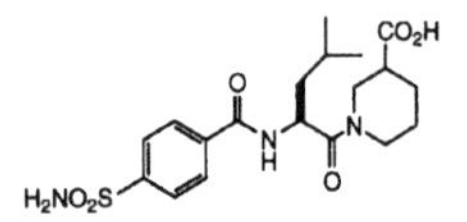 Enzyme: Carbonic anhydrase II Activity: K_D = 4.0 nM
Library 2.14: Dihydrobenzopyran **Size**: 1 143 members **Affiliation**: Pharmacopeia, Inc. [29] **Year**: 1995	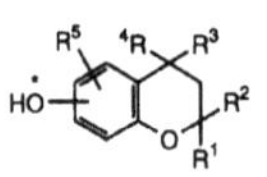	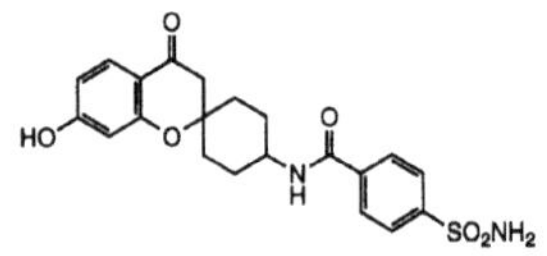 Enzyme: Carbonic anhydrase II Activity: K_D = 15.0 nM

Table 2. (continued)

Other enzymes			
Library 2.15: Peptidyl carboxysulfonamide **Size:** 545 members **Affiliation:** Whitesides, G. M. et al. [32] **Year:** 1996	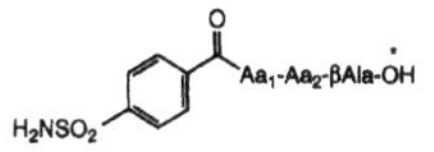	→	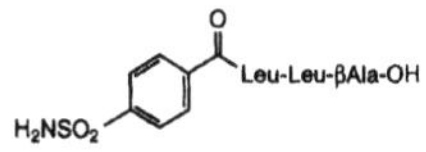 **Enzyme:** Carbonic anhydrase II **Activity:** K_b = 14.0 nM
Library 2.16: Thiazolidinone **Size:** 375 members **Affiliation:** Affymax [35] **Year:** 1996	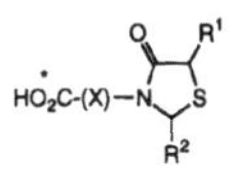	→	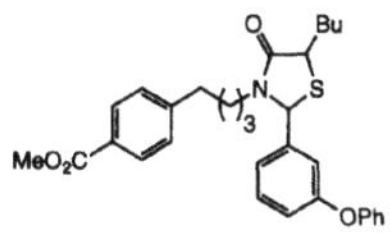 **Enzyme:** Cyclooxygenase I **Activity:** IC_{50} = 3.7 μM
Library 2.17: Tetrapeptide **Size:** 2 000 000 members **Affiliation:** Wallace, A. et al. [43] **Year:** 1996	H-Aa$_1$-Aa$_2$-Aa$_3$-Aa$_4$-NH$_2$ *	→	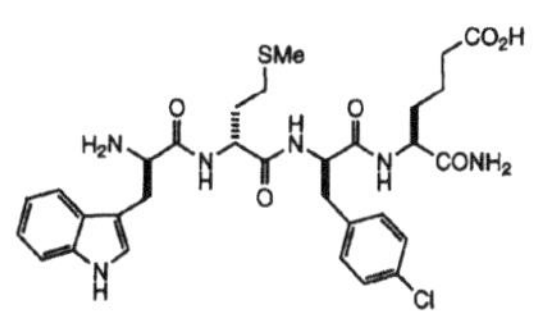 **Enzyme:** Farnesyl-protein transferase (FPTase; human) **Activity:** K_i = 2.0 nM
Library 2.18: Pentapeptide **Size:** 537 824 members **Affiliation:** Samson, I. et al. [41] **Year:** 1997 **Note:** On-bead assay.	H-D-Aa$_1$-D-Aa$_2$-D-Aa$_3$-D-Aa$_4$-D-Aa$_5$-●	→	H-D-Lys-D-(2-Thienyl)alanine-D-Lys-D-Ala-D-Arg-NH$_2$ **Enzyme:** Fructose-1,6-bisphosphate Aldolase (*Trypanosoma brucei*) **Activity:** IC_{50} = 200.0 μM
Library 2.19: 9-(2-Hydroxy-3-alkyl)adenine **Size:** 8 members **Affiliation:** Venton, D. L. et al. [45] **Year:** 1997 **Note:** Solution mixture synthesis.	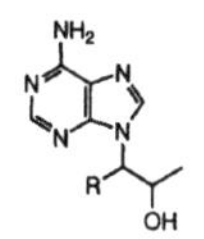	→	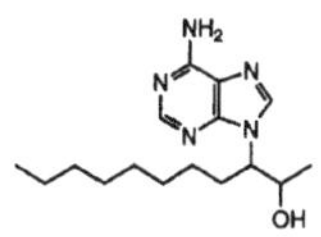 **Enzyme:** Adenosine deaminase **Activity:** K_d = ca. 3.5 nM

* Represents the point of attachment to the solid support.

ICE-inhibitor (Ac-Tyr-Val-Ala-Asp-H) complex. The substrate library 1.27 was also used to define the optimal P_1–P_4 specificity for the family of caspases and granzyme B [23b].

The only authentic description of a library active against a cysteine protease was that disclosed by Chiron some five years ago, when their N-substituted glycine library 1.26 was screened for inhibitory activity against the viral cysteine protease, hepatitis A 3C protease [27].

In one of the very few examples using natural products as a template for enzyme inhibitor discovery, Wipf and co-workers described a library of putative protein serine/threonine phosphatase (PSTPase) inhibitors [44]. There are a number of natural products including okadaic acid, tautomycin, nodularin, motuporin, microcystin LR **10**, and calyculin A **11** (Figure 2) which possess potent inhibitory activity against PSTPases. These natural products were collectively analyzed to generate a common phamacophore model for enzyme inhibition. Using this model, library 2.6 was designed and synthesized. Screening library 2.6 against protein phosphatase-1 and -2A, two of three known PSTPase subtypes, furnished weak micromolar inhibitors (50% inhibition at 100 μM) of the enzymes.

Pulsed ultrafiltration, when coupled to electrospray mass spectrometry, may be used to analyze ligand/biomolecule interactions [45]. As proof of principle, a small library of adenosine deaminase

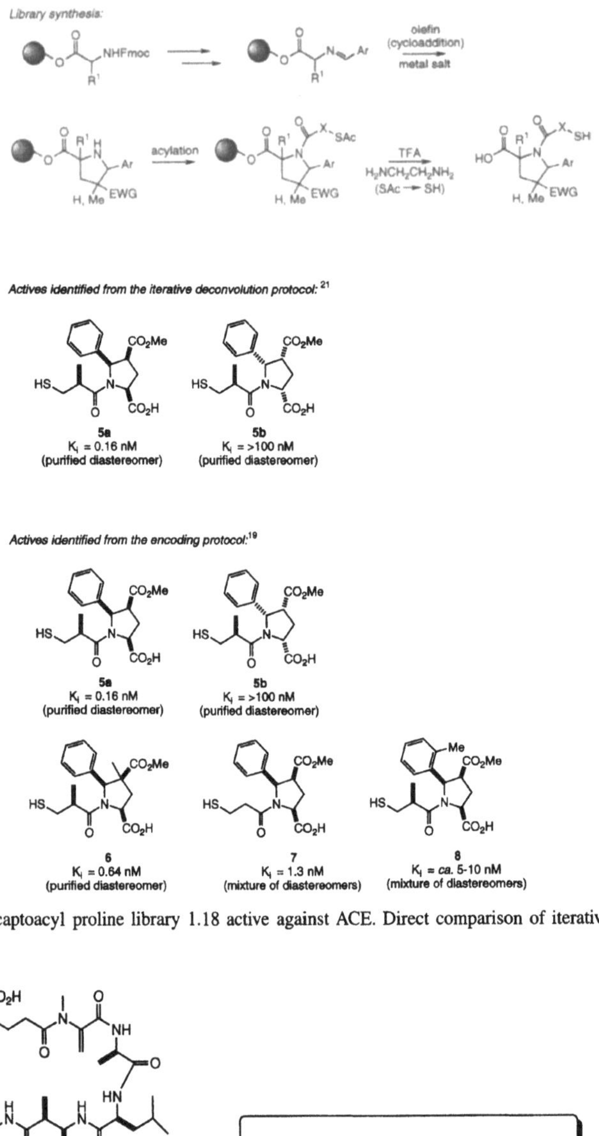

Scheme 1. The Affymax mercaptoacyl proline library 1.18 active against ACE. Direct comparison of iterative deconvolution and encoding protocols [19,21].

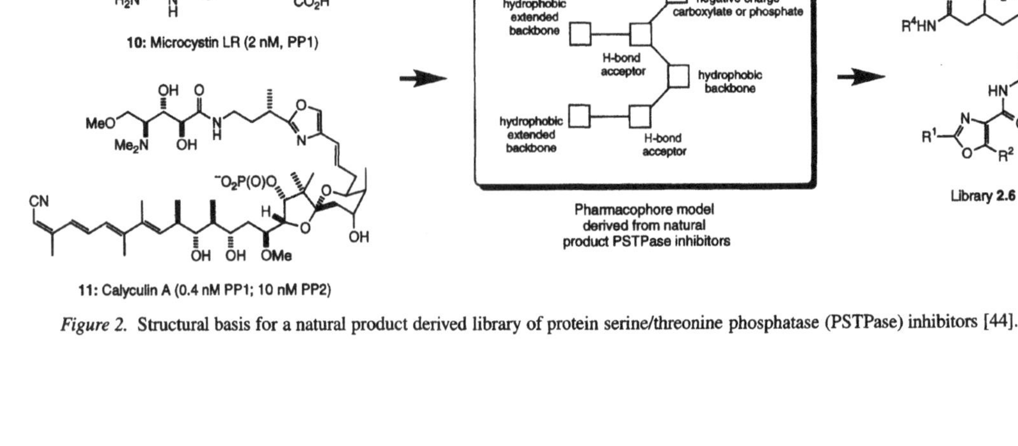

Figure 2. Structural basis for a natural product derived library of protein serine/threonine phosphatase (PSTPase) inhibitors [44].

Table 3. Chemical libraries targeted for G-protein coupled receptors (GPCRs)

Opiate receptors		
Library 3.1: Hexapeptide Size: 52 128 400 members Affiliation: Houghten, R. A. et al. [54b,c, 55] Year: 1993 Note: Dual positional scanning.	H-Aa_1-Aa_2-Aa_3-Aa_4-Aa_5-Aa_6-NH_2 •	H-Tyr-Gly-Gly-Phe-Met-Ala-NH_2 Receptor: μ-Opiate (rat brain homogenate) Activity: IC_{50} = 28.0 nM (μ-selective agonist) H-Trp-Trp-Pro-Lys-His-Gly-NH_2 Receptor: μ-Opiate Activity: IC_{50} = 9.0 nM (μ-selective agonist)
Library 3.2: Hexapeptide Size: 52 128 4000 members Affiliation: Dooley, C. T. et al. [52, 55] Year: 1993 Note: Dual positional scanning	Ac-Aa_1-Aa_2-Aa_3-Aa_4-Aa_5-Aa_6-NH_2 •	Ac-Arg-Phe-Met-Trp-Met-Arg-NH_2 Receptor: Opiate (guinea pig brain suspension) Activity: IC_{50} = 1.1 nM (μ); 32.4 nM (δ); 494.0 nM (κ_1); >10 000 nM (κ_2); 2.6 nM (κ_3) (μ-selective antagonist) Ac-Phe-Arg-Trp-Trp-Tyr-Met-NH_2 Receptor: μ-opiate (rat brain homogenate) Activity: IC_{50} = 33.0 nM (μ-selective agonist)
Library 3.3: Hexapeptide Size: ca. 204 073 344 members Affiliation: Houghten, R. A. et al. [54a] Year: 1993 Note: Iterative positional scanning requiring synthesis of six libraries of 34 million members each.	H-Aa_1-Aa_2-Aa_3-Aa_4-Aa_5-Aa_6-NH_2 •	H-Tyr-Gly-Gly-Phe-Met-Tyr-NH_2 Receptor: Opiate (rat brain) Activity: IC_{50} = 17.0 nM
Library 3.4: N-Substituted glycine Size: 5 000 members Affiliation: Chiron Corp. [68] Year: 1993 Note: Cross-reference: library	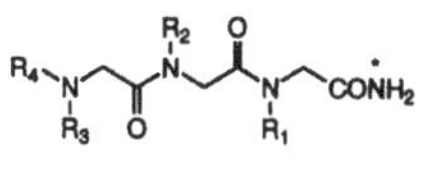	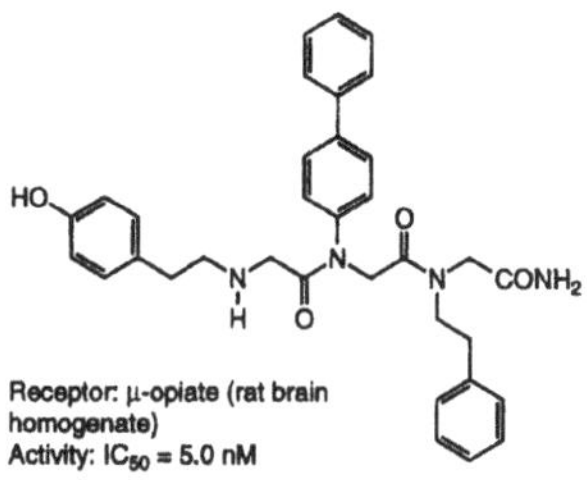 Receptor: μ-opiate (rat brain homogenate) Activity: IC_{50} = 5.0 nM
Library 3.5: D-Hexapeptide Size: 52 128 400 members Affiliation: Dooley, C. T. et al. [53] Year: 1995 Note: D- Amino acids only; dual positional scanning.	Ac-D-Aa_1-D-Aa_2-D-Aa_3-D-Aa_4-D-Aa_5-D-Aa_6-NH_2 •	Ac-D-Arg-D-Phe-D-Trp-D-Ile-D-Asn-D-Lys-NH_2 Receptor: μ-opiate (rat brain homogenate) Activity: K_i = 16.0 nM (μ_1); 41.0 nM (μ_2); (μ-selective agonist)
Library 3.6: Glycopeptide Size: 4 members Affiliation: Drouillat, B. et al. [56] Year: 1997	• (Glucuronate)$_n$-Aa_1-Phe-Gly-Gly-Tyr-H	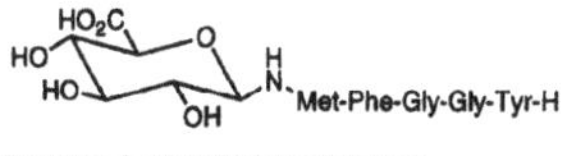 Receptor: δ-opiate (guinea pig ileum) Activity: IC_{50} = 110.0 nM

Table 3. (continued)

Tachykinin receptors			
Library 3.7: Undecapeptide **Size:** 512 members **Affiliation:** Chiron Mimotopes [67] Year: 1993	H–Aa_1-Aa_2-.....-Aa_7-Aa_8-Gly-Aa_{10}-Met-NH_2	→	H-Arg-Pro-Lys-Pro-Gln-Gln-Phe-Pro-Gly-Leu-Met-NH_2 (substance P) Receptor: NK1 (rat brain synaptosome preparation) Activity: IC_{50} = 1.0 nM
Library 3.8: Dipeptide Size: 256 members Affiliation: Parke-Davis [47, 59] Year: 1994	R-Aa_2-Aa_1-NH_2	→	Boc-Phe-Phe-NH_2 Receptor: NK-3 (human receptor expressed in CHO cells) Activity: IC_{50} = 15.5 μM
Library 3.9: Cyclic hexapeptide Size: 4 members Affiliation: Hirschmann, R. et al. [58] Year: 1996 Note: Solution mixture synthesis and HPLC separation.	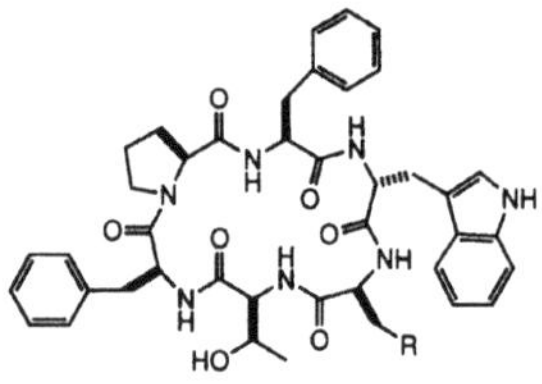	→	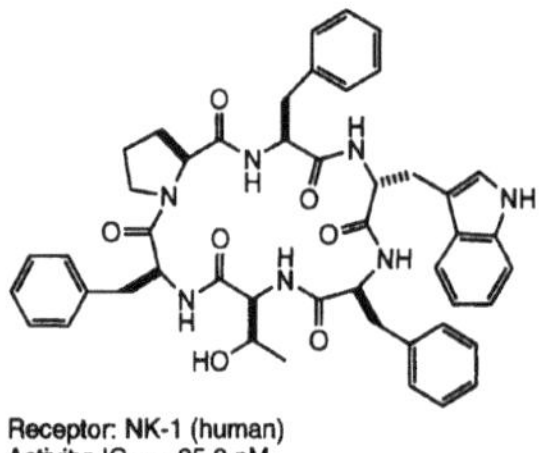 Receptor: NK-1 (human) Activity: IC_{50} = 95.0 nM
Library 3.10: 3,4-Dichlorophenylbutyldiamine Size: 1 180 members Affiliation: Pfizer [42] Year: 1996 Note: Solution-phase synthesis. Most active library member was the original lead.	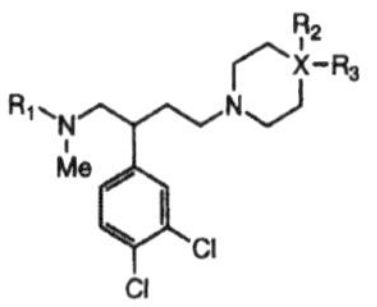	→	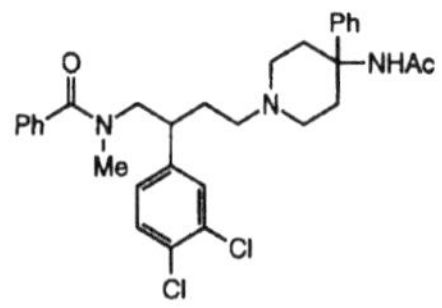 Receptor: NK-2 (rat duodenum membrane preparation) Activity: IC_{50} = 2.4 nM
Library 3.11: Peptidomimetics Size: several thousand members Affiliation: Pfizer [66] Year: 1997	R-X-Aa-NH_2	→	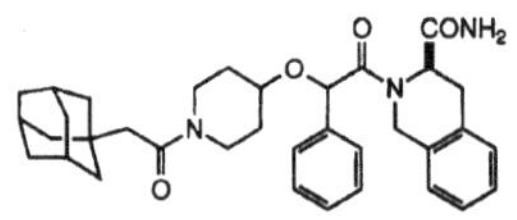 Receptor: NK-2 Activity: IC_{50} = 59.0 nM
Library 3.12: Cyclopeptide Size: 60 members Affiliation: Bitan, G. et al. [46] Year: 1997	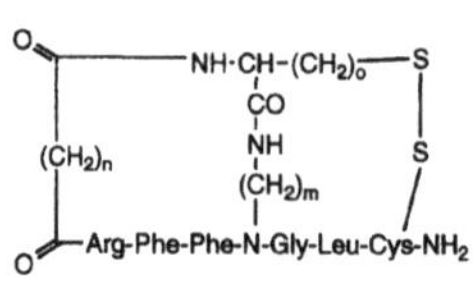	→	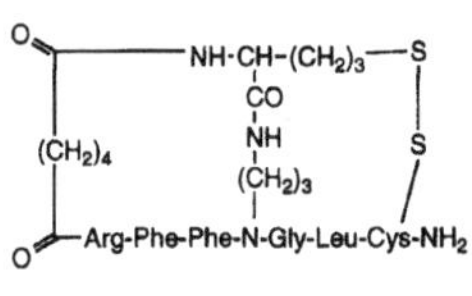 Receptor: NK-1 Activity: K_i = 4.0 μM (agonist)

Table 3. (continued)

Endothelin A (ET_A) receptors		
Library 3.13: Hexapeptide Size: ca. 64 members Affiliation: Chiron Corp. [64] Year: 1993	Ac-Aa_1-Aa_2-Aa_3-Aa_4-Aa_5-Aa_6-OH →	Ac-D-Phe-D-Orn-D-Asp-D-Ile-Ile-D-Trp-OH Receptor: ET_A (murine 3T3 cells) Activity: IC_{50} = 250.0 nM
Library 3.14: Hexapeptide Size: 300 members Affiliation: Chiron Corp. [63] Year: 1993	Ac-Aa_1-Aa_2-Aa_3-Aa_4-Aa_5-Aa_6-OH →	Ac-D-Phe-(N-benzyl)Gly-Asp-Ile-Ile-Trp-OH Receptor: ET_A (murine 3T3 cells) Activity: IC_{50} = 12.0 nM
Library 3.15: Hexapeptide Size: ca. 300 members Affiliation: Schering-Plough [61] Year: 1995	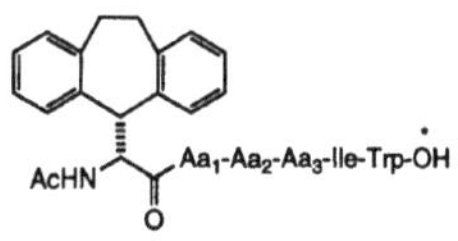 →	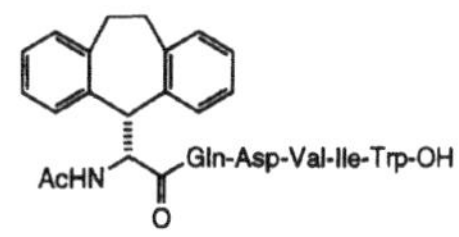 Receptor: ET_A (porcine heart membrane preparation) Activity: IC_{50} = 1.7 nM
Library 3.16: Tripeptide Size: 30 752 members Affiliation: Pfizer [65] Year: 1995	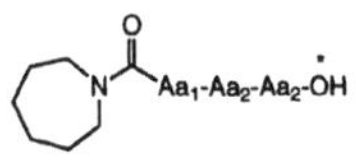 →	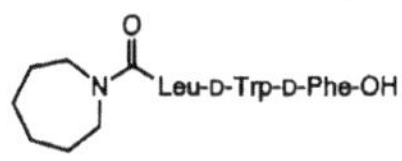 Receptor: ET_A (dog spleen) Activity: IC_{50} = 2.5 nM
Other G-protein coupled receptors		
Library 3.17: N-Substituted glycine Size: 5 000 members Affiliation: Chiron Corp. [71] Year: 1993 Note: Cross reference: library 3.3.	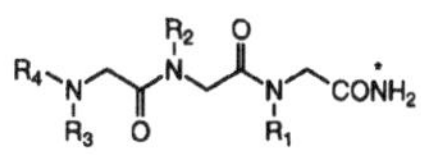 →	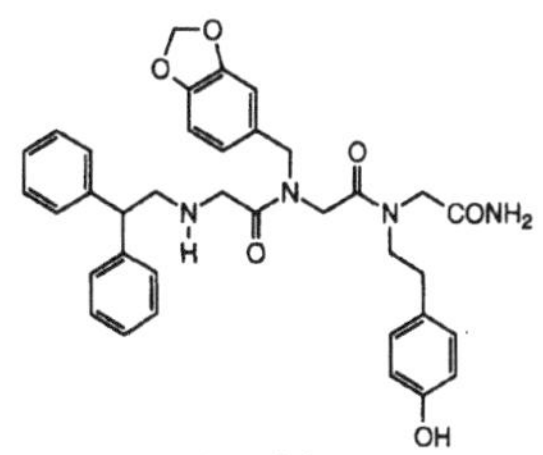 Receptor: α_1-adrenergic (rat brain homogenate) Activity: IC_{50} = 6.0 nM
Library 3.18: 1,4-Benzodiazepine Size: 192 members Affiliation: Ellman, J. A. et al. [48] Year: 1994	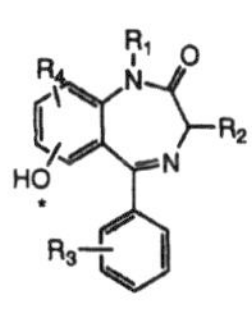 →	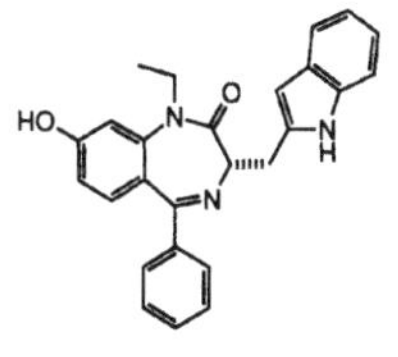 Receptor: CCK-A (rat pancreas membrane preparation) Activity: IC_{50} = 80.0 nM
Library 3.19: Tripeptide Size: 221 184 members Affiliation: Quillan, J. M. et al. [62] Year: 1995	H-Aa_1-Aa_2-Aa_3-NH_2 →	H-D-Trp-Arg-Leu-NH_2 Receptor: melanocyte-stimulating hormone (MSH; endogenous amphibian and transfected human receptors) Activity: IC_{50} = 0.62 μM

Table 3. (continued)

Other G-protein coupled receptors		
Library 3.20: N-Substituted glycinamide Size: 328 509 members Affiliation: Heizmann, G. et al. [57] Year: 1995	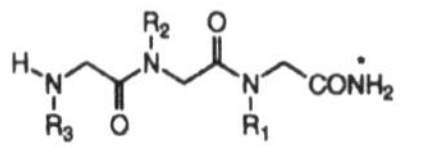	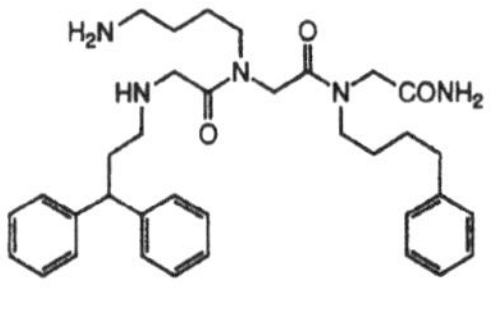 Receptor: melanocyte-stimulating hormone (MSH; human HBL melanoma cells) Activity: K_D = 1.6 nM
Library 3.21: Tripeptide Size: 15 625 members Affiliation: Glaxo [51] Year: 1995	H-Aa$_1$-Aa$_2$-Aa$_3$-OH*	H-D-Tic-D-Tyr-D-(4-nitro)-Phe-OH Receptor: Vasopressin-2 (LLCPK1 cells) Activity: IC_{50} = 63.0 nM
Library 3.22: Oligo-nucleoside phosphate Size: 20 736 members Affiliation: Isis Pharm. [30] Year: 1995 Note: Cross-reference: library 2.10.	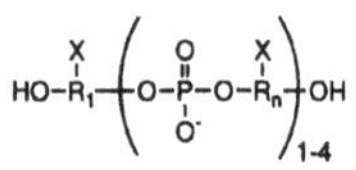	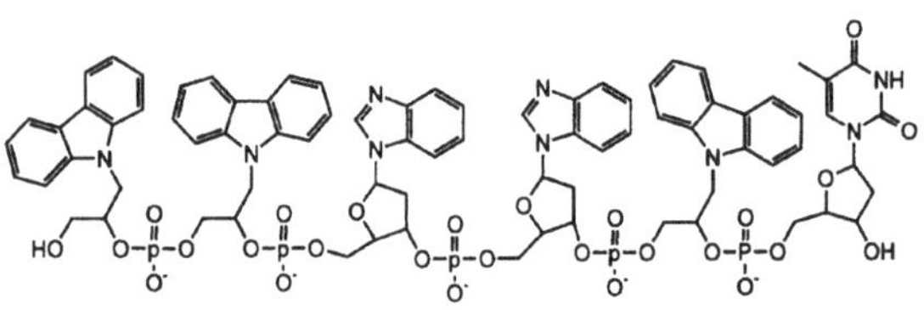 Receptor: LTB_4 (guinea pig spleen) Activity: IC_{50} = 0.7 μM
Library 3.23. Diaminotriazine Size: 350 members Affiliation: Neurocrine Biosciences [50] Year: 1996	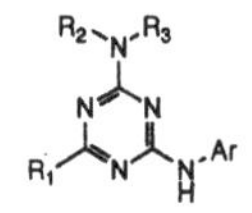	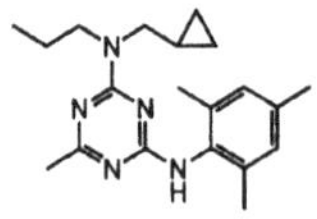 Receptor: CRF-1 (membrane preparation of cells transfected with human receptor) Activity: IC_{50} = 57.0 nM
Library 3.24: Peptidomimetic Size: 146 members Affiliation: Scios Nova Inc. [49] Year: 1996	H-D-Arg-Arg-X$_1$-X$_2$-Arg-OH*	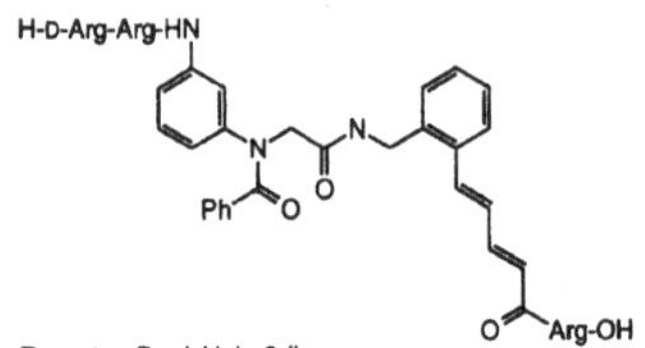 Receptor: Bradykinin-2 (human receptor expressed in CHO cells) Activity: IC_{50} = 80.0 nM
Library 3.25: Thiazolylethenylanilide Size: 700 members Affiliation: Hoffmann-La Roche [60] Year: 1997 Note: Most active compound was the original lead.	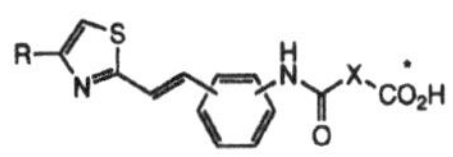	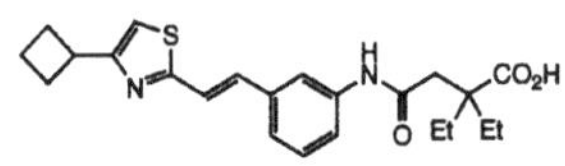 Receptor: LTD_4 (guinea pig tracheal smooth muscle) Activity: IC_{50} = 6.4 nM

* Represents the point of attachment to the solid support.

Library 3.1 (dual positional scanning SPCL)[54b,c,55]

H-O_1-O_2-X-X-X-X-NH_2

- one library composed of 400 separate sublibraries defined by the first two amino acids (O_1 and O_2)
- 130 321 members per sublibrary
- L-amino acids only
- free N-terminus
- total number of peptides: 52 128 400

↓ opioid receptor screening

H-Tyr-Gly-Gly-Phe-Met-Ala-NH_2
12: IC_{50} = 28 nM (μ-agonist)
- sequence related to Met-enkephalin

H-Trp-Trp-Pro-Lys-His-Gly-NH_2
13: IC_{50} = 9 nM (μ-agonist)
- novel hexapeptide sequence

Library 3.2 (dual positional scanning SPCL)[52,55]

Ac-O_1-O_2-X-X-X-X-NH_2

- one library composed of 400 separate sublibraries defined by the first two amino acids (O_1 and O_2)
- 130 321 members per sublibrary
- L-amino acids only
- acetylated N-terminus
- total number of peptides: 52 128 400

↓ opioid receptor screening

Ac-Arg-Phe-Met-Trp-Met-Arg-NH_2
15: IC_{50} = 1.1 nM (μ-antagonist)
- active against κ_3 (IC_{50} = 2.6 nM)

H-Phe-Arg-Trp-Trp-Tyr-Met-NH_2
17: IC_{50} = 33 nM (μ-agonist)

Library 3.3 (iterative positional scanning SPCL)[54a]

H-O_1-X-X-X-X-X-NH_2 ....H-X-X-X-X-X-O_6-NH_2

- library composed of 6 separate libraries, each defined by a single amino acid (O_1 to O_6)
- 34 012 224 members per library
- L-amino acids only
- free N-terminus
- total number of peptides: 204 073 344

↓ opioid receptor screening

18: H-Tyr-Gly-Gly-Phe-Met-Tyr-NH_2
IC_{50} = 17 nM (μ-agonist)

Library 3.5 (dual positional scanning SPCL)[53]

Ac-O_1-O_2-X-X-X-X-NH_2

- one library composed of 400 separate sublibraries defined by the first two amino acids (O_1 and O_2)
- 130 321 members per sublibrary
- D-amino acids only
- acetylated N-terminus
- total number of peptides: 52 128 400

↓ opioid receptor screening

19: Ac-D-Arg-D-Phe-D-Trp-D-Ile-D-Asn-D-Lys-NH_2
IC_{50} = 16 nM (μ-agonist)

Figure 3. Synthetic peptide combinatorial libraries (SPCLs) of Houghten, Dooley and co-workers [52–55].

inhibitors (library 2.19), analogs of the known potent inhibitor *erythro*-9-(2-hydroxy-3-nonyl)adenine, was constructed and analyzed by this method [45]. Homologs of *erythro*-9-(2-hydroxy-3-nonyl)adenine were identified and the results were in good agreement with the corresponding bioassay data.

Libraries yielding G-protein coupled receptor agonists and antagonists

Table 3 delineates the libraries in from which GPCR inhibitors have been identified. The table is subdivided into several receptor classes: opiate receptors (libraries 3.1–3.6), tachykinin receptors (libraries 3.7–3.12), endothelin A receptors (libraries 3.13–3.16) and other GPCRs (libraries 3.17–3.25).

Some of the early examples of libraries yielding receptor agonists and antagonists were peptide-based. Houghten, Dooley and co-workers introduced synthetic peptide combinatorial libraries (SPCLs) in which defined mixtures of peptides were synthesized, cleaved from resin and assayed free in solution against either soluble (enzyme) or membrane associated (receptor) molecular targets [52–55]. Three hexapeptide SPCLs (Libraries 3.1, 3.2 and 3.5) containing ca. 52 million members each were synthesized and evaluated against the μ opiate receptor (Figure 3). Library 3.1 was composed of L-amino acid amides having a free N-terminus (-NH_2). Peptide **12** was identified as μ-selective opiate agonist, with the first five residues corresponding to Leu- or Met-enkephalin [54b,c]. Subsequently, two novel agonist series, unrelated to the enkephalins, were identified from the same library: H-Trp-Trp-Pro-Lys-His-Gly-NH_2 **13** (IC_{50} = 9 nM) and H-Tyr-Pro-Phe-Gly-Phe-Arg-NH_2 **14** (IC_{50} = 13 nM) [55]. Library 3.2 was identical to library 3.1 except that the N-terminus was acetylated (-NHAc; Figure 3). The triply active opiate inhibitors, Ac-Arg-Phe-Met-Trp-Met-Arg-NH_2 **15** [52] and Ac-Arg-Trp-Ile-Gly-Trp-Arg-NH_2 **16** [55] were identified from the library. The peptides, termed 'acetalins', demonstrated high affinity for the μ and κ_3 opioid receptors, somewhat lower affinity for δ receptors, weak affinity for κ_1 receptors, and no affinity for κ_2 receptors. Each were found to be potent μ receptor antagonists in the

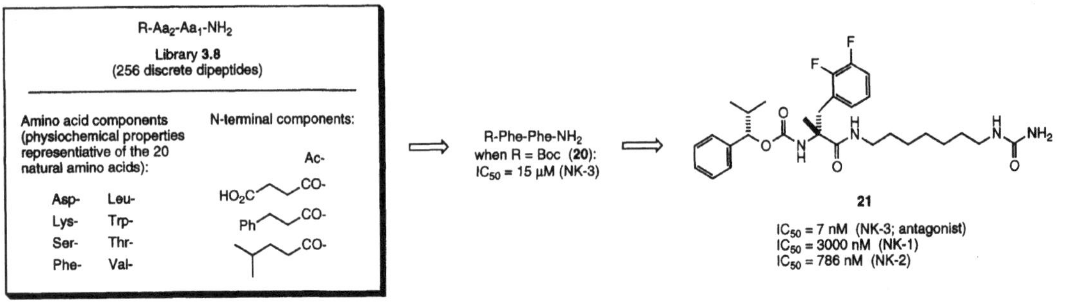

Figure 4. The Parke-Davis dipeptide library [47,59].

guinea pig ileum assay and relatively weak antagonists in the mouse vas deferens assay. Further examination of the library revealed Ac-Phe-Arg-Trp-Trp-Tyr-Met-NH_2 **17** as a μ selective agonist [55].

Houghten also produced a 6-part mega library (library 3.3) of hexapeptides containing all possible hexameric combinations of 18 of the 20 natural L-amino acids [54a]. The six positional SPCLs totaled $34\,012\,224 \times 6 = 204\,073\,344$ hexapeptides. Although much larger in size than library 3.1, the peptides obtained from this library corresponded to the highly homologous enkephalins identified in library 3.1 (e.g., **18**, Figure 3). This observation, following the screening of some 200 million peptides, would seem to suggest that nature has largely perfected Leu- and Met-enkephalin as the optimal peptide ligands for the opioid receptor.

Perhaps less well known is the preparation of library 3.5, an all D-amino acid version of library 3.1 (Figure 3) [53]. Iterative deconvolution afforded the all D-peptide **19**, Ac-D-Arg-D-Phe-D-Trp-D-Ile-D-Asn-D-Lys-NH_2. This agent exhibits high selectivity for the μ receptors (μ_1 = 16 nM; μ_2 = 41 nM) in a rat brain homogenate assay. The D-hexapeptide **19** is a full agonist as determined in a guinea pig ileum assay, IC_{50} = 440 nM, only 2-fold less active than Met-enkephalin in the same assay. Modifications to the agonist were made, including individual replacement of each D-amino acid with its L-antipode, resulting in a 2- to 200-fold decrease in activity.

Tachykinins have been the subject of library design and screening. A library of 512 substance P analogs (library 3.7) was prepared by researchers at Chiron Mimotopes [67]. Highlighting their multipin peptide synthesis technology, all possible single and multiple L- to D-amino acid replacements in nine of the residues in substance P was achieved. The group confirmed that single D-amino acid chiral changes at positions Arg^1 to Gln^5 result in moderate decrease in binding, whereas substantial decreases in activity are observed in chiral changes from Gln^6 to Leu^{10}.

The Parke-Davis group at Cambridge prepared the 256 member dipeptide library 3.8, combinatorializing eight amino acids with four N-capping groups ($8 \times 8 \times 4$ matrix; Figure 4) [59]. The eight amino acids used in the library were judiciously selected to span a diverse set of calculated physiochemical properties for the twenty natural amino acids. In this way, it was thought that the 265 member library was representative of the corresponding virtual library of 1600 compounds (as theoretically derived from all possible combinations of twenty natural amino acids and four capping groups; $20 \times 20 \times 4$ matrix). Although the screening and biological activity of library 3.8 was not disclosed *per se*, it was a library of this type which afforded Boc-Phe-Phe-NH_2 **20** as a ca. 15 μM NK-3 lead. This lead was subjected to traditional medicinal chemistry optimization generating the potent antagonist **21** (Figure 4) [47]. This research represents one of the rare examples of the derivation of a high-affinity nonpeptide antagonist from a synthetic peptide library.

A remarkable application of combinatorial chemistry is the *conversion of a selective, high affinity somatostatin receptor agonist to a selective, high affinity NK-1 receptor antagonist* as reported by Hirschmann and co-workers (Figure 5) [58]. During the course of their investigations on the design of novel peptidomimetics as receptor antagonists, it was discovered that certain saccharide-based peptidomimetics bearing functionality to interact with the somatostatin receptor possessed unexpectedly high affinity for the NK-1 receptor. This observation led to a hypothesis that the somatostatin and NK-1 GPCRs may share some measure of common topology. To test this hypothesis, a four component mixture of cyclic hexapeptides, derived from the known potent somatostatin

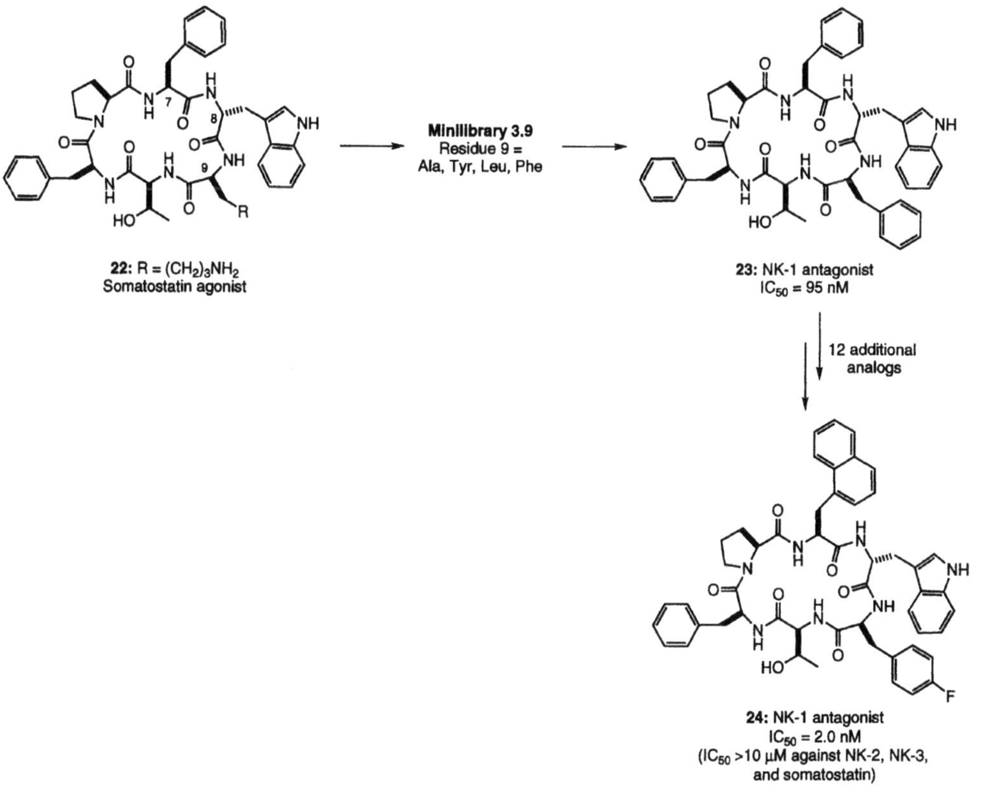

Figure 5. Conversion of a somatostatin agonist to an NK-1 antagonist [58].

receptor agonist **22**, was synthesized and evaluated for NK-1 receptor binding (library 3.9; Figure 5). The rationale for selecting the hydrophobic residues, Ala, Leu, Tyr and Phe, as replacements for the lysine side chain at residue 9 in **22** was based upon the presence of multi-aromatic functionality present in established high affinity NK-1 receptor antagonists. The cyclic hexapeptide analog **23** containing the Phe replacement for Lys, possessed good affinity (IC_{50} = 95 nM) against NK-1, and was devoid of any somatostatin receptor affinity. The corresponding Ala (IC_{50} = 4.1 μM), Tyr (IC_{50} = 1.0 μM), and Leu (IC_{50} = 2.7 μM) replacements were less active. Further modification of **23** resulted in **24** (IC_{50} = 2 nM), a selective, high affinity NK-1 antagonist.

Pfizer's contributions to the discovery of novel NK-2 receptor antagonists include Libraries 3.10 and 3.11 [42,66]. Library 3.10, composed of several parts and totaling 1180 members, is based on the 3,4-dichlorophenylalkyldiamine pharmacophore found in Sanofi Winthrop's neurokinin-2 antagonist **25** (Figure 6) [42]. The SAR information obtained from screening the total library suggested that only conservative structural changes in the lead result in active analogs (e.g., **26–28**).

Library 3.11 is presumably a more diverse peptidomimetic-oriented library, although the actual components of the library were not disclosed (Figure 7) [66]. In this instance, library 3.11 furnished potent NK-2 antagonists including **29**, which are structurally distinct from the antagonists of library 3.10. Antagonist **29** was submitted to optimization by traditional medicinal chemistry (Figure 6). The carboxamide 'handle' was removed (**29** to **30**) resulting in a <2-fold loss in activity. Reducing one of the amide carbonyls (**29** to **31**) gave a slight improvement in potency, suggesting that neither the carbonyl nor the primary carboxamide functional groups contribute significantly to the binding affinity of **29** to the NK-2 receptor.

Three research groups independently reported on peptide-based libraries targeted for the endothelin A

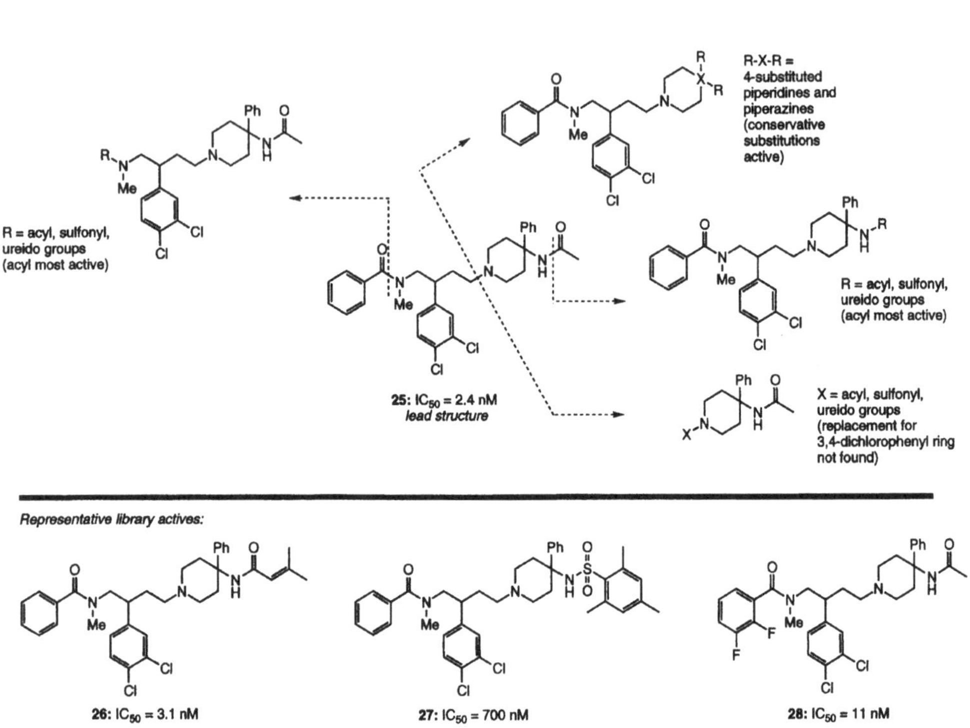

Figure 6. Library 3.10 designed from a known NK-2 antagonist [42].

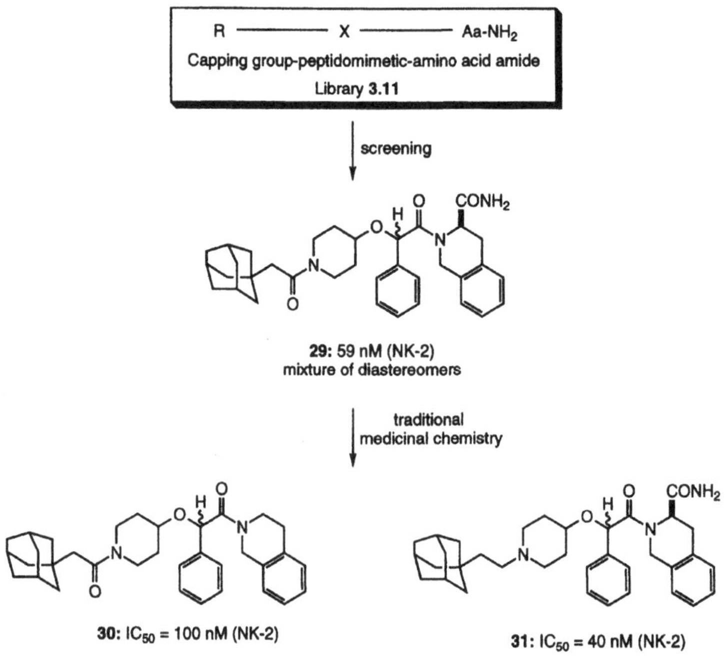

Figure 7. NK-2 receptor antagonist from a peptidomimetic library [66].

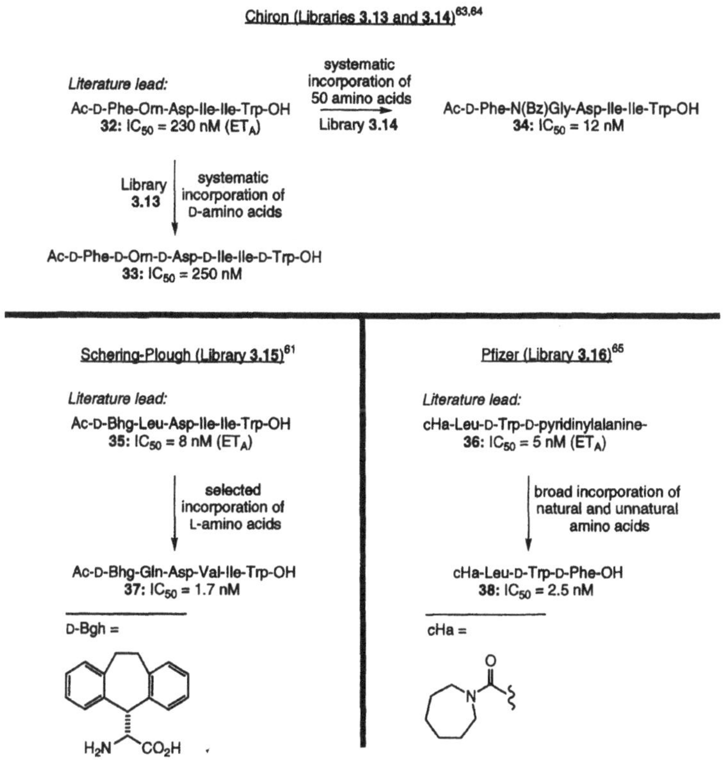

Figure 8. Endothelin receptor A (ET_A) antagonists [61,63–65].

(ET_A) receptor (libraries 3.13–3.16; Figure 8). In the earliest study, Chiron constructed two different libraries based on the lead **32** via multipin peptide synthesis (Figure 8). For library 3.13, all possible single and multiple D-amino acid replacements were introduced into hexapeptide **32** [64]. This exercise furnished **33**, an equipotent diastereomer of **32** containing four new D-amino acids. In the second library (library 3.14), a set of twenty standard L-amino acids, twenty isomeric D-amino acids, four N-substituted glycines, and six other unnatural amino acids (total number of amino acid replacements = 50) was systematically introduced at each residue in the same lead hexapeptide **32** [62]. The net result was the discovery of N-benzyl glycine as a replacement for ornithine giving rise to **34** with a 20-fold enhancement in binding affinity relative to peptide **32**.

Both Schering Plough [61] and Pfizer [65] also started with literature ET_A peptide antagonist leads, **35** and **36**, respectively, with the intent of designing optimization libraries to either enhance potency or to discovery surrogates for a portion of the amino acids present in the peptide leads. In both instances, compounds similar in structure and potency to the original leads were found (**35** versus **37**, and **36** versus **38**; Figure 8).

In an effort to move away from peptide libraries, so-called peptoid libraries of oligomeric N-substituted glycine amides have been created (Figure 9) [27,57,71]. Peptoid leads may have a potential metabolic stability advantage versus pure peptides and may represent a better starting point for drug discovery. Zuckerman was the first to prepare libraries of this class which yielded enzyme inhibitors **39** and **40** (Libraries 1.26 and 2.12) [27] and receptor antagonists **41** and **42** (Libraries 3.4 and 3.17) [71]. A ca. 330 000 member trimeric N-substituted glycine library was independently prepared by Heizmann [57] (library 3.20) affording **43**, which possesses high affinity (K_D = 1.6 nM) for human melanocyte-stimulating hormone receptor.

Triazine **44** (IC_{50} = 2.1 μM) was identified as a lead in a screening program directed toward the discovery of antagonists of corticotropin releasing

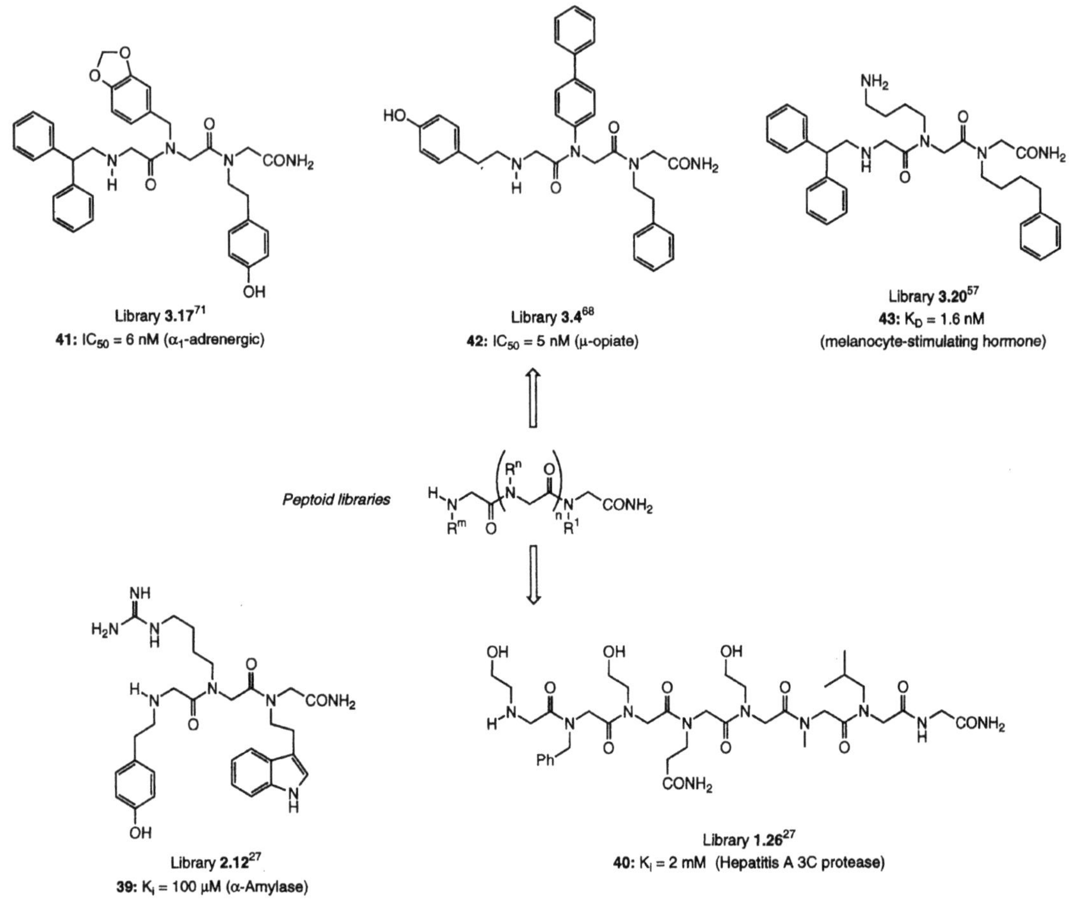

Figure 9. N-substituted glycine amide (peptoid) libraries.

factor-1 (CRF_1) [50]. In an effort to optimize the lead structure, a 350 member library of triazine analogs (library 3.23; Figure 10) was synthesized. Chemical synthesis was carried in solution via 'rapid microscale synthesis', a semi-automated technique that utilizes a reconfigured commercially available robotic system (Hewlett-Packard 7686 PrepStation) to assist in reagent addition and reaction work-up. Triazine **45**, IC_{50} = 57 nM emerged as an antagonist with enhanced affinity for CRF_1. Analysis of the library SAR data indicated a rather stringent requirement for a trisubstituted aniline and the cyclopropylmethylpropyl-amine functional groups. Subsequent examination of the corresponding pyrimidine analogs of **45** and exchange of the trimethyl- for the trichloro-substituted aniline ring system, demonstrated that the pyrimidine series was essentially equipotent with the triazine series (**46a,b** versus **45**), provided that the appropriate pyrimidine regioisomer was presented to the receptor (Figure 10). Introducing a second methyl group into the pyrimidine ring of **46a** led to a further increase in potency, as exemplified by the identification of **47** with an IC_{50} = 2 nM, a 1000-fold increase in affinity relative to the original lead **44**.

Tripeptide libraries have shown activity against certain G-protein coupled receptors. For example, screening a library of ca. 221 000 tripeptides (library 3.19) furnished the potent and selective melanocyte-stimulating hormone agonist, H-D-Trp-Arg-Leu-NH_2 (IC_{50} = 620 nM) [62]. A vasopressin-2 receptor antagonist, H-D-Tic-D-Tyr-D-(4-nitro)-Phe-OH (IC_{50} = 63 nM), was revealed in a second independently prepared tripeptide library construct (library 3.21; 15 625 members) [51].

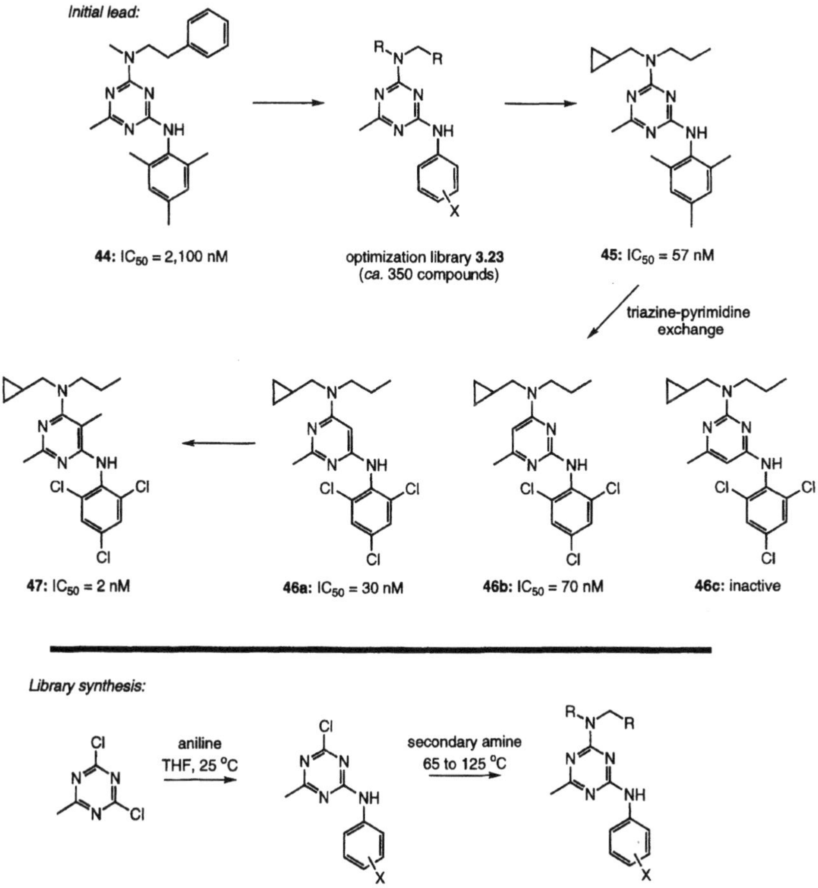

Figure 10. Triazine- and pyrimidine-based CRF_1 receptor antagonists [50].

Libraries yielding non-GPCR binding agents

Libraries yielding active structures against molecular targets not classified as G-protein coupled receptors are found in Table 4. The table is sub-divided into mammalian and non-mammalian targets. The mammalian targets include the serotonin cotransporter (tripeptide, library 4.1), the nuclear hormone receptors (hydroxystilbene, library 4.2; thiazolidinedione, library 4.3), the novel human neutrophil receptor, fMLF (β-turn mimetic, library 4.4), and the integrins (peptidomimetic, libraries 4.5–4.7). The non-mammalian targets include viral proteins (libraries 4.8, 4.10, and 4.12) and microorganisms (libraries 4.9, 4.11, 4.13 and 4.14).

The antidiabetic activity associated with thiazolidinedione **48** is effected through binding to the gamma sub-type of the peroxisome proliferator-activated receptor, PPARγ (Scheme 2) [80]. Fatty acids, e.g., **49**, are the natural ligands for PPARγ. Since ligands **48** and **49** adopt turn structures in solution as suggested by ^{1}H NOESY and ROESY spectra analysis, it was believed that certain hybrid structures of **48** and **49** should be active against the receptor. As a result, library 4.3 incorporated aliphatic dicarboxylic acid monoamides as replacements for the pyridine ring in **48** (Scheme 2). Long chain aliphatic acids (C5-C8 alkyl and alkenyl chains **50** and **51**) were tolerated as surrogates for the 2-(N-methyl)pyridinyl group in **48** validating the original hypothesis.

The -Arg-Gly-Asp- tripeptide sequence is a common recognition motif of the integrins. A 27 member peptidomimetic library (library 4.5) based on this motif was constructed, containing surrogates for both the aspartic acid β-carboxylate and the arginine guanidinyl group which are essential for integrin recogni-

Table 4. Chemical libraries targeted for non-GPCR binding agents

Mammalian targets			
Library 4.1: Tripeptide **Size:** 10 000 members **Affiliation:** Lilly [77] **Year:** 1995	$H\text{-}Aa_3\text{-}Aa_2\text{-}Aa_1\text{-}\dot{N}H_2$	→	H-D-Trp-Phe-D-Tyr-NH_2 **Protein:** Serotonin cotransporter (rat basophil leukemia cells) **Activity:** IC_{50} = 3.2 μM
Library 4.2: Hydroxystilbene **Size:** 23 members **Affiliation:** Williard, R. et al. [70] **Year:** 1995	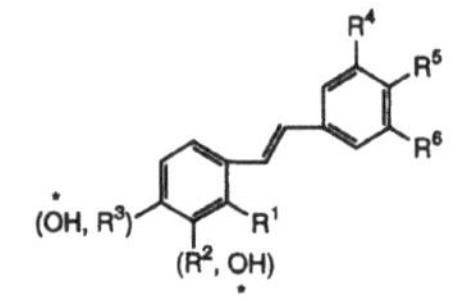	→	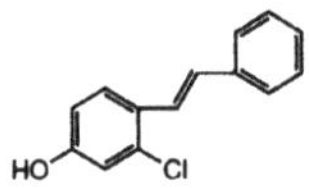 **Assay:** nuclear estrogen receptor (ERC1 cells expressing human estrogen receptors) **Activity:** EC_{50} = ca. 5.0 μM
Library 4.3: Thiazolidinedione **Size:** 22 members **Affiliation:** Glaxo Wellcome [80] **Year:** 1997	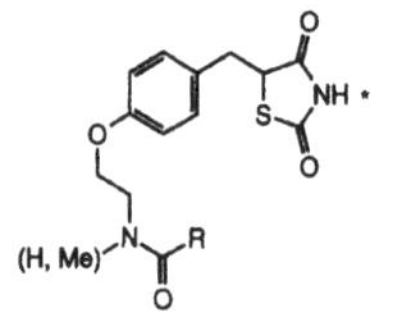	→	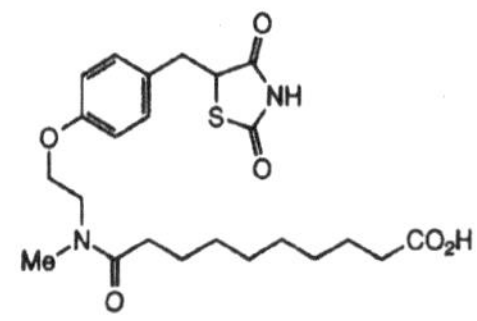 Receptor: peroxisome proliferator-activated receptor (PPARγ, human) Activity: K_i = 18.0 nM
Library 4.4: β-Turn mimetic **Size:** 1 152 members **Affiliation:** Ellman, J. A. et al. [69] **Year:** 1997	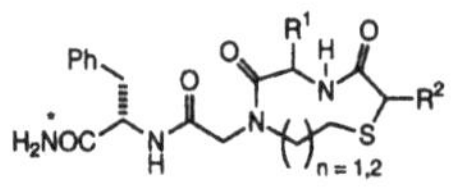	→	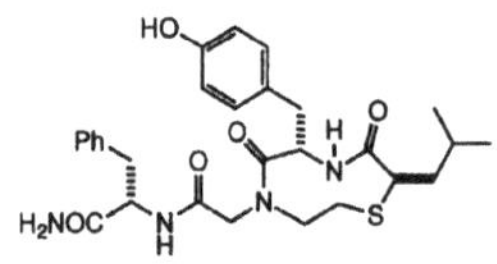 **Receptor:** fMLF (human neutrophils) **Activity:** IC_{50} = 10.0 μM
Library 4.5: Peptidomimetic **Size:** 27 members **Affiliation:** Life Science Div., Nippon Steel Corp. [75] **Year:** 1997	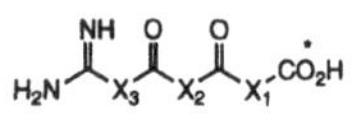	→	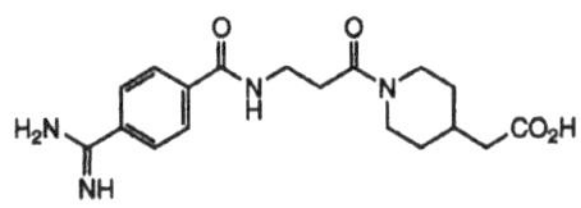 **Assay:** platelet aggregation (human platelet-rich plasma; $\alpha_{IIb}\beta_3$ integrin) **Activity:** IC_{50} = 2.5 μM
Library 4.6: Peptidomimetic **Size:** 18 members **Affiliation:** Life Science Div., Nippon Steel Corp. [75] **Year:** 1997 **Note:** Based on active found in library 4.5.	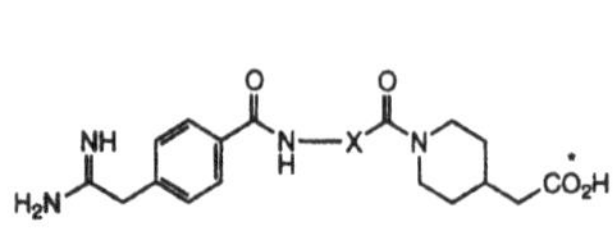	→	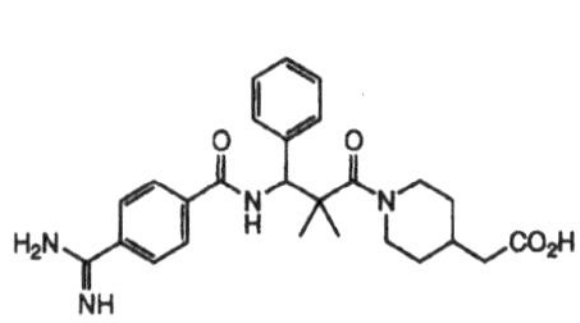 **Assay:** platelet aggregation (human platelet-rich plasma; $\alpha_{IIb}\beta_3$ integrin) **Activity:** IC_{50} = 0.19 μM

Table 4. (continued)

Library	Library structure	Example hit
Mammalian targets		
Library 4.7: Peptidomimetic Size: ca. 112 members Affiliation: DuPont Merck [73] Year: 1997	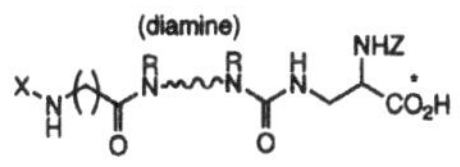	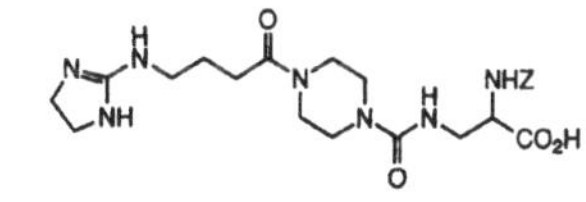 Protein: $\alpha_v\beta_3$ Integrin (human placenta) Activity: IC_{50} = 1.1 nM ($\alpha_v\beta_3$); 660.0 nM ($\alpha_5\beta_1$); 420.0 nM ($\alpha_v\beta_5$); 20,000.0 nM ($\alpha_{IIb}\beta_3$)
Non-mammalian targets and anti-infectives		
Library 4.8: N-Substituted glycine Size: >10 000 members Affiliation: Chiron Corp. [27] Year: 1992 Note: Cross reference: library 1.26.	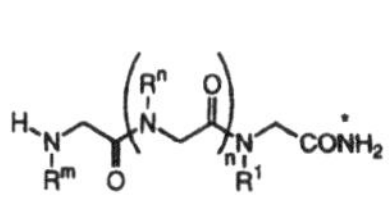	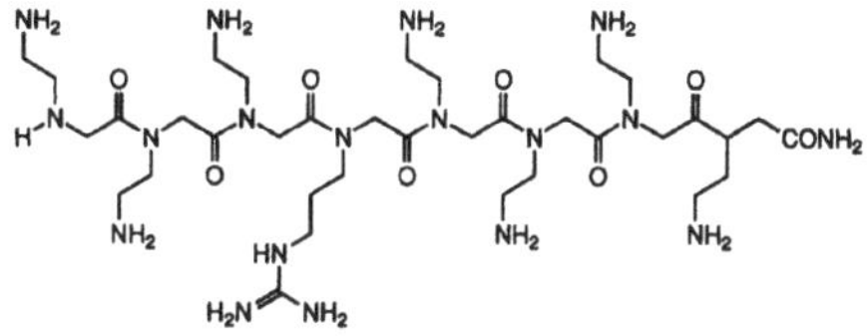 Protein: HIV Tat RNA-binding domain Activity: IC_{50} = ca. 2.0 μM
Library 4.9: Urea Size: 4 000 members Affiliation: Lily [76] Year: 1996 Note: Solution synthesis using scavanger resin.	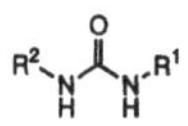	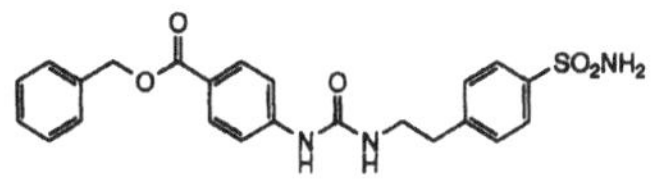 Assay: Human rhinovirus-14 (cell culture) Activity: IC_{50} = 0.8 μM
Library 4.10: Neomycin B mimetics Size: 52 members Affiliation: Wong, C.-H. et al. [78] Year: 1996 Note: Library constructed on soluble polymer, PEG Me ether (MW ca. 5000).	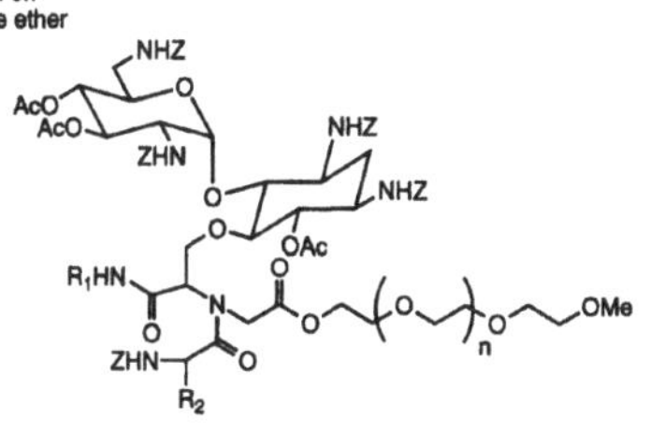	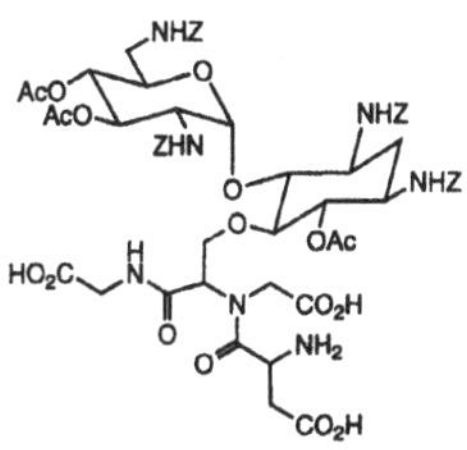 Protein: Rev (HIV transactivator protein Rev)-RRE (Rev response element) Activity: 85% inhibition @ 200 μM
Library 4.11: Hexapeptide Size: >1 000 000 members Affiliation: Reed, J. D. et al. [79] Year: 1997	H-Aa$_1$-Aa$_2$-Aa$_3$-Aa$_4$-Aa$_5$-Aa$_6$-OH*	H-D-Phe-D-Arg-D-Leu-D-Lys-D-Phe-D-His-OH Assay: Phytopathogenic fungi Activity: MIC = 36 μg/mL (*Pythium ultimum*)

Table 4. (continued)

Non-mammalian targets and anti-infectives			
Library 4.12: Biaryl carboxylic acid **Size**: 28 members **Affiliation**: Abbott Labs. [74] **Year**: 1997 **Note**: SAR by NMR method used to derive a 15 μM inhibitor from the library active (see text).	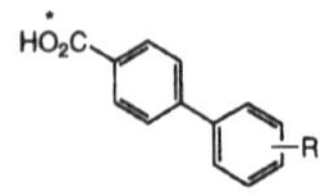	→	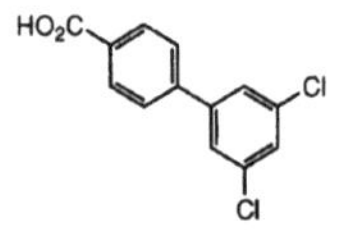 Assay: DNA-binding domain of human papillomavirus E2 protein Activity: IC_{50} = 150.0 μM
Library 4.13: Pyridinopolyamine **Size**: 1 638 members **Affiliation**: Isis Pharm. [72] **Year**: 1997 **Note**: Solution synthesis of the template.	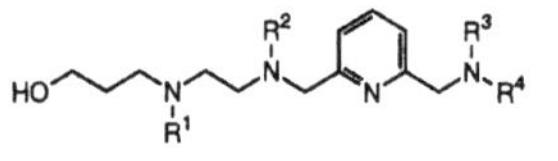	→	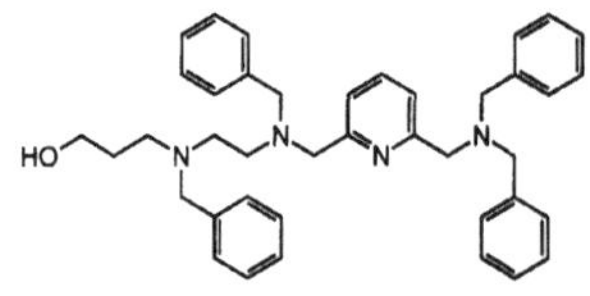 Assay: Bacteria and fungi growth inhibition Activity: MIC = 1-5 μg/mL (C. *albicans*); MIC = 5-25 μg/mL (S. *Pyogenes*; E. *cole imp*⁻)
Library 4.14: Polyazapyridinophanes **Size**: 1 600 members **Affiliation**: Isis Pharm. [71] **Year**: 1997 **Note**: Solution synthesis of the template.	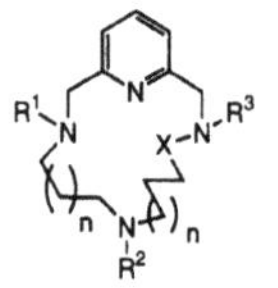	→	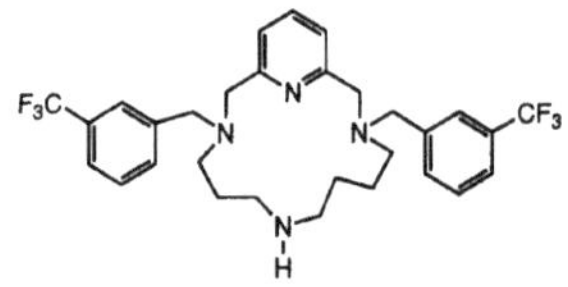 Assay: Bacteria growth inhibition Activity: MIC = 3-6 μg/mL (K. *pneumonie*); MIC = 2.5-5 μg/mL (E. *cole imp*⁻); MIC = <2.5 μg/mL (S. *Pyogenes*)

* Represents the point of attachment to the solid support.

tion [75]. Library 4.5 was screened against the $\alpha_{IIb}\beta_3$ integrin, which is abundantly present on the surface of activated platelets and binds to fibrinogen to cause platelet aggregation. Non-α-amino acid synthons for the library were selected to included piperazine-4-carboxylic acid derivatives and guanidine or benzamidine functional groups separated by a conformationally flexible spacer (Figure 11). The most active antagonist identified from the library was **52** (IC_{50} = 2.5 μM). The corresponding analogs **53** and **54** were not active, demonstrating a preferred spatial separation of the charged functionalities in **52**–**54**. An optimization library was then generated (library 4.6) wherein the C- and N-termini of 52 were held constant and 18 variations of the spacer group were surveyed. The 2,2-dimethyl-3-amino-3-phenyl-propionic acid spacer was found to be superior to the original 3-aminopropionic acid spacer: **55** (IC_{50} = 0.19 μM) versus **52** (IC_{50} = 2.5 μM). Resolution of **55** afforded one active enantiomer ((**+)-55**; IC_{50} = 0.09 μM), showing that the binding to the integrin was stereospecific. The absolute stereochemistry of the active enantiomer **(+)-55** was not disclosed.

The integrin $\alpha_v\beta_3$ is an important receptor for mediating the attachment of osteoclast to bone during bone resorption. Selective inhibitors of the integrin may represent a therapeutically relevant class of anti-osteoporoetic agents. In this regard, a research group at DuPont Merck [74] constructed library 4.7, a peptidomimetic library consisting of amine or guanidinyl N-terminal groups and a C-terminal α-N-Z-(2-amino)alanine (known to impart $\alpha_v\beta_3$ specificity) separated by diamine spacers of varying length and rigidity (Figure 12). The inhibitory action of library members against $\alpha_v\beta_3$ was assessed using an ELISA assay and their specificity was measured against the platelet integrin $\alpha_{IIb}\beta_3$ and related integrins $\alpha_5\beta_1$and $\alpha_v\beta_5$. The most active and selective agents contained the piperazine or N,N′-dimethylethylenediamine spacer coupled to ei-

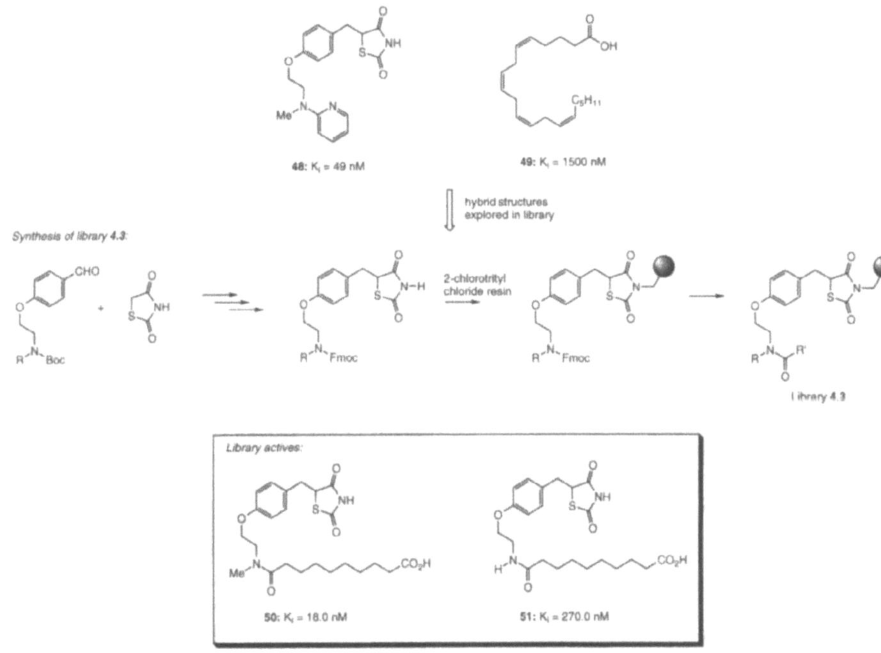

Scheme 2. Hybrid thiazolidinedione-fatty acid PPARγ ligands [80].

ther the butyryl guanidine or butyryl cyclic guanidine functionalities (**56–58**; Figure 12).

A recently introduced NMR-based approach to lead discovery known as ‘SAR by NMR’ was used in conjunction with combinatorial chemistry to discovery inhibitors that block DNA binding of human papillomavirus E2 protein (Figure 13). SAR by NMR was used to screen for initial ligands for the E2 protein by monitoring changes in the $^{15}N/^{1}H$ amide chemical shifts of the protein upon the addition of compounds. This approach led to the identification of the 4-phenylbenzoic acid **59** and the biphenyl ether **60** as weak ligands (K_D = 2.5 mM and 1.9 mM, respectively). The parallel library 4.12 of biaryl carboxylic acids was constructed as a means to enhance the affinity of **59** (Figure 13). This was accomplished with the identification of the 4-(3,5-dichlorophenyl)benzoic acid **61** (K_D = 0.06 mM). Optimization of **60** to **62** (K_D = 0.35 mm) was achieved by further screening of commercially available biaryl ethers and evaluation of a limited set of analogs. Both the biphenyl **59** and biphenyl ether **60** cause the same set of amide cross-peaks in the NMR spectrum of the E2 protein to shift, suggesting a common binding site for the two compounds. Merging the two lead structures **61** and **62** into a single hybrid structure furnished the biaryl ether **63**, possessing an IC_{50} = 10 μM against the HPV-E2 DNA binding protein.

A series of ureas demonstrating anti-rhinoviral activity in an whole cell assay was obtained from a library (4.9) prepared in solution using a covalent amine scavenger [76]. For library synthesis, an equimolar mixture of primary and secondary amines in a solution of THF was treated with a slight excess of an isocyanate (Figure 14). After a predetermined reaction time, aminomethylpolystyrene resin was added to covalently bind the excess isocyanate reagent. Filtration of the reaction mixture and evaporation of solvent afforded an equimolar mixture of ten ureas. In this fashion, the 4000 member library 4.9 was created and screened for antirhinoviral activity furnishing **64** and **65** with comparable potency (IC_{50} = ca. 1 μM).

Several structurally distinct library motifs have been evaluated for anti-infective activity. The first library of this type was peptide-based, displaying antibacterial and antifungal activity as described by Houghten in 1991 [82]. Reed and coworkers [79] have since prepared a hexapeptide library 4.11 of >1 million members from which a hexapeptide H-D-Phe-D-Arg-D-Leu-D-Lys-D-Phe-D-His-OH was found with potent activity against certain phytopathogenic fungi. Isis Pharmaceuticals described the solution synthe-

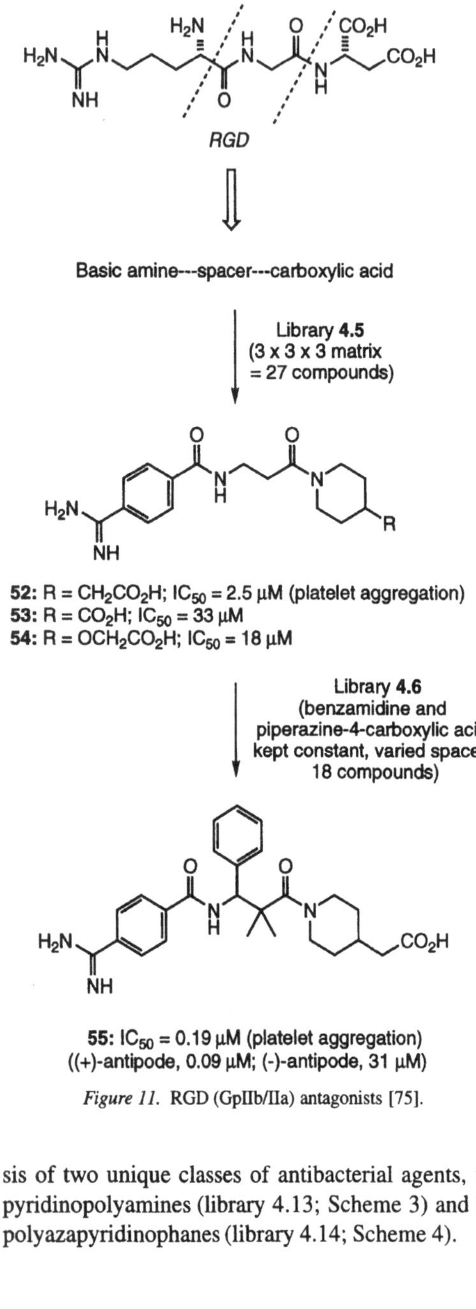

Figure 11. RGD (GpIIb/IIa) antagonists [75].

sis of two unique classes of antibacterial agents, the pyridinopolyamines (library 4.13; Scheme 3) and the polyazapyridinophanes (library 4.14; Scheme 4).

Library statistics and summary

A total of 86 biologically active libraries is presented in Tables 1–4. Approximately 50% of the libraries contain up to 1000 members, spanning a range in size from as few as 4 members (library 3.9) to 200 million members (library 3.3). There are 47 citations for libraries yielding enzyme inhibitors, 25 citations for libraries yielding G-protein coupled receptor agonists or antagonists, and 14 citations for libraries yielding miscellaneous biological activities including antibacterial and antifungal activity. As gleaned from the tables, 60% of the contributions have come from pharmaceutical and biotechnology laboratories. It is interesting to note that ca. 50% of all the libraries are peptide- or peptidomimetic-based, ca. 70% of the libraries incorporate at least one α-amino acid, and 85% contain at least one amide bond.

As suggested by these statistics, a recurring library theme throughout this time period was the synthesis and evaluation of oligomeric peptide libraries. First described by Houghten and Dooley in the discovery of enzyme inhibitors and opiate receptor binding agents (libraries 1.1, 1.3, 2.7, 3.1, 3.2, 3.3, and 3.5) [52–55], some 18 independent laboratories have reported the preparation and screening of either cyclic or acyclic peptide libraries (acyclic peptide libraries: 1.4, 1.5, 2.1, 2.17, 2.18, 3.7, 3.8, 3.13, 3.14, 3.15, 3.16, 3.19, 3.21, and 4.1; cyclic peptide libraries: 1.10, 2.9, 3.9, and 3.12). This may be a reflection of (1) the facility with which these libraries may be prepared, (2) a lingering belief that peptides represent a viable starting point for non-peptide ligand design, or as in many cases, (3) a proof of principle exercise for research groups as they embrace the new technology for the first time resulting in an outlet for publication. A few of the notable discoveries which have resulted from this activity include the potent all D-hexapeptide μ-selective opiate agonist (library 3.5), the all D-tripeptide vasopressin-2 inhibitor (library 3.21), the tripeptide MSH agonist (library 3.19), and the conversion of a cyclic hexapeptide somatostatin agonist to a potent and selective NK-1 antagonist via a single amino acid change (library 3.9).

Library constructs based on a known pharmacophore as a means to obtain potent and selective biological agents are a largely successful design concept. Overall, approximately 50% of the tabulated libraries incorporate a structural bias directed toward a particular molecular target. Specifically, ca. 70–80% of the enzyme directed libraries are pharmacophore-biased, while about half that number, ca. 35–40%, of the receptor directed libraries are likewise constructed utilizing a known pharmacophore. A successful application of this strategy is the peptidyl phosphinic acid libraries of Dive and coworkers [13,14] (libraries 1.19 and 1.20), where a phosphinic acid functional group imbedded into a peptide backbone served as

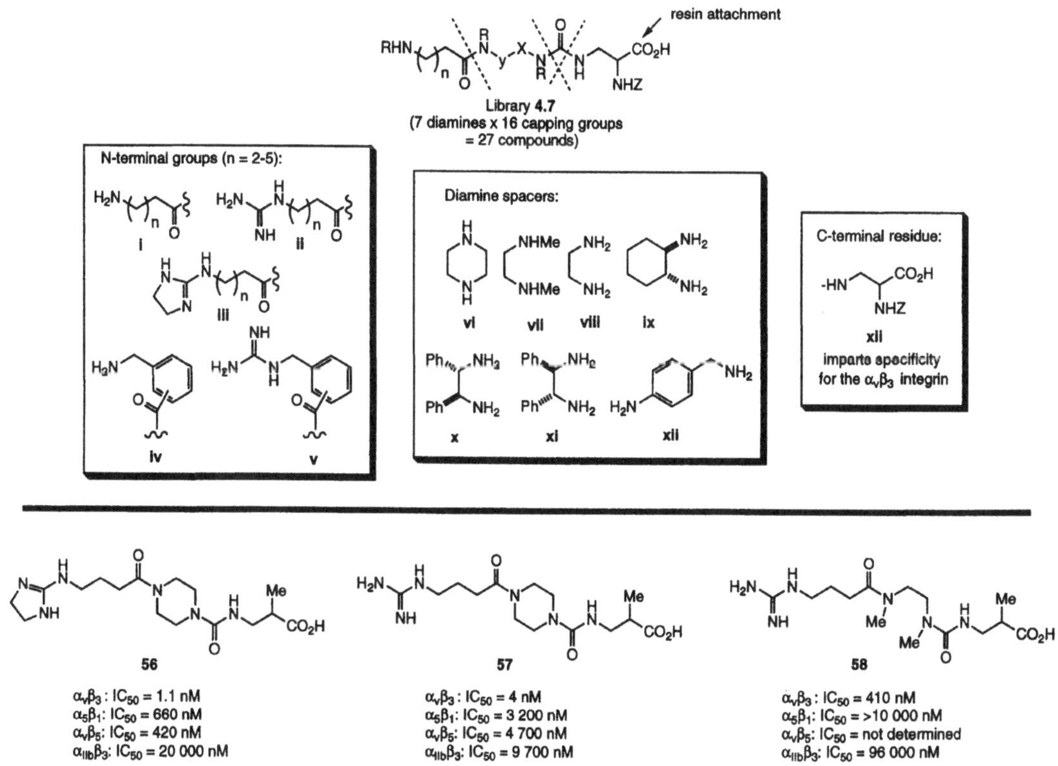

Figure 12. $\alpha_v\beta_3$ Integrin antagonists [73].

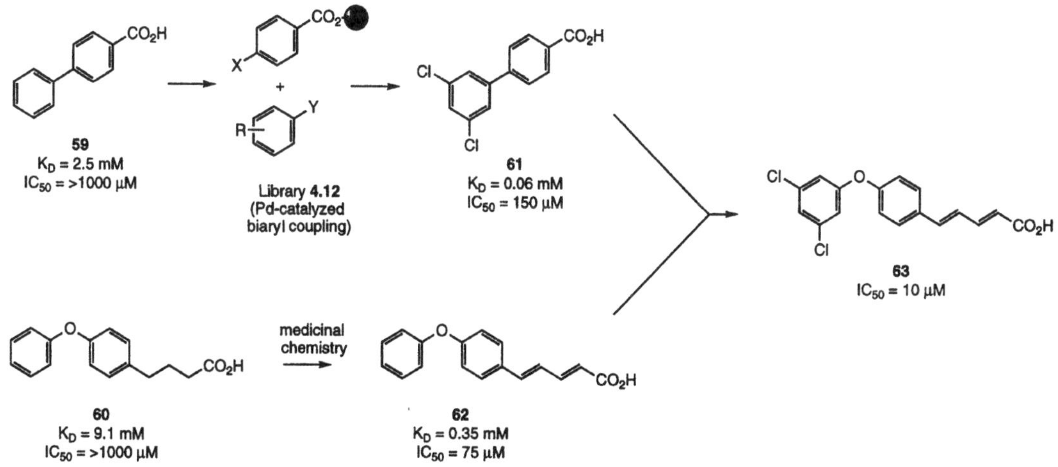

Figure 13. Optimization of ligands for the DNA binding of human papillomavirus E2 protein [74].

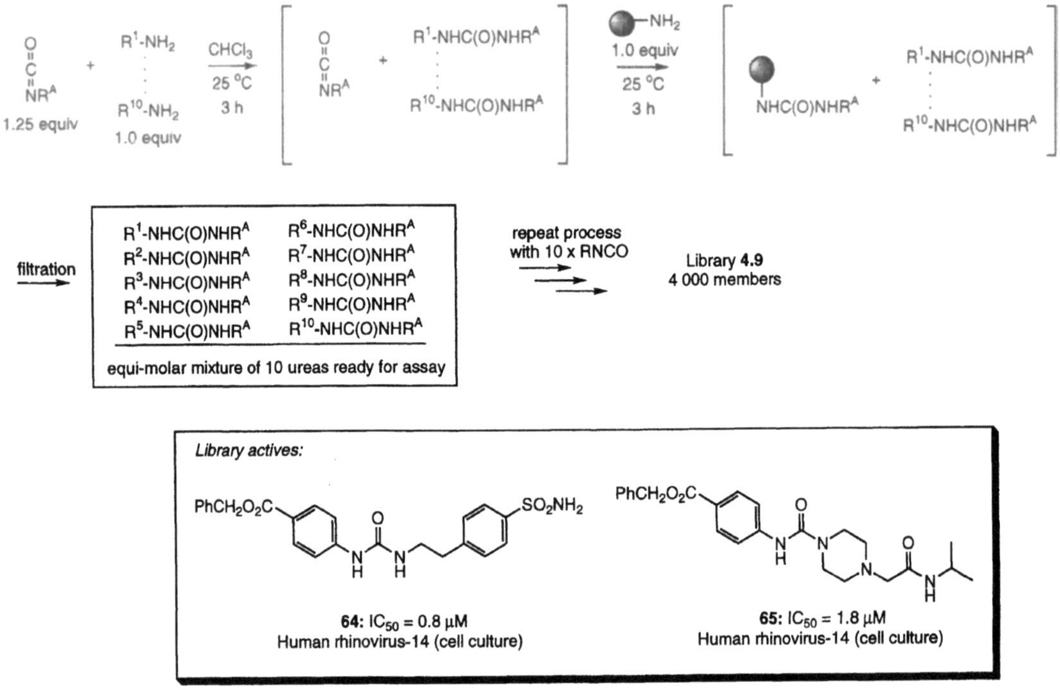

Figure 14. Anti-rhinoviral ureas produced in solution using a scavenger resin [76].

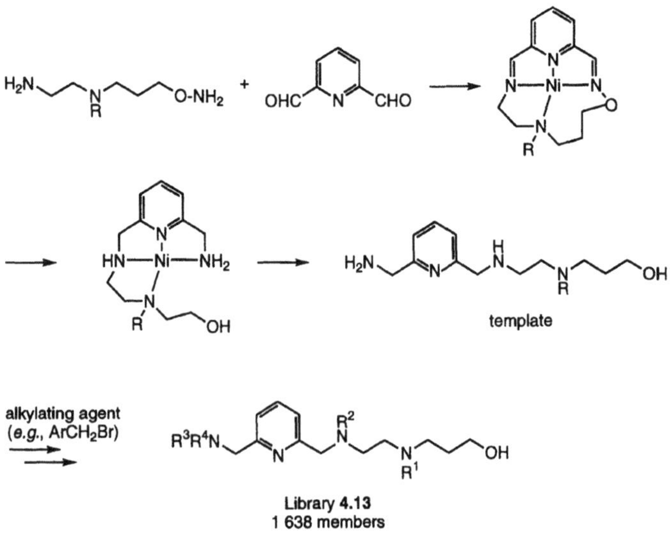

Scheme 3. Antibacterial pyridinopolyamine [72].

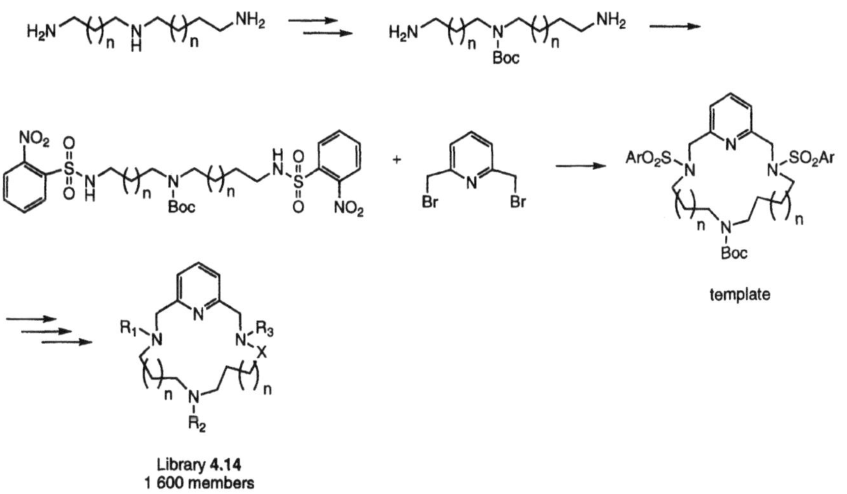

Scheme 4. Antibacterial polyazapyridinophanes [71].

a zinc endopeptidase transition state mimetic. In this case, the libraries produced sub-nanomolar and >300-fold selective inhibitors of thimet oligopeptidase and neurolysin. They also provided a new definition of the enzyme's P_1' and P_2' specificity determinants, which was contrary to prevailing literature precedent. Success, however, is not always guaranteed as in the case of the 1180 member library of 3,4-dichlorophenylbutyldiamines (library 3.10); the most active compound identified from the library was the original lead itself.

Related to the incorporation of known pharmacophores into libraries, are those libraries utilizing biologically active natural products as design anchors. Library 2.6, based on the protein phosphatase inhibitor, calyculin-A, and library 4.10, derived from the antibacterial, neomycin B, illustrate this strategy.

Template-based approaches to combinatorial libraries have been explored by a few research groups. The macrocyclic lysine-derived template of Houghten (library 1.2) is an early example of this library concept. More recently, less peptidic scaffolds have been employed in libraries including the Kemp triacid (library 1.6), a xanthene tetracarboxylic acid (library 1.7), 2,4-dichloro-6-methylpyrimidine (library 3.23), pyridinotriamines (library 4.13) and the polyazapyridinophanes (library 4.14).

Although there is much interest in the design and synthesis of low molecular weight (<600 daltons), nonpeptide drug-like molecules, descriptions of such libraries with assay screening results are at present rather sparse. A library of 1,4-benzodiazepines active against CCK (library 3.18) and the CRF_1 antagonists derived from the diaminotriazine (library 3.23) are notable exceptions. In general, these types of libraries may be regarded as more proprietary in nature relative to simple peptide-based libraries, and hence, there may be some reluctance on the part of drug discovery companies to release the details of the library synthesis and screening results to the scientific community. It is expected that disclosure of these will be forthcoming in light of the increased number of conference presentations on the design and synthesis of non-peptide libraries.

Finally, the integration of combinatorial chemistry with other drug discovery tools is occurring. The identification of hydroxyethylamine-based inhibitors of cathepsin D by Ellman (libraries 1.14a,b and 1.15), and the penultimate derivation of selective MMP-3 inhibitors from a library of N-carboxyalkyl amino acid by researchers at DuPont Merck (library 1.23), represent two illustrations of the successful merger of structure-based ligand design and library synthesis. The recent introduction of SAR by NMR methodology for lead discovery has also been used in conjunction with a combinatorial library (library 4.12).

The number of libraries yielding biologically active agents is impressive and it is a testimony to the rapid acceptance of this technology in drug discovery. In the future, one will likely witness greater attention applied to the selection of synthons and the physiochemical properties of libraries (MW, logP, pharmacologically acceptable functionality, etc.) to enhance the overall performance of a library and the quality of the compounds emerging from combinatorial lead discovery and optimization programs. Full integration of the technology with more traditional discovery tools is expected. Hopefully, one will observe the discovery of leads active against the more perplexing molecular targets, i.e., large protein ligand-large protein receptors.

References

1. a. Thompson, L.A. and Ellman, J.A., *Synthesis and application of small molecule libraries*, Chem. Rev., 96 (1996) 555–600.
   b. Gallop, M.A., Barrett, R.W., Dower, W.J., Fodor, S.P. and Gordon, E.M., *Applications of combinatorial technologies to drug discovery. 1. Background and peptide combinatorial libraries*, J. Med. Chem., 37 (1994) 1233–1251.
   c. Gordon, E.M., Gallop, M.A. and Patel, D.V., *Strategy and tactics in combinatorial organic synthesis. Applications to drug discovery*, Acc. Chem. Res., 29 (1996) 144–154.
   d. Terrett, N., Gardner, M., Gordon, D.W., Kobylecki, R.J. and Steele, J., *Combinatorial synthesis – The design of compound libraries and their application to drug discovery*, Tetrahedron, 51 (1995) 8135–8173.
   g Wilson, S.R. and Czarnik, A.W., Combinatorial chemistry. Synthesis and application, John Wiley & Sons, New York, 1997.
2. a. Balkenhohl, F., von dem Bussche-Hunnefeld, C., Lansky, A. and Zechel, C., *Combinatorial synthesis of small organic molecules*, Angew. Chem. Intl. Ed. Engl., 35 (1996) 228–2337.
   b. Ellman, J.A., *Design, synthesis, and evaluation of small-molecule libraries*, Acc. Chem. Res., 29 (1996) 132–143.
   c. Hermkens, P.H., Ottenheijm, H.C. and Rees, D.C., *Solid-phase organic reactions. II: A review of the literature*, Tetrahedron, 53 (1997) 5643–5678.
   d. Hermkens, P.H., Ottenheijm, H.C. and Rees, D.C., *Solid-phase organic reactions: A review of the recent literature*, Tetrahedron, 52 (1996) 4527–4554.
   e. Flynn, D.L., Crich, J.Z., Devraj, R.V., Hockerman, S.L., Parlow, J.J., South, M.S. and Woodward, S., *Chemical library purification strategies based on principles of complementary molecular reactivity and molecular recognition*, J. Am. Chem. Soc., 119 (1997) 4874–4881.
   f. Parlow, J.J., Naing, W., South, M.S. and Flynn, D.L., *In situ chemical tagging: Tetrafluorophthalic anhydride as a 'Sequestration enabling reagent' (SER) in the purification of solution-phase combinatorial libraries*, Tetrahedron Lett., 38 (1997) 7959–7962.
3. a. Dolle, R., *Discovery of enzyme inhibitors through combinatorial chemistry*, Mol. Div., 2 (1996) 223–236.
   b. Cowley, P.M. and Rees, D., *Application of solid-phase synthesis to drug discovery*, Curr. Med. Chem., 4 (1997) 211–227.
4. Bastos, M., Maeji, N.J. and Abeles, R.H., *Inhibitors of human heart chymase based on a peptide library*, Proc. Natl. Acad. Sci. USA, 92 (1995) 6738–6742.
5. Campbell, D.A., Bermak, J.C., Burkoth, T.S. and Patel, D.V., *A transition state analogue inhibitor combinatorial library*, J. Am. Chem. Soc., 117 (1995) 5381–5382.
6. Carell, T., Wintner, E.A., Sutherland, A.J., Rebek, J., Jr., Dunayevskiy, Y.M. and Vouros, P., *New promise in combinatorial chemistry: synthesis, characterization, and screening of small-molecule libraries in solution*, Chem. Biol., 2 (1995) 171–183.
7. a. Chen, C.L., Strop, P., Lebl, M. and Lam, K.S., *One bead-one compound combinatorial peptide library: Different types of screening*, Combinatorial Chem., 267 (1996) 211–219.
   b. Seligmann, B., Abdul-Latif, F., Al-Obeidi, F., Flegelova, Z., Issakova, O., Kocis, P., Krchnak, V., Lam, K.S., Lebl, M., Ostrem, J., Safar, P., Sepetov, N., Stierandova, A., Strop, P. and Wildgoose, P., *The construction and use of peptide and non-peptidic combinatorial libraries to discover enzyme inhibitors*, Eur. J. Med. Chem. 30 (supplement) (1995) 319s–335s.
   c. Kocis, P., Issakova, I., Sepetov, N.F. and Lebl, M., *Kemp's triacid scaffolding for synthesis of combinatorial nonpeptide uncoded libraries*, Tetrahedron Lett., 37 (1995) 6623–6626.
8. a. Eichler, J. and Houghten, R.A., *Identification of substrate-analog trypsin inhibitors through the screening of synthetic peptide combinatorial libraries*, Biochemistry, 32 (1993) 11035–11041.
   b. Coombs, G.S., Hazzard, J. and Corey, D.R., *Kinetic characterization of a peptide inhibitor of trypsin isolated from a synthetic peptide combinatorial library*, Bioorg. Med. Chem. Lett., 5 (1995) 611–614.
   c. Pinilla, C., Appel, J.R., Blondell, S.E., Dooley, C.T., Eichler, J., Ostresh, J.M. and Houghten, R.A., *Versatility of positional scanning synthetic combinatorial libraries for the identification of individual compounds*, Drug Dev. Res., 33 (1994) 133–145.
   d. Pinilla, C., Appel, J.R., Blondell, S.E., Dooley, C.T., Dorner, B., Eichler, J., Ostresh, J. and Houghten, R.A., *A review of the utility of soluble peptide combinatorial libraries*, Biopolymers (Pep. Sci.) 37 (1995) 221–240.
9. Eichler, J., Lucka, A.W. and Houghten, R.A., *Cyclic peptide template combinatorial libraries: Synthesis and identification of chymotrypsin inhibitors*, Peptide Res., 7 (1994) 300–307.
10. Esser, C.K., Kevin, N.J., Yates, N.A. and Chapman, K.T., *Solid-phase synthesis of a N-carboxyalkyl tripeptide combinatorial library*, Bioorg. Med. Chem. Lett., 20 (1997) 2639–2644.
11. Ferry, G., Boutin, J.A., Atassi, G., Fauchere, J.L. and Tucker, G.C., *Selection of a histidine-containing inhibitor of gelatinases through deconvolution of combinatorial tetrapeptide libraries*, Mol. Div., 2 (1996) 135–146.
12. Foley, M.A., Hassman, A.S., Drewry, D.H., Greer, D.G., Wagner, C.D., Feldman, P.L., Berman, J., Bickett, D.M., McGeeham, G.M., Lambert, M.H., and Green, M., *Rapid synthesis of novel dipeptide inhibitors of human collagenase and gelatinase using solid phase chemistry*, Bioorg. Med. Chem. Lett., 6 (1996) 1905–1910.
13. Jiracek, J., Yiotakis, A. Vincent, B., Checler, F. and Dive, V., *Development of the first potent and selective inhibitor of the zinc endopeptidase neurolysin using a systematic approach based on combinatorial chemistry of phosphinic peptides*, J. Biol. Chem., 271 (1996) 19606–19611.
14. Jiracek, J., Yiotakis, A., Vincent, B., Lecoq, A., Nicolaou A., Checler, F. and Dive, V., *Development of highly potent and selective phosphinic peptide inhibitors of zinc endopeptidase 24-15 using combinatorial chemistry*, J. Biol. Chem., 270 (1995) 21701–21706.

15. Kick, E.K. and Ellman, J.A., *Expedient method for the solid-phase synthesis of aspartic acid protease inhibitors directed toward the generation of libraries*, J. Med. Chem., 38 (1995) 1427–1430.
16. Kick, E.K., Roe, D.C., Skillman, A.G., Liu, G., Ewing, T.J.A., Sun, Y., Kuntz, I.D. and Ellman, J.A., *Structure-based design and combinatorial chemistry yield low nanomolar inhibitors of cathepsin D*, Chem. Biol., 4 (1997) 297–307.
17. Lloyd, J., Schmidt, J.B., Hunt, J.T., Barrish, J.C., Little, D.K. and Tymiak, A.A., *Solid phase synthesis of phosphinic acid endothelin converting enzyme inhibitors*, Bioorg. Med. Chem. Lett., 6 (1996) 1323–1326.
18. McBride, J.D., Freeman, N., Domingo, G.J. and Leatherbarrow, R.J., *Selection of chymotrypsin inhibitors from a conformationally-constrained combinatorial peptide library*, J. Mol. Biol., 259 (1996) 819–827.
19. Maclean, D., Schullek, J.R., Murphy, M.M., Ni, Z.J., Gordon, E.M. and Gallop, M.A., *Encoded combinatorial chemistry: Synthesis and screening of a library of highly functionalized pyrrolidines*, Proc. Natl. Acad. Sci. USA, 94 (1997) 2805–2810.
20. Meldal, M. and Svendsen, I., *Direct visualization of enzyme inhibitors using a portion mixing inhibitor library containing a quenched fluorogenic peptide substrate. Part 1. Inhibitors for subtilisin carlsberg*, J. Chem. Soc., Perkin Trans. 1, 12 (1995) 1591–1596.
21. Murphy, M.M., Schullek, J.R., Gordon, E.M. and Gallop, M.A., *Combinatorial organic synthesis of highly functionalized pyrrolidines: Identification of a potent angiotensin converting enzyme inhibitor from a mercaptoacyl proline library*, J. Am. Chem. Soc., 117 (1995) 7029–7030.
22. Owens, R.A., Gesellchen, P.D., Houchins, B.J. and DiMarchi, R.D., *The rapid identification of HIV protease inhibitors through the synthesis and screening of defined peptide mixtures*, Biochem. Biophys. Res. Commun., 181 (1991) 402–408.
23. a. Rano, T.A., Timkey, T., Peterson, E.P., Rotonda, J., Nicholson, D.W., Becker, J.W., Chapman, K.T. and Thornberry, N.A., *A combinatorial approach for determining protease specificities: Application to interleukin-1β converting enzyme (ICE)*, Chem. Biol., 4 (1997) 149–155.
b. Thornberry, N.A., Rano, T.A., Peterson, E.P., Rasper, D.M., Timkey T., Garcia-Calvo, M., Houtzager, V.M., Nordstrom, P.A., Roy, S., Vaillancourt, J.P., Chapman, K.T. and Nicholson, D.W., *A combinatorial approach defines specificities of members of the caspase family and granzyme B*, J. Biol. Chem., 29 (1997) 17907–17911.
24. Rockwell, A., Melden, M., Copeland, R.A., Hardman, K., Decicco, C.P. and DeGrado, W.F., *Complementarity of combinatorial chemistry and structure-based ligand design: application to the discovery of novel inhibitors of matrix metalloproteinases*, J. Am. Chem. Soc., 118 (1996) 10337–10338.
25. Rutenber, E.E., McPhee, F., Kaplan, A.P., Gallion, S.L., Hogan, J.C., Jr., Craik, C.S. and Stroud, R.M., *A New class of HIV-1 protease inhibitors: The crystallographic structure, inhibition and chemical synthesis of an aminimide peptide isostere*, Bioorg. Med. Chem. Lett., 4 (1996) 1545–1558.
26. Schullek, J.R., Butler, J.H., Ni, Z.J., Chen, D. and Yuan, Z., *A high-density screening format for encoded combinatorial libraries: Assay miniaturization and its application to enzymatic reactions*, Anal. Biochem., 246 (1997) 20–29.
27. Simon, R.J., Kania, R.S., Zuckermann, R.N., Huebner, V.D., Jewell, D.A., Banville, S., Ng, S., Wang, L., Rosenberg, S., Marlowe, C.K., Spellmeyer, D.C., Tan, R., Frankel A.D., Santi, D.V., Cohen, F.E. and Bartlett, P.A., *Peptoids: A modular approach to drug discovery*, Proc. Natl. Acad. Sci. USA, 89 (1992) 9367–9371.
28. Wang, G.T., Li, S., Wideburg, N., Krafft, G.A. and Kempf, D.J., *Synthetic chemical diversity: Solid phase synthesis of libraries of C(2) symmetric inhibitors of HIV protease containing diamino diol and diamino alcohols cores*, J. Med. Chem., 38 (1995) 2995–3002.
29. a. Chabala, J.C., Baldwin, J.J., Burbaum, J.J., Chelsky, D., Dillard, L.W., Henderson, I., Li, G., Ohlmeyer, M.H.J., Randle, T.L., Reader, J.C., Rokosz, L. and Sigal, N.H., *Binary encoded small-molecular libraries in drug discovery and optimization*, Perspectives in Drug Discovery Design, 2 (1994) 305–318.
b. Baldwin, J. J., Burbaum, J.J., Henderson, I. and Ohlmeyer, M.H., *Synthesis of a small molecule combinatorial library encoded with molecular tags*, J. Am. Chem. Soc., 117 (1995) 5588–5589.
c. Barbaum, J.J., Ohlmeyer, M.H.J., Reader, J.C., Henderson, I., Dillard, L.W., Li, G., Randle, T.L., Sigal, N.H., Chelsky, D. and Baldwin, J.J., *A pardigm for drug discovery employing encoded combinatorial libraries*, Proc. Natl. Acad. Sci. USA, 92 (1995) 6027–6031.
30. Davis, P.W., Vickers, T., Wilson-Lingardo, L., Wyatt, J.R., Guinosso, C.J., Sanghvi, Y.S., DeBaets, E.A., Acevedo, O., Cook, P.D. and Ecker, D.J., *Drug leads from combinatorial phosphodiester libraries*, J. Med. Chem., 38 (1995) 4363–4366.
31. Eichler, J., Lucka, A.W., Pinilla, C. and Houghten, R.A., *Novel a-glucosidase inhibitors identified using multiple cyclic peptide combinatorial libraries*, Mol. Div., 1 (1995) 233–240.
32. Gao, J., Cheng, R.C., Sigal, G.B., Bruce, J.E., Schwartz, B.L., Hofstadler, S.A., Anderson, G.A., Smith, R.D. and Whitesides, G.M., *Screening derivatized peptide libraries for tight binding inhibitors to carbonic anhydrase II by electrospray ionization-mass spectrometry*, J. Med. Chem., 39 (1996) 1949–1955.
33. Green, J., *Solid Phase Synthesis of Lavendustin A and Analogues*, J. Org. Chem., 60 (1995) 4287–4290.
34. Lingardo, L.W., Davis, P.W., Ecker, D.J., Herbert, N., Acevedo, O., Sprankle, K., Brennan, T., Schwarcz, L., Freier, S.M. and Wyatt, J.R., *Decovolution of combinatorial libraries for drug discovery: experimental comparison of pooling strategies*, J. Med. Chem., 39 (1996) 2720–2726.
35. Look, G.C., Schullek, J.R., Holmes, C.P., Chinn, J.P., Gordon, E.M. and Gallop, M.A., *The identification of cyclooxygenase-1 inhibitors from 4-thiazolidinone combinatorial libraries*, Bioorg. Med. Chem. Lett., 6 (1996) 707–712.
36. Lutzke, R.A.P., Eppens, N.A., Weber, P.A., Houghten, R.A. and Plasterk, R.H.A., *Identification of a hexapeptide inhibitor of the human immunodeficiency virus integrase protein by using a combinatorial chemical library*, Proc. Natl. Acad. Sci. USA, 92 (1995) 11456–11460.
37. Meyers, H.V., Dilley, G.J., Durgin, T.L., Powers, T.S., Winssinger, N.A., Zhu, H. and Pavia, M.R., *Multiple simultaneous synthesis of phenolic libraries*, Mol. Div., 1 (1995) 13–20.
38. Moran, E.J., Sarshar, S., Cargill, J.F., Shahbaz, M.M., Lio, A., Mjalli, A.M. and Armstrong, R.W., *Radio frequency tag encoded combinatorial library method for the discovery of tripeptide-substituted cinnamic acid inhibitors of the protein tyrosine phosphotase PTP1B*, J. Am. Chem. Soc., 117 (1995) 10787–10788.

39. Norman, T.C., Gray, N.S., Koh, J.T. and Schultz, P.G., *A structure-based library approach to kinase inhibitors*, J. Am. Chem. Soc., 118 (1996) 7430–7431.
40. Samson, I., Kerremans, L., Rozenski, J., Samyn, B., Van Beeumen, J., Van Aerschot A. and Herdewijn, P., *Identification of a peptide inhibitor against glycosomal phosphoglycerate kinase of Trypanosoma brucei by a synthetic peptide library approach*, Bioorg. Med. Chem. Lett., 3 (1995) 257–265.
41. Samson, I., K., Rozenski, J., Samyn, B., Van Aerschot A., Van Beeumen, J. and Herdewijn, P., *Screening a random pentapeptide library, composed of 14 D-amino acids, against the COOH-terminal sequence of fructose-1,6-bisphosphate aldolase from Trypanosoma brucei*, J. Biol. Chem., 272 (1997) 11378–11383.
42. Selway, C.N. and Terrett, N.K., *Parallel-compound synthesis: Methodology for accelerating drug discovery*, Bioorg. Med. Chem., 4 (1996) 645–654.
43. Wallace, A., Koblan, K.S., Hamilton, K., Marquis-Omer, D.J., Miller, P.J., Mosser, S.D., Omer, C.A., Schaber, M.D., Cortese, R., Oliff, A., Gibbs, J.B. and Pessi, A., *Selection of potent inhibitors of farnesyl-protein transferase from a synthetic tetrapeptide combinatorial library*, J. Biol. Chem., 271 (1996) 31306–31311.
44. Wipf, P., Cunningham, A., Rice, R.L. and Lazo, J.S., *Combinatorial synthesis and biological evaluation of library of small-molecule Ser/Thr-protein phosphatase inhibitors*, Bioorg. Med. Chem., 5 (1997) 165–177.
45. Zhao, Y.Z., Van Breemen, R.B., Nikolic, D., Huang, C.R., Woodbury, C.P., Schilling, A. and Venton, D.L., *Screening solution-phase combinatorial libraries using pulsed ultrafiltration/electrospray mass spectrometry*, J. Med. Chem., 40 (1997) 4006–4012.
46. Bitan, G., Sukhohotinsky, I., Mashriki, Y., Hanaini, M., Selinger, Z. and Gilon, C., *Synthesis and biological activity of novel backbone-bicyclic substance-P analogs containing lactam and disulfide bridges*, J. Pept. Res., 49 (1997) 421–426.
47. Boden, P., Eden, J.M., Hodgson, J., Horwell, D.C., Howson, W., Hughes, J., McKnight, A.T., Meecham, K., Pritchard, M.C., Raphy, J., Ratcliffe, G.S., Suman-Chauhan, N.S. and Woodruff, G.N., *The rational development of small molecule tachykinin NK_3 receptor selective antagonists-the utilization of a dipeptide chemical library in drug design*, Bioorg. Med. Chem. Lett., 4 (1994) 1679–1684.
48. Bunin, B., Plunkett, M. and Ellman, J., *The combinatorial synthesis and chemical and biological evaluation of a 1,4-benzodiazepine library*, Proc. Natl. Acad. Sci. USA, 91 (1994) 4708–4712.
49. Chakravarty, S., Mavunkel, B.J., Goehring, R.R. and Kyle, D.J., *Novel bradykinin receptor antagonists from a structurally directed non-peptide combinatorial library*, Immunopharm., 33 (1996) 61–67.
50. a. Chen, C., Dagnino, R., Jr., De Souza, E.B., Grigoriadis, D.E., Huang, C.Q., Kim, K.I., Liu, Z., Moran, T., Webb, T.R., Whitten, J.P., Xie, Y.F. and McCarthy, J.R., *Design and synthesis of a series of non-peptide high-affinity human corticotropin-releasing $factor_1$ receptor antagonists*, J. Med. Chem., 39 (1996) 4358–4360.
b. Whitten, J.P., Xie, Y.F., Erickson, P.E., Webb, T.R., De Souza, E.B., Grigoriadis, D.E. and McCarthy, J.R., *Rapid microscale synthesis, a new method for lead optimization using robotics and solution phase chemistry: Application to the synthesis and optimization of corticotropin-releasing $factor_1$ receptor antagonists*, J. Med. Chem., 39 (1996) 4354–4357.
51. Deprez, B., Willard, X., Bourel, L., Coste, H., Hyafil, F. and Tartar, A., *Orthogonal Combinatorial Libraries*, J. Am. Chem. Soc., 117 (1995) 5405–5406.
52. Dooley, C.T., Chung, N.N., Schiller, P.W. and Houghten, R.A., *Acetalins: Opioid receptor antagonists determined through the use of synthetic peptide combinatorial libraries*, Proc. Natl. Acad. Sci. USA, 90 (1993) 10811–10815.
53. Dooley, C.T., Chung, N.N., Wilkes, B.C., Schiller, P.W., Bidlack, J.M., Pastrnack, G.W. and Houghten, R.A., *An all D-amino opioid peptide with central analgesic activity from a combinatorial library*, Science, 266 (1994) 2019–2022.
54. a. Dooley, C. and Houghten, R., *The use of positional scanning synthetic peptide combinatorial libraries for the rapid determination of opioid receptor ligands*, Life Sci., 52 (1993) 1509–1517.
b. Houghten, R. and Dooley, C., *The use of synthetic peptide combinatorial libraries for the determination of peptide ligands in radio-receptor assays: Opioid peptides*, Bioorg. Med. Chem. Lett., 3 (1993) 405–412.
(c) Dooley, C., Prochera, A.M. and Houghten, R.A., *A priori determination of peptide ligands for opiate receptors using synthetic peptide combinatorial libraries,* Peptides, (1992) 649–650.
55. Dooley, C.T., Kaplan, R.A., Schiller, P.W., Bidlack, J.M. and Houghten, R.A., *Six highly active Mu-selective opioid peptides identified from two synthetic combinatorial libraries*, Pep. Res., 3 (1995) 124–137.
56. Drouillant, B., Kellam, B., Dekany, G., Starr, M.S. and Toth, I., *Solid phase synthesis of C-terminal carbohydrate modified enkephalins*, Bioorg. Med. Chem. Lett., 7 (1997) 2247–2250.
57. Heizmann, G., Tanner, H. and Eberle, A.N., *New ligands for the human melanoma MSH receptor identified by a peptoid library (oligo n-substituted glycines)*, Innovation Perspect. Solid Phase Synth. Comb. Libr. Collect. Pap., 4th International Symposium, Meeting Date 1995, 391–394.
58. Hirshmann, R., Yao, W., Cascieri, M., Strader, C., Maechler, L., Cichy-Knight, M., Hynes, J., Jr., van Rijn, R., Spengeler, P. and Smith, A., *Synthesis of potent cyclic hexapeptide NK-1 antagonists. Use of a minilibrary in transforming a peptidal somatostatin receptor ligand into an NK-1 receptor ligand via a polyvalent peptidomimetic*, J. Med. Chem., 39 (1996) 2441–2448.
59. a. Horwell, D.C., Howson, W., Ratcliffe, G.S., and Rees, D.C.,*The design of a dipeptide library for screening at peptide receptor sites,* Bioorg. Med. Chem. Lett., 3 (1993) 799–802.
b. Horwell, D.C., Pritchard, M., Raphy, J. and Ratcliffe, G.S., *Targeted molecular diversity: Design and development of non-peptide antagonists for cholecystokinin and tachykinin receptors*, Immunopharm., 33 (1996) 68–72.
60. a. Maher, H. and Yang, R., *Structure optimization of a leukotriene D_4 antagonist by combinatorial chemistry in solution*, Bioorg. Med. Chem. Lett., 5 (1997) 493–496.
b. O'Donnell, M., Crowley, H.J., Yaremko, B., O'Neill, N. and Welton, A.F., *Pharmacologic actions of Ro 24-5913, a novel antagonist of leukotriene D_4*, J. Pharm. Exp. Ther., 259 (1991) 751–758.
61. Neustadt, B., Wu, A., Smith, E.M., Nechuta, T., Fawzi, A., Zhang, H. and Ganguly, A.K., *A case study of combinatorial libraries: Endothelin receptor antagonist hexapeptides*, Bioorg. Med. Chem. Lett., 5 (1995) 2041–2044.
62. Quillian, J.M., Jayawickreme, C.K. and Lerner, M.R., *Combinatorial diffusion assay used to identify topically active melanocyte-stimulating hormone receptor antagonists*, Proc. Natl. Acad. Sci. USA, 92 (1995) 2894–2898.

63. Spellmeyer, D.C, Brown, S. and Stauber, G., *Endothelin receptor ligands. Replacement net approach to SAR determination potent hexapeptides*, Bioorg. Med. Chem. Lett., 3 (1993) 519–524.
64. Spellmeyer, D.C., Brown, S. and Stauber, G., *Endothelin receptor ligands. Multiple* D-*amino acid replacement net approach*, Bioorg. Med. Chem. Lett., 3 (1993) 1253–1256.
65. Terrett, N.K., Bojanic, D., Brown, D., Bungay, P., Gardner, M., Gordon, D.W., Mayers, C.J. and Steele, J.,*The combinatorial synthesis of a 30,752-compound library: Discovery of SAR around the endothelin antagonist, FR-139, 317*. Bioorg. Med. Chem. Lett., 5 (1995) 917–922.
66. Terrett, N.K., Gardner, M., Gordon, D.W., Kobylecki, R.J. and Steele, J., *Drug discovery by combinatorial chemistry-The development of a novel method for the rapid synthesis of single compounds*, J. Eur. Chem., 3 (1997) 1917–1920.
67. a. Wang, J.X., Bray, A.M., DiPasquale, A.J., Maeji, N.J. and Geysen, H.M., *Systematic study of substance P analogs: Evaluation of peptides synthesized by the multipin method for quantitative receptor binding assay*, Int. J. Peptide Protein Res., 42 (1993) 384–391.
b. Wang, J.X., DiPasquale, A.J., Bray, A.M., Maeji, N.J. and Geysen, H.M., *Study of stereo-requirements of substance P binding to NK1 receptors using analogues with systematic d-amino acid replacements*, Bioorg. Med. Chem. Lett., 13 (1993) 451–456.
c. Wang, J.X., Bray, A.M., DiPasquale, A.J., Maeji, N.J. and Geysen, H.M., *Application of the multipin peptide synthesis technique for peptide receptor binding studies: Substance P as a model system*, Bioorg. Med. Chem. Lett., 3 (1993) 447–450.
68. Zuckerman, R.N., Martin, E.J., Spellmeyer, D.C., Stauber, G.B., Shoemaker, K.R., Kerr, J.M., Figliozzi, G.M., Goff, D.A., Siani, D., Simon, R., Banville, S.C., Brown, E.G., Wang, L., Richter, L.S. and Moos, W., *Discovery of nanomolar ligands for 7-transmembrane G-protein-coupled receptors from a diverse N-(substituted) glycine peptoid library*, J. Med. Chem. 37 (1994) 2678–2685.
69. Virgilio, A.A., Bray, A.A., Zhang, W., Trinh, L., Snyder, M., Morrissey, M.M. and Ellman, J.A., *Synthesis and evaluation of a library of peptidomimetics based upon the B-turn*, Tetrahedron, 53 (1997) 6635–6644.
70. Williard, R., Jammalamadaka, V., Zava, D., Bens, C.C., Hunt, C.A., Kushner, P.J. and Scanlan, T.S., *Screening and characterization of estrogenic activity from a hydroxystilbene library*, Chem. Biol., 2 (1995) 45–51.
71. An, H., Cummins, L.L., Griffey, R.H., Bharadwaj, R., Haly, B.D., Fraser, A.S., Wilson-Lingardo, L., Risen, L.M., Wyatt, J.R. and Cook, P.D., *Solution phase combinatorial chemistry: Discovery of novel polyazapyridinophanes with potent antibacterial activity by a solution phase simultaneous addition of functionalities approach*, J. Am. Chem. Soc., 119 (1997) 3696–3708.
72. An, H., Haly, B.D., Fraser, A.S., Guinosso, C.J. and Cook, P.D., *Solution phase combinatorial chemistry. Synthesis of novel linear pyridinopolyamine libraries with potent antibacterial activity*, J. Org. Chem., 62 (1997) 5156–5164.
73. Corbett, J.W., Graciani, N.R., Mousa, S.A. and DeGrado, W.E., *Solid-phase synthesis of a selective $\alpha_v\beta_3$ integrin antagonist library*, Bioorg. Med. Chem. Lett., 7 (1997) 1371–1376.
74. Hajduk, P.J., Dinges, J., Miknis, G.F., Merlock, M., Middleton, T., Kempf, D.J., Egan, D.A., Walter, K.A., Robins, T.S., Shuker, S.B., Holzman, T.F. and Fesik, E., *NMR-based discovery of lead inhibitors that block DNA binding of the human papillomavirus E2 protein*, J. Med. Chem., 40 (1997) 3144–3150.
75. Harada, T., Katada, J., Tachiki, A., Asari, T., Iijima, K., Uno, I., Ojima, I. and Hayashi, Y., *Development of the new potent non-peptide GpIIb/IIIa antagonist NSL-95301 by utilizing combinatorial technique*, Bioorg. Med. Chem. Lett., 7 (1997) 209–212.
76. Kaldor, S., Fritz, J.E., Tang, J. and McKinney, E.R., *Discovery of antirhinoviral leads by screening a combinatorial library of ureas prepared using covalent scavengers*, Bioorg. Med. Chem. Lett., 6 (1996) 3041–3044.
77. Koppel, G., Dodds, C., Houchins, B., Hunden, D., Johnson, D., Owens, R., Chaney, M., Usdin, T., Hoffman, B. and Brownstein, M., *Use of peptide combinatorial libraries in drug design: The indication of a potent serotonin reuptake inhibitor derived from a tripeptide cassette library*, Chem. Biol., 2 (1995) 483–487.
78. Park, W.C., Auer, M., Jaksche, H. and Wong, C.H., *Rapid combinatorial synthesis of amino glycoside antibiotic mimetics: Use of a polythylene glycol-linked amine and a neamine-derived aldehyde in multiple component condensation as a strategy for the discovery of new inhibitors of the HIV RNA rev responsive element*, J. Am. Chem. Soc., 118 (1996) 10150–10155.
79. Reed, J.D., Edwards, D.L. and Gonzalez, C.F., *Synthetic peptide combinatorial libraries: A method for the identification of bioactive peptides against phytopathogenic fungi*, Am. Path. Soc. 10 (1997) 537–549.
80. Tomkinson, N.C., Sefler, A.M., Plunket, K.D., Blanchard, S.G., Parks, D.J. and Willson, T.M., *Solid-phase synthesis of hybrid thiazolidinedione-fatty acid PPARγ ligands*, Bioorg. Med. Chem. Lett., 19 (1997) 2491–2496.
81. Houghten, R.A., Pinilla, C., Blondell, S.E., Appel, J.R., Dooley, C.T. and Cuervo, J.H., *Generation and use of synthetic peptide combinatorial libraries for basic research and drug discovery*, Nature, 354 (1991) 84–86.

M.R. Pavia and W.H. Moos (eds.), Annual Reports in Combinatorial Chemistry and Molecular Diversity, Vol. 2, 129–161.
KLUWER/ESCOM

A compendium of solid-phase chemistry publications

Ian W. James
Chiron Technologies Pty. Ltd., 11 Duerdin St., Clayton, Victoria 3168, Australia
E-mail: Ian_James@cc.Chiron.com

Key words: review, solid-phase chemistry, solid-phase inorganic chemistry, solid-phase organic chemistry

Abstract

The solid-phase chemistry literature from the 1960s to 1997 has been reviewed with an emphasis on organic chemistry. Reactions have been classified according to the products formed on the solid phase, the types of chemistry performed, names of chemical reactions, asymmetric reactions and transition metal chemistry. Examples of inorganic chemistry performed on the solid phase have also been included.

Introduction

The explosion in publications in the field of solid phase organic chemistry (SPOC) has continued. Over 200 new references have been added to last year's collection of over 300 references (Jam1997). The main impetuous for this resurgence in SPOC continues to be combinatorial chemistry and its application to drug discovery. This is typified by the large number of heterocycle syntheses on the solid phase that have been published. Another example is the significant amount of work published recently on the solid phase preparation of hydroxamic acids, potential metalloproteinase inhibitors. Apart from application to drug discovery, there has also been an increase in the number of SPOC papers in the application of combinatorial chemistry to new fields such as the development of new catalysts and solid phase reagents.

This review covers both recent literature, and literature going back to the 1960s. References have been sorted according to the type of compounds prepared, functional group transformations, and the type of chemistry performed on the solid phase. Multiple references are often given. These may be used to assess which are the most commonly used conditions, or to select which conditions most suit the chemist's requirements such as molecule or linker stability.

Note though that the reported conditions may not be readily amenable to large numbers, or a different polymeric support may have been used. The reported conditions, however, may still be useful as the basis for optimization studies of the reaction. Variables such as solvent, temperature, catalyst, and reaction time may all be investigated to optimize the reaction. Modern parallel handling techniques may be used so that a number of conditions may be investigated to rapidly find the optimal conditions.

This compilation, although extensive, is not exhaustive. Apologies to any who find their papers omitted! My thanks to those who have notified me of their upcoming and recently published papers. Comments to improve future editions are most welcome.

Solid-phase organic chemistry (SPOC)

Reviews

Jam1997, Lam1997, Nef1997a, Choo1996, Ell1996, Fru1996, Herm1996, Patl1996, Ter1995, Des1994, Goro1994, Las1987, Frec1981, Hod1980, Lez1978, Crow1976, Patc1975

The following issues are devoted to combinatorial chemistry and contain some interesting SPOC reviews:

Acc. Chem. Res., **1996**, *29*(3)
Chemtracts-Org. Chem., **1995**, *8*(1)

For historical perspective, the following paper is strongly recommended: Mer1963

Products

Acetal	Alt1997, Aur1995, Dou1995, Kic1995, Liu1995, Mel1994, Tho1994, Blo1990, Veen1987, Frec1984, Hod1983, Lez1977a, Frec1975c, Hane1974, Dum1973, Lez1973, Won1973
Acyl Phosphate	Reb1975a
Alcohol	Alt1997, Fral1997, Furt1997, Fur1997, Gay1997, Kan1997, Kat1997, LeH1997, Nic1997b, Nic1997c, Park1997, Rod1997a, Rod1997b, Rou1997, Syl1997, Boe1996, Regg1996, Rot1996, Aja1995, Dou1995, Liu1995, Meye1995, Kur1994, Shap1994b, Ale1992, Dor1984, Gir1984, Hod1983, Xu1983, Lez1980, Frec1979b, Far1976, Frec1975b, Won1974, Lez1973, Zeh1973, Frec1971, Dor1969
Aldehyde	Cheg1997, Marx1997, Nic1997a, Nic1997b, Pea1997, Syl1997, Regg1996, Rot1996, Che1994, Bee1992, Frec1984, Tay1982, Far1976, Frec1975a, Dum1973, Frec1971, Aye1965
Alkene	Bro1997b, Chin1997, Cun1997, Kro1997, Nic1997a, Nic1997b, Rout1997b, Ruh1997, Wip1997, Bro1996, Cer1996, Gor1996a, Gos1996, Pop1996, Rot1996, Sch1996, Tie1996a, Bee1995, Hiro1995, John1995, Wil1995, Che1994, Yu1994, Chan1985, Svi1984, Fyl1978, Fyl1977, Lez1977a, Lez1977d, Nieu1976, Won1974, Lez1973, Frec1971
Alkyl Iodide	Wor1979
Alkyne	Col1997, Yu1994, Svi1979, Fyl1977, Lez1977d, Lez1976
3° Amide	Aru1997, Bha1997, Bro1997a, Bun1997, Fit1997, Gari1997, Hoe1997, Kan1997, Krc1997a, Marx1997, Nef1997b, Nef1997c, Plu1997, Raj1997a, Rous1997, Sara1997, Woo1997, Bic1996, Gof1996, Hoe1996, Koh1996, Nor1996, Roe1996, Vir1996, Yun1996, Booj1995, Kic1995, Ran1995, Ric1995, Ang1994, Simo1994, Zam1994, Zuc1994a, Zuc1994b, Simo1992, Zuc1992, Gol1978, Lez1977c
Amidine	Fur1997
Amidoxime	Rous1997

Enamine	Hird1997, Zar1996b, Gor1996b, MacD1996
Enol Ether	Fral1997
Enol Triflate	Fral1997
Epoxide	LeH1997, See1997, Cer1996, Rot1996, Rob1995, Dani1994, Dani1993
Ether	Alt1997, Dee1997, Du1997, Furt1997, Fur1997, Gari1997, Gay1997, Hei1997, Hern1997, Kan1997, Kat1997, Nic1997a, Nic1997b, Rod1997b, Rous1997, Rout1997a, Rout1997b, Rou1997, Sara1997, Schl1997, Stra1997, Vla1997, All1996, Boe1996, Bro1996, Hamp1996, Hane1996, Kur1996, Loh1996, Roe1996, Tie1996c, Tor1996, Val1996a, Val1996b, Bee1995, Booj1995, Chen1995, Krc1995, Ran1995, Che1994, Moo1994, Ric1994a, Tho1994, Yan1994, Bee1992, Its1987, Sieb1987, Frec1984, Gir1984, Its1984, Hall1982, McA1982, Frec1980, Lez1977b, Lez1977d, Lez1972
Furan (Tetrahydro)	Rout1997b
Guanidine	Drew1997, Wan1997b, Kow1996
Hydantoin	Kim1997, Xia1997, Sho1996
Hydrazide	Hol1997
Hydrazine	Kli1996, Sem1995
Hydrazone	Li1997, Tie1997, Hut1996a, Tie1996b, Shao1995, Blo1990
Hydroxamate	Bau1997, Che1997, Mell1997, Ngu1997a, Ric1997, Flo1996
Hydroxylamine ether	Bau1997, Mel1997b, Ric1997, Flo1996
Hydroxyquinone	Tem1997
Imidazole	Sar1996, Zha1996
Imidazopyrimidine (Tetrahydro)	Hut1996b
Imide	Gow1997, Whit1997, Reb1975b
Imine	Bun1997, Ede1997, Gor1997a, Katr1997, Krc1997a, Pea1997, Pei1997, Plu1997, Ruh1997, Sara1997, Bic1996, Boyd1996, Cole1996, Loo1996, Ni1996, O'D1996, Ruh1996, Sch1996, Booj1995, Bra1995, Chan1995, Chen1995, Goro1995, Gre1995, Loo1995, Mur1995, Plu1995b, Plu1995a, Bun1994a, Bun1994b, DeW1993, Bun1992, McA1982
Iminophosphorane	Drew1997, Wan1997b
Indazole	Yan1996b
Indole	Cheg1997, Col1997, Fag1997, Zha1997a, Zha1997b, Hut1996a, Yun1996
Indoline	Cheg1997
Iodide	Ngu1997b, Nic1997a, Hun1996, Wor1979
Isocyanide	Zha1996
Isoquinoline	
Dihydro	Meu1995
Tetrahydro	Kis1997, Hut1996b, Roe1996, Meu1995

Isoquinolinone	Gof1995b
Isothiocyanate	Step1997
Isoxazole	Kan1997, Marz1996, Pei1994, Bee1992, Yed1980
Isoxazoline	Kur1996, Lor1996
Ketal	Clar1997
β-Ketoester	Tie1997, Gor1996a, Chag1981
Ketone	Furt1997, Nic1997a, Plu1997, Rout1997a, Wall1997, Boe1996, Aja1995, Bra1995, Plu1995b, Zik1995, Blo1990, Gir1984, DeG1980, Lez1980, Rol1975, Crow1970
Lactam	Ni1996, Booj1995, Gof1995a
β-Lactam	Pei1997, Ruh1997, Ruh1996
Lactone	Moo1994, Moo1992
Mesylate	Bha1997, Kan1997, Kro1997, Ngu1997b, Nug1997b, Ric1997, Roe1996, Vir1996, Fyl1978, Fyl1977, Lez1977d, Lez1976, Hane1974
4-Metathiazanone	Hol1995
Nitrile	Dup1989, Chem1980, Frec1980, Frec1979b, Gav1979, Aye1965
Oligosaccharide	Ary1997, Nic1997c, Rod1997a, Rod1997b, See1997, Sei1997, Adi1996, Hun1996, Rad1996, Shi1996, Sof1996, Dou1995, Rob1995, Dani1994, Halc1994, Sch1994, Yan1994, Dani1993, Ver1993, Dou1991, Veen1987, Chiu1976, Exc1975, Gut1973, Zeh1973, Frec1972a, Frec1972b, Frec1971
Oligosaccharide Mimetic	Mue1995
Oxathioacetal	Rav1994
Oxazaborolidine	Frac1995
Oxazaborolidinone	Kam1996, Kam1995
Oxazoline	Ede1997
2-Oxindole	Aru1997
Oxime	Shao1995, Fin1989, Gir1984, DeG1980, Far1976, Lez1973
Ozonide	Syl1997
Phenol	Fan1997, Boe1996, Frec1980
Phosphate Ester	Cao1996, Wij1996, Berg1995a, Tan1995, Robl1994, Dor1984, Yip1971
Phosphine	Gil1996, Bern1983, Mas1978, Far1976, Gru1976, Pit1974, Rell1974, Str1974, Hei1972
Phosphine Dichloride	Rell1974
Phosphine Oxide	Rell1974
Phosphine Sulfide	Gil1996
Phosphinic Acid	Boyd1996
Phosphite Ester	Tan1995, Dor1984
Phosphonate	Fat1996, LeB1996, Zha1996, Camb1995, Camb1994, Hane1974
Phosphonium Salt	Nic1997a, Nic1997b, Hug1996, Frec1984, Bern1983, Fyl1978, Tun1978, Hei1972, Frec1971

Named reactions

Aldol Addition	Nic1997a, Nic1997b, Kob1996a, Regg1996, Kur1994
Aldol Condensation	Lez1973
Arndt-Eistert Homologation	Mari1997
Aza-Wittig (Staudinger) Reaction	Drew1997, Wan1997b, Gof1995a
Biginelli Dihydroprimidine Synthesis	Wip1995
Bischler-Napieralski Reaction	Roe1996, Meu1995
Claisen Condensation	Marz1996
Dieckmann Cyclization	Crow1980, Crow1970
Diels-Alder	Pan1996, Tie1996a, Yed1980, Gav1979
Ene	
intramolecular	Tie1996
Finkelstein	Hun1996, Wor1979
Fischer Indole Synthesis	Cheg1997, Hut1996a
Friedel-Crafts	Aja1995, Zik1995, Gir1984, DeG1980
Grieco Three Component Condensation	Kis1997
Grignard Reaction	Alt1997, Fral1997, Rou1997, van1997, Chen1995, Hau1995, Liu1995, Xu1983, Nieu1976, Frec1975b, Kaw1974, Lez1973, Kaw1972, Cra1968
to imine	Katr1997
Hantzsch Heterocyclization	Gor1996a
Henry Reaction	Bee1995, Bee1992
Hofmann Rearrangement	Kow1996
Knoevenagel Condensation	Gor1996a, Tie1996a, Tie1996c
Mannich Type Reaction	Yman1997, Kob1996b
Michael Addition	
enolate	
intermolecular	Aru1997, Ley1995
intramolecular	Ley1995
RNH_2	Hei1997, Hol1997, Kro1997, Wan1997b, Gof1996, Chem1980, Frec1980
RO^-	Hei1997, Bee1995
RS^-	
intramolecular	Shar1996
intermolecular	Aru1997, Che1994, Mae1992
secondary amines	Kol1996, Morp1996
Mitsunobu Reaction	Fag1997, Fan1997, Gay1997, Gray1997, Hei1997, Krc1997b, Pis1997, Rous1997, Wan1997a, Bro1996, Hamp1996, Nor1996, Val1996a, Val1996b, Camb1995, Krc1995, Ran1995, Ric1994a
on sulfonamide nitrogen	Yang1997
intramolecular, amine formation	Swa1997
Pauson-Khand Cycloaddition	Bol1996, Scho1990

Palladium	
allylic nucleophilic substitution	Fle1996
alkyne amination	Col1997
aryl amination	War1996, Will1996
arylhalide-alkyne coupling	Col1997, Fag1997, Fan1997, Zha1997a, Nel1996
arylhalide-arylzinc coupling	Marq1996
deallylation	Krc1997a, Rous1997, Thi1997, Val1996b, Berg1995a, Vet1995, Kate1994, Shap1994a
Heck	
intermolecular	Ruh1997, Pop1996, Hiro1995, You1994, Yu1994
intramolecular	Aka1997, Aru1997, Fag1997, Fan1997, Zha1997a, Zha1997b, Koh1996, Yun1996, Gof1995b, Hiro1995
Stille	Plu1997, Form1995, Plu1995b, Plu1995a, Desh1994
Suzuki	Bro1997b, Fral1997, Gari1997, Piet1997, Raj1997b, Ruh1997, Tem1997, Woo1997, Yoo1997, Bro1996, Gui1996, Koh1996, Lar1996, Chen1995, Bac1994, Fren1994
Platinum	
Suzuki type reaction	Bro1996
Rhodium	
carbene insertion (reaction with solid support in tagging)	Bal1995, Nes1994
[3+2] dipolar cycloaddition, diazo with alkyne	Whit1997
Ruthenium	
metathesis	Cun1997, Sch1996
Scanadium	
Aldol reaction	Kob1996a
Silver	
ketene formation	Mari1997
Vanadium	
oxidative coupling	ApS1982
Zinc	
arylhalide-arylzinc coupling	Marq1996
Ene reaction	Tie1996c
Lewis acid	See1997, Hut1996a, Dor1984

Solid phase inorganic chemistry (SPIC)

Aluminum	
1,1′-bi-2-naphthol complex	Hu1996
phthalocyanine complexes	Lez1993
Arsenic	
aryl arsonic acid	Jac1979

Chromium	
aryl complex	Pit1972
Cobalt	
aniline complex	Das1997, De1997
phosphine complex	Reg1977, Eva1974
porphyrin complex	Rol1975, Coll1974
Copper	
amine complex	Oht1995
Iridium	
carbonyl cluster	Raf1978
phosphine complex	Coll1972
Iron	
cyclobutadiene complex	Reb1975b, Reb1974
phosphine complex	Eva1974, Str1974
porphyrin complex	Coll1973
Lithium	
naphthalenide	van1997
Manganese	
amine complex	Rao1997
Magnesium	
organomagnesium halide	van1997, Burl1978, Wei1975
Mercury	
organomercury compounds	Jano1982, Tay1982, Burl1978, Reb1975b
thiol complex	Gir1997
Nickel	
amine complex	Pet1995
phosphine complex	Pet1995, Bem1982, Pit1975a, Pit1975b, Eva1974
Osmium	
pyridine complex	Nan1997
Sn-Os compound	Burl1975
Palladium	
bipyridyl complex	Card1979
phosphine complex	Jang1997, Uoz1997, Wan1994, Bem1982, Kin1979
Platinum	
phosphine complex	Bem1982
Rhodium	
anthranilic acid complex	Holy1979
amine complex	Pet1995
ammonium ion pair	Set1994
cyclopentadienyl complex	Dyg1996
1,3-diketone complex	Cap1982
phosphine complex	Gil1996, Ach1978, Mas1978, Gru1976, Pit1975b, Pit1975c, Pit1974, Str1974, Dum1973, Coll1972, Cap1971

Ruthenium	
phosphine complex	Nguy1995, Pit1975b, Pit1975c, Str1974
Scandium	
sulfonamide complex	Kob1996a, Kob1996c
Sodium	
naphthalenide	van1997
Tin	
distannane	Jun1997
tin chloride	Burl1978
tin hydride	Blan1991, Gerl1991, Geri1990, Wei1975
Titanium	
alkoxy	Alt1997
titanocene	Gru1977, Bond1975, Gru1973
Vanadium	
imine complex	Das1995

Acknowledgements

The assistance of Susan N. Eagle and Beata M. Krywalt with entering the data used in this compilation is gratefully acknowledged.

References

Ach1978 Achiwa, K., Catalytic Asymmetric Hydrogenations with Polymer Supported Chiral Pyrrolidinephosphine-Rhodium Complexes, Chem. Lett., (1978) 905–908.

Adi1996 Adinolfi, M., Barone, G., De Napoli, L., Iadonisi, A. and Piccialli, G., Solid Phase Synthesis of Oligosaccharides, Tetrahedron Lett., 37 (1996) 5007–5010.

Aja1995 Ajayaghosh, A. and Rajasekharan Pillai, V.N., Solid-Phase Synthesis and C-Terminal Amidation of Peptides Using a Photolabile o-Nitrobenzhydrylaminopolystyrene Support, Tetrahedron Lett., 36 (1995) 777–780.

Aka1997 Akaji, K. and Kiso, Y., Macrocyclization on Solid Support using Heck Reaction, Tetrahedron Lett., 38 (1997) 5185–5188.

Aki1985 Akiyama, M., Shimizu, K., Aiba, S. and Katoh, H., N-Hydroxy Amides. III. Active Esters of Polystyrene-bound 1-Hydroxy-2-pyrrolidinone and Their Use in Peptide Synthesis, Bull. Chem. Soc. Jpn, 58 (1985) 1421–1425.

Ale1992 Alewood, P.F., Brinkworth, R.I., Dancer, R.J., Garnham, B., Jones, A. and Kent, S.B.H., Solid Phase Synthesis of Hydroxyethylamine Peptide Bond Isosteres: Synthesis of the Potent HIV-1 Protease Inhibitor JG365, Tetrahedron Lett., 33 (1992) 977–980.

All1996 Allin, S.M. and Shuttleworth, S.J., The Preparation and First Application of a Polymer-Supported 'Evans' Oxazolidinone, Tetrahedron Lett., 37 (1996) 8023–8026.

Als1994 Alsina, J., Rabanal, F., Giralt, E. and Albericio, F., Solid-Phase Synthesis of 'Head-to-Tail' Cyclic Peptides via Lysine Side-Chain Anchoring, Tetrahedron Lett., 35 (1994) 9633–9636.

Alt1997 Altava, B., Burguette, M.I., Escuder, B., Luis, S.V. and Salvador, R.V., Polymer-Grafted Ti-TADDOL Complexes. Preparation and Use as Catalysts in Diels-Alder Reactions, J. Org. Chem., 62 (1997) 3126–3134.

Ang1994 Angell, Y.M., García-Echeverría, C. and Rich, D.H., Comparative Studies of the Coupling of N-Methylated, Sterically Hindered Amino Acids During Solid-Phase Synthesis, Tetrahedron Lett., 35 (1994) 5981–5984.

Anu1995 Anuradha, M.V. and Ravindranath, B., Ultrasound in Peptide Synthesis. 3: Zinc-Salt Assisted Anchoring of Carboxylic Acids to Merrifield Resin, Tetrahedron, 51 (1995) 5671–5674.

ApS1982 ApSimon, J.W. and Dixit, D.M., Biogenetic Type Solid Phase Phenol Coupling Synthesis of Alkaloids. Prelycorane Species, Can. J. Chem., 60 (1982) 368–370.

Aru1997 Arumugam, V., Routledge, A., Abell, C. and Balasubramarian, S., Synthesis of 2-Oxindole Derivatives via the Intramolecular Heck Reaction on Solid Support, Tetrahedron Lett., 38 (1997) 6473–6476.

Ary1997 Arya, P. and Ben, R.N., Combinatiorial Chemistry for the Synthesis of Carbohydrate Libraries, Angew. Chem., Int. Ed. Engl., 36 (1997) 1281–1282.

Aur1995 Aurell, M.J., Boix, C., Ceita, M.L., Llopis, C., Tortajada, A. and Mestres, R., Polymer Supported o-Nitrophenylethylene Glycols for Photoremovable Protection of Aldehydes, J. Chem. Res. (M), (1995) 2569–2583.

Aye1965 Ayers, J.T. and Mann, C.K., Some Chemical Reactions of Poly(p-chloromethylstyrene) Resin in Dimethylsulfoxide, Polymer Lett., 3 (1965) 505–508.

Bac1994 Backes, B.J. and Ellman, J.A., Carbon-Carbon Bond-Forming Methods on Solid Support. Utilization of Kenner's 'Safety-Catch' Linker, J. Am. Chem. Soc., 116 (1994) 11171–11172.

Bac1996 Backes, B.J., Virgilio, A.A. and Ellman, J.A., Activation Method to Prepare a Highly Reactive Acylsulfonamide 'Safety-Catch' Linker for Solid-Phase Synthesis, J. Am. Chem. Soc., 118 (1996) 3055–3056.

Bal1995 Baldwin, J.J., Burbaum, J.J., Henderson, I. and Ohlmeyer, M.H.J., Synthesis of a Small Molecule Combinatorial Library Encoded with Molecular Tags, J. Am. Chem. Soc., 117 (1995) 5588–5589.

Bar1996 Barn, D.R., Morphy, J.R. and Rees, D.C., Synthesis of an Array of Amides by Aluminum Chloride Assisted Cleavage of Resin-Bound Esters, Tetrahedron Lett., 37 (1996) 3213–3216.

Bau1997 Bauer, U., Ho, W.B. and Koskinen, A.M.P., A Novel Linkage for the Solid-Phase Synthesis of Hydroxamic Acids, Tetrahedron Lett., 38 (1997) 7233–7236.

Bea1996 Beaver, K.A., Siegmund, A.C. and Spear, K.L., Application of the Sulfonamide Functional Group as an Anchor for Solid Phase Organic Synthesis (SPOS), Tetrahedron Lett., 37 (1996) 1145–1148.

Bee1992 Beebe, X., Schore, N.E. and Kurth, M.J., Polymer-Supported Synthesis of 2,5-Disubstituted Tetrahydrofurans, J. Am. Chem. Soc., 114 (1992) 10061–10062.

Bee1995 Beebe, X., Schore, N.E. and Kurth, M.J., Polymer-Supported Synthesis of Cyclic Ethers: Electrophilic Cyclization of Isoxazolines, J. Org. Chem., 60 (1995) 4196–4203.

Bem1982 Bemi, L., Clark, H.C., Davies, J.A., Fyfe, C.A. and Wasylishen, R.E., Studies of Phosphorus(III) Ligands and Their Complexes of Ni(II), Pd(II), and Pt(II) Immobilized on Insoluble Supports by High-Resolution Solid-State ^{31}P NMR Using Magic-Angle Spinning Techniques, J. Am. Chem. Soc., 104 (1982) 438–445.

Berg1995a Bergmann, F. and Bannwarth, W., Solid Phase Synthesis of Directly Linked Peptide-Oligodeoxynucleotide Hybrids Using Standard Synthesis Protocols, Tetrahedron Lett., 36 (1995) 1839–1842.

Berg1995b Bergmann, F., Kueng, E., Iaiza, P. and Bannwarth, W., Allyl as Internucleotide Protecting Group in DNA Synthesis to be Cleaved Off by Ammonia, Tetrahedron, 51 (1995) 6971–6976.

Bern1983 Bernard, M. and Ford, W.T., Wittig Reagents Bound to Crosslinked Polystyrenes, J. Org. Chem., 48 (1983) 326–332.

Bha1997 Bhalay, G., Blaney, P., Palmer, V.H. and Baxter, D., Solid-Phase Synthesis of Diverse Tetrahydro-1,4-Benzodiazepine-2-ones, Tetrahedron Lett., 38 (1997) 8375–8378.

Bic1996 Bicknell, A.J. and Hird, N.W., Synthesis of a Highly Functionalized Rigid Template by Solid Phase Azomethine Ylide Cycloaddition, Bioorg. Med. Chem. Lett., 6 (1996) 2441–2444.

Blan1991 Blanton, J.R. and Salley, J.M., Reducing Alkyl Halides Using a Polymer-Bound Crown Ether/Tin Hydride Cocatalyst, J. Org. Chem., 56 (1991) 490–491.

Blo1990 Blossey, E.C., Cannon, R.G., Ford, W.T., Periyasamy, M. and Mohanraj, S., Synthesis, Reactions and Carbon-13 FT NMR Spectroscopy of Polymer-Bound Steroids, J. Org. Chem., 55 (1990) 4664–4668.

Boe1996 Boehm, T.L. and Showalter, H.D.H., Development of a Novel Silyl Ether Linker for Solid-Phase Organic Synthesis, J. Org. Chem., 61 (1996) 6498–6499.

Bol1996 Bolton, G.L., Solid Phase Synthesis of Azabicyclo[4.3.0]nonen-8-one Amino Acid Derivatives via Intramolecular Pauson-Khand Cyclization, Tetrahedron Lett., 37 (1996) 3433–3436.

Bond1975 Bonds, W.D., Jr., Brubaker, C.H., Chandrasekaran, E.S., Gibbons, C., Grubbs, R.H. and Kroll, L.C., Polystyrene Attached Titanocene Species. Preparation and Reactions, J. Am. Chem. Soc., 97 (1975) 2128–2132.

Booj1997 Boojamra, C.G., Burow, K.M., A., T.L. and Ellman, J.A., Solid-Phase Synthesis of 1,4-Benzodiazepine-2,5-diones. Library Preparation and Demonstration of Synthesis Generality, J. Org. Chem., 62 (1997) 1240–1256.

Booj1995 Boojamra, C.G., Burow, K.M. and Ellman, J.A., An Expedient and High-Yielding Method for the Solid-Phase Synthesis of Diverse 1,4-Benzodiazepine-2,5-diones, J. Org. Chem., 60 (1995) 5742–5743.

Boyd1996 Boyd, E.A., Chan, W.C. and Loh, V.M., Jr., Multiple Solid Phase Synthesis of (RS)-1-Aminophosphinic Acids, Tetrahedron Lett., 37 (1996) 1647–1650.

Bra1995 Bray, A.M., Chiefari, D.S., Valerio, R.M. and Maeji, N.J., Rapid Optimization of Organic Reactions on Solid Phase Using the Multipin Approach: Synthesis of 4-Aminoproline Analogs by Reductive Amination, Tetrahedron Lett., 36 (1995) 5081–5084.

Bro1997a Brown, E.G. and Nuss, J.M., Alkylation of Rink's Amide Linker on Polystyrene Resin: A Reductive Amination Approach to Modified Amine-Linkers for the Solid Phase Synthesis of N-Substituted Amide Derivatives, Tetrahedron Lett., 38 (1997) 8457–8460.

Bro1996 Brown, J.M. and Ramsden, J.A., Supports for Solid-Phase Synthesis with High Fluidity and High Functionality, J. Chem. Soc., Chem. Commun., (1996) 2117–2118.

Bro1997b Brown, S.D. and Armstrong, R.W., Parallel Synthesis of Tamoxifen and Derivatives on Solid Support via Resin Capture, J. Org. Chem., 62 (1997) 7076–7077.

Buc1996 Buckman, B.O. and Mohan, R., Solid-Phase Synthesis of 1,3-Dialkylquinazoline-2,4-diones, Tetrahedron Lett., 37 (1996) 4439–4442.

Bun1992 Bunin, B.A. and Ellman, J.A., A General and Expedient Method for the Solid-Phase Synthesis of 1,4-Benzodiazepine Derivatives, J. Am. Chem. Soc., 114 (1992) 10997–10998.

Bun1994a Bunin, B.A. and Ellman, J.A., Increasing the Diversity of a 1,4-Benzodiazepine Library Through Side-Chain Functionalization, Polymer Preprints, 35 (1994) 983–984.

Bun1994b Bunin, B.A., Plunkett, M.J. and Ellman, J.A., The Combinatorial Synthesis and Chemical and Biological Evaluation of a 1,4-Benzodiazepine Library, Proc. Natl. Acad. Sci. USA, 91 (1994) 4708–4712.

Bun1997 Bunin, B.A., Plunkett, M.J., Ellman, J.A. and Bray, A.M., The Synthesis of a 1680 Member 1,4-Benzodiazepine Library, New J. Chem., 21 (1997) 125–130.

Burd1993 Burdick, D.J., Struble, M.E. and Burnier, J.P., Solid Phase Synthesis of Peptide para-Nitroanilides, Tetrahedron Lett., 34 (1993) 2589–2592.

Burg1997a Burgess, K., Ibarzo, J., Linthicum, D.S., Russell, D.H., Shin, H., Shitangkoon, A., Totani, R. and Zhang, A.J., Solid Phase Syntheses of Oligoureas, J. Am. Chem. Soc., 119 (1997) 1556–1564.

Burg1997b Burgess, K., Lim, D., Bois-Choussy, M. and Zhu, J., Rapid and Efficient Solid Phase Syntheses of Cyclic Peptides with Endocyclic Biaryl Ether Bonds, Tetrahedron Lett., 38 (1997) 3345–3348.

Burg1995 Burgess, K., Linthicum, D.S. and Shin, H., Solid-Phase Synthesis of Unnatural Biopolymers Containing Repeating Urea Units, Angew. Chem., Int. Ed. Engl., 34 (1995) 907–909.

Burl1975 Burlitch, J.M. and Winterton, R.C., Solid Phase Organometallic Synthesis. Bis(di-n-butylchlorotin)tetracarbonylosmium, J. Am. Chem. Soc., 97 (1975) 5605–5606.

Burl1978 Burlitch, J.M. and Winterton, R.C., Macroreticular Copolymer Bound Analogs of Phenylmercuric Chloride, Phenyllithium, and Phenyldi-n-Butyltin Chloride, J. Organomet. Chem., 159 (1978) 299–316.

Byk1997 Byk, G., Frederic, M. and Scherman, D., One Pot Synthesis of Unsymmetrically Functionalized Polyamines by a Solid Phase Strategy Starting from their Symmetrical polyamine-Counterparts, Tetrahedron Lett., 38 (1997) 3219–3222.

Camb1994 Campbell, D.A. and Bermak, J.C., Solid-Phase Synthesis of Peptidylphosphonates, J. Am. Chem. Soc., 116 (1994) 6039–6040.

Camb1995 Campbell, D.A., Bermak, J.C., Burkoth, T.S. and Patel, D.V., A Transition State Analog Inhibitor Combinatorial Library, J. Am. Chem. Soc., 117 (1995) 5381–5382.

Campl1971a Camps, F., Castells, J., Ferrando, M.J. and Font, J., Organic Synthesis with Functionalized Polymers: I. Preparation of Polymeric Substrates and Alkylation of Esters, Tetrahedron Lett., (1971) 1713–1714.

Campl1971b Camps, F., Castells, J., Font, J. and Vela, F., Organic Syntheses with Functionalized Polymers: II. Wittig Reaction with Polystyryl-p-Diphenylphosphoranes, Tetrahedron Lett., (1971) 1715–1716.

Can1995 Canne, L.E., Walker, S.M. and Kent, S.B.H., A General Method for the Synthesis of Thioester Resin Linkers for Use in the Solid Phase Synthesis of Peptide-α-Thioacids, Tetrahedron Lett., 36 (1995) 1217–1220.

Cao1996 Cao, X. and Mjalli, A.M.M., Combinatorial Method for the Synthesis of α-Hydroxy Phosphonates on Wang Resin, Tetrahedron Lett., 37 (1996) 6073–6076.

Cap1982 Capka, M., Czakoová, M., Urbaniak, W. and Schubert, U., Hydrogenation and Hydrosilylation Activity of Homogeneous and Immobilized Dicarbonyl(2,4-pentanedionato)rhodium complexes, J. Mol. Cat., 74 (1982) 335–334.

Cap1971 Capka, M., Svoboda, P., Cerny, M. and Hetfleje, J., Hydrogenation, Hydrosilation and Hydroformylation of Olefins Catalysed by Polymer-Supported Rhodium Complexes, Tetrahedron Lett., (1971) 4787–4790.

Card1979 Card, R.J., Liesner, C.E. and Neckers, D.C., Poly(styryl)bipyridinepalladium Complexes as Heterogeneous Catalyst for Hydrogenation of Alkanes and Alkynes, J. Org. Chem., 44 (1979) 1095–1098.

Cer1996 Cernerud, M., Reina, J.A., Tegenfeldt, J. and Moberg, C., Chiral Polymers via Asymmetric Epoxidation and Asymmetric Dihydroxylation, Tetrahedron: Asymmetry, 7 (1996) 2863–2870.

Chan1985 Chan, T.-H. and Huang, W.-Q., Polymer-Anchored Organosilyl Protecting Group in Organic Synthesis, J. Chem. Soc., Chem. Commun., (1985) 909–911.

Chan1995 Chan, W.C. and Mellor, S.L., Reductive Alkylation of 9-Amino-xanthen-3-yloxymethylpoly(styrene): a Novel Procedure for the Synthesis of Peptidyl N-Alkyl Amides by Fmoc/But Chemistry, J. Chem. Soc., Chem. Commun., (1995) 1475–1477.

Chag1981 Chang, Y.H. and Ford, W.T., Use of Highly Cross-Linked Polystyrene Supports for Acylation and Alkylation of an Ester Enolate at Room Temperature, J. Org. Chem., 46 (1981) 5364–5371.

Chap1975 Chapman, P.H. and Walker, D., Amino Acids and Peptides. Preparation and Reactions of a Polymer Diazomethylene, J. Chem. Soc., Chem. Commun., (1975) 690–691.

Chem1980 Cheminat, A., Benezra, C., Farrall, M.J. and Frechet, J., Use of Polymeric Nucleotides for the Selective Binding and Removal of α-Methylene-γ-Butyrolactone Allergens from Complex Mixtures, Tetrahedron Lett., 21 (1980) 617–618.

Che1994 Chen, C., Randall, L.A.A., Miller, R.B., Jones, A.D. and Kurth, M.J., 'Analogous' Organic Synthesis of Small-Compound Libraries: Validation of Combinatorial Chemistry in Small-Molecule Synthesis, J. Am. Chem. Soc., 116 (1994) 2661–2662.

Che1997 Chen, J.J. and Spatola, A.F., Solid Phase Synthesis of Peptide Hydroxamic Acids, Tetrahedron Lett., 38 (1997) 1511–1514.

Chen1995 Chenera, B., Finkelstein, J.A. and Veber, D.F., Protodetachable Arylsilane Polymer Linkages for Use in Solid Phase Organic Synthesis, J. Am. Chem. Soc., 117 (1995) 11999–12000.

Cheg1997 Cheng, Y. and Chapman, K.T., Solid Phase Synthesis of Spiroindoline, Tetrahedron Lett., 38 (1997) 1497–1500.

Chin1997 Chin, J., Fell, B., Shapiro, M.J., Tomesch, J., Waring, J.R. and Bray, A.M., Magic Angle Spinning NMR for Reaction Monitoring and Structure Determination of Molecules Attached to Multipin Crowns, J. Org. Chem., 62 (1997) 538–539.

Chiu1976 Chiu, S.-H.L. and Anderson, L., Oligosaccharide Synthesis by the Thioglycoside Scheme on Soluble and Insoluble Polystyrene Supports, Carbohydr. Res., 50 (1976) 227–238.

Cho1993 Cho, C.Y., Moran, E.J., Cherry, S.R., Stephans, J.C., Fodor, S.P.A., Adams, C.L., Sundaram, A., Jacobs, J.W. and Schultz, P.G., An Unnatural Biopolymer, Science, 261 (1993) 1303–1305.

Choo1996 Choong, I.C. and Ellman, J.A., Solid-Phase Synthesis: Applications to Combinatorial Libraries, Annu. Rep. Med. Chem., 31 (1996) 309–318.

Chu1995 Chu, S.S. and Reich, S.H., NPIT: A New Reagent for Quantitatively Monitoring Reactions of Amines in Combinatorial Synthesis, Bioorg. Med. Chem. Lett., 5 (1995) 1053–1058.

Clar1997 Clarke, I.D. and Hodge, P., Partial Separation of Enantiomeric 1,2-Diols via Ketal Formation with a Polymer-Supported Chiral Ketone, J. Chem. Soc., Chem. Commun., (1997) 1395–1396.

Coh1977 Cohen, B.J., Kraus, M.A. and Patchornik, A., Organic Synthesis Involving Multipolymer Reactions. Polymeric Trityllithium, J. Am. Chem. Soc., 99 (1977) 4165–4167.

Cole1996 Cole, B.M., Shimizu, K.D., Krueger, C.A., Harrity, J.P.A., Snapper, M.L. and Hoveyda, A.H., Discovery of Chiral Catalysts Through Ligand Diversity: Ti-Catalyzed Enantioselective Addition of TMSCN to Meso Epoxides, Angew. Chem., Int. Ed. Engl., 35 (1996) 1668–1671.

Col1997 Collini, M.D. and Ellingboe, J.W., The Solid Phase Synthesis of Tri-Substituted Indoles, Tetrahedron Lett., 38 (1997) 7963–7966.

Coll1974 Collman, J.P., Gagne, R.R., Kouba, J. and Ljusberg-Wahren, H., Reversible Oxygen Adduct Formation in Cobalt(II) 'Picket Fence Porphyrins', J. Am. Chem. Soc., 96 (1974) 6800–6802.

Coll1972 Collman, J.P., Hegedus, L.S., Cooke, M.P., Norton, J.R., Dolcetti, G. and Marquardt, D.N., Resin-Bound Transition Metal Complexes, J. Am. Chem. Soc., 94 (1972) 1789–1790.

Coll1973 Collman, J.P. and Reed, C.A., Syntheses of Ferrous-Porphyrin Complexes. A Hypothetical Model for Deoxymyoglobin, J. Am. Chem. Soc., 95 (1973) 2048–2049.

Coy1988 Coy, D.H., Hocart, S.J. and Sasaki, Y., Solid Phase Reductive Alkylation Techniques in Analogue Peptide Bond and Side-Chain Modification, Tetrahedron, 44 (1988) 835–841.

Cra1968 Cramer, F. and Köster, H., Synthesis of Oligonucleotides on a Polymeric Carrier, Angew. Chem., Int. Ed. Engl., 7 (1968) 473–474.

Craw1997 Crawshaw, M., Hird, N.W., Irie, K. and Nagai, K., Synthesis of 3,4,5-Trisubstituted Cyclohexanones by Cycloaddition

To Solid Phase 2-Aminobutadienes, Tetrahedron Lett., 38 (1997) 7115–7118.

Cros1977 Crosby, G.A. and Kato, M., The Utilization of a Polymeric Phenylthiomethyllithium Reagent for the Homologation of Alkyl Iodides and Its Application for the Study of Intraresin Reactions of Polymer-Bound Functional Groups, J. Am. Chem. Soc., 99 (1977) 278–280.

Cros1975 Crosby, G.A., Weinshenker, N.M. and Uh, H.S., Polymeric Reagents. III. Synthesis of an Insoluble Polymeric Thioanisole and Its Utilization for the Oxidation of Alcohols, J. Am. Chem. Soc., 97 (1975) 2232–2235.

Crow1970 Crowley, J.I. and Rapoport, H., Cyclization via Solid Phase Synthesis. Unidirectional Dieckmann Products from Solid Phase and Benzyl Triethylcarbinyl Pimelates, J. Am. Chem. Soc., 92 (1970) 6363–6365.

Crow1976 Crowley, J.I. and Rapoport, H., Solid-Phase Organic Synthesis: Novelty or Fundamental Concept?, Acc. Chem. Res., 9 (1976) 135–144.

Crow1980 Crowley, J.I. and Rapoport, H., Unidirectional Dieckmann Cyclizations on a Solid Phase and in Solution, J. Org. Chem., 45 (1980) 3215–3227.

Cun1997 Cuny, G.D., Cao, J. and Hauske, J.R., Ring Opening Cross-Metathesis on Solid Support, Tetrahedron Lett., 38 (1997) 5237–5240.

Dani1993 Danishefsky, S.J., McClure, K.F., Randolph, J.T. and Ruggeri, R.B., A Strategy for the Solid-Phase Synthesis of Oligosaccharides, Science, 260 (1993) 1307–1309.

Dani1994 Danishefsky, S.J., Randolph, J.T. and Roberge, J.Y., Application of the Glycal Assembly Method in the Solid Phase: a New Strategy for the Synthesis of Oligosaccharides and Glycoconjugates, Polym. Prepr., 35 (1994) 977–978.

Dank1995 Dankwardt, S.M., Newman, S.R. and Krstenansky, J.L., Solid Phase Synthesis of Aryl and Benzylpiperazines and Their Application in Combinatorial Chemistry, Tetrahedron Lett., 36 (1995) 4923–4926.

Das1997 Das, B.C. and Iqbal, J., Polyaniline Supported Cobalt(II) Catalyst: Oxidation of Alkenes with Molecular Oxygen, Tetrahedron Lett., 38 (1997) 1235–1238.

Das1995 Das, S.K., Kumar, A., Jr., Nandrajog, S. and Kumar, A., Polymer Supported VO^{2+} Schiff Base Catalyst for Hydroxylation of Benzene, Tetrahedron Lett., 36 (1995) 7909–7912.

De1997 De, A., Basak, P. and Iqbal, J., Polyaniline Supported Cobalt Catalysed One Pot Stereoselective Synthesis of the Structural Analogues of Aminopeptidase Inhibitor Bestatin, Tetrahedron Lett., 68 (1997) 8383–8386.

Dee1997 Deegan, T.L., Gooding, O.W., Baudart, S. and Porco, J.A., Jr, Non-Acidic Cleavage of Wang Derived Ethers From Solid Support: Utilization of a Mixed-Bed Scavenger for DDQ, Tetrahedron Lett., 38 (1997) 4973–4976.

DeG1980 DeGrado, W.F. and Kaiser, E.T., Polymer-Bound Oxime Esters as Supports for Solid-Phase Peptide Synthesis. The Preparation of Protected Peptide Fragments, J. Org. Chem., 45 (1980) 1295–1300.

Dele1995 Deleuze, H. and Sherrington, D.C., Poly(p-acetoxystyrene) Resin: A Prospective New Support for Combinatorial Synthesis, J. Chem. Soc., Perkin Trans. 2, (1995) 2217–2221.

Des1994 Desai, M.C., Zuckermann, R.N. and Moos, W.H., Recent Advances in the Generation of Chemical Diversity Libraries, Drug Dev. Res., 33 (1994) 174–188.

Desh1994 Deshpande, M.S., Formation of Carbon-Carbon Bond on Solid Support: Application of the Stille Reaction, Tetrahedron Lett., 35 (1994) 5613–5614.

DeW1993 DeWitt, S.H., Kiely, J.S., Stankovic, C.J., Schroeder, M.C., Cody, D.M.R. and Pavia, M.R., 'Diversomers': An Approach to Nonpeptide, Nonoligomeric Chemical Diversity, Proc. Natl. Acad. Sci. USA, 90 (1993) 6909–6913.

Dix1978 Dixit, D.M. and Leznoff, C.C., The Use of Polymer Supports in Organic Synthesis. 17. The Synthesis of Unsymmetrical Diamides and Monoamide Monotosylamides from Symmetrical Diamines, Isr. J. Chem., 17 (1978) 248–252.

Dor1969 Dorman, L.C. and Love, J., Esterification of Neutral and Basic Dialkyl(arylmethylene)sulfonium Resins with N-Protected Amino Acids for Use in Solid Phase Peptide Synthesis, J. Org. Chem., 34 (1969) 158–164.

Dor1984 Dorman, M.A., Noble, S.A., McBride, L.J. and Caruthers, M.H., Synthesis of Oligodeoxynucleotides and Oligodeoxynucleotide Anologs Using Phosphoramidite Intermediates, Tetrahedron, 40 (1984) 95–102.

Dou1991 Douglas, S.P., Whitfield, D.M. and Krepinsky, J.J., Polymer-Supported Solution Synthesis of Oligosaccharides, J. Am. Chem. Soc., 113 (1991) 5095–5097.

Dou1995 Douglas, S.P., Whitfield, D.M. and Krepinsky, J.J., Polymer-Supported Solution Synthesis of Oligosaccharides Using a Novel Versatile Linker for the Synthesis of D-Mannopentose, a Structural Unit of D-Mannans of Pathogenic Yeasts, J. Am. Chem. Soc., 117 (1995) 2116–2117.

Dres1996 Dressman, B.A., Spangle, L.A. and Kaldor, S.W., Solid Phase Synthesis of Hydantoins Using a Carbamate Linker and a Novel Cyclization/Cleavage Step, Tetrahedron Lett., 37 (1996) 937–940.

Drew1997 Drewry, D.H., Gerritz, S.W. and Linn, J.A., Solid-Phase Synthesis of Trisubstituted Guanidines, Tetrahedron Lett., 38 (1997) 3377–3380.

Dro1997 Drouillat, B., Kellam, B., Dekany, G., Starr, M.S. and Toth, I., Solid Phase Synthesis of C-Terminal Carbohydrate Modified Enkephalins, Bioorg. Med. Chem. Lett., 7 (1997) 2247–2250.

Du1997 Du, X. and Armstrong, R., Synthesis of Benzofuran Derivatives on Solid Support via SmI_2-Mediated Radical Cyclisation, J. Org. Chem., 62 (1997) 5678–5679.

Dum1973 Dumont, W., Poulin, J.C., Dang, T.P. and Kagan, H.B., Asymmetric Catalytic Reduction with Transition Metal Complexes. II. Asymmetric Catalysis by a Supported Chiral Rhodium Complex, J. Am. Chem. Soc., 95 (1973) 8295–8299.

Dup1989 Dupas, G., Decormeille, A., Bourguignon, J. and Quéguiner, G., Optimization of the Reaction Conditions Using a NADH Model Grafted on a Merrifield Type Resin, Tetrahedron, 45 (1989) 2579–2590.

Dyg1996 Dygutsch, D.P. and Eilbracht, P., Synthesis of Cyclopentanone Derivatives with Polystyrene-Supported Cyclopentadienyl Rhodium Catalysts, Tetrahedron, 52 (1996) 5461–5468.

Ede1997 Ede, N.J. and Bray, A.M., A Simple Linker for the Attachment of Aldehydes to the Solid Phase. Application to Solid Phase Synthesis by the Multipin™ Method, Tetrahedron Lett., 38 (1997) 7119–7122.

Ell1996 Ellman, J.A., Design, Synthesis, and Evaluation of Small-Molecule Libraries, Acc. Chem. Res., 29 (1996) 132–143.

Ess1997 Esser, C.K., Kevin, N.J., Yates, N.A. and Chapman, K.T., Solid-Phase Synthesis of a N-Carboxyalkyl Tripeptide Combinatorial Library, Bioorg. Med. Chem. Lett., 7 (1997) 2639–2644.

Eva1974 Evans, G.O., Pittman, C.U., Jr., McMillan, R., Beach, R.T. and Jones, R., Synthetic and Catalytic Studies of Polymer-Bound Metal Carbonyls, J. Organomet. Chem., 67 (1974) 295–314.

Exc1975 Excoffier, G., Gagnaire, D., Utille, J.P. and Vignon, M., Synthesis of Oligosaccharides on Polymer Supports. IV. Reactions on a Popcorn Polymer, Tetrahedron, 31 (1975) 549–553.

Fag1997 Fagnola, M.C., Candiana, I., Visentin, G., Cabri, W. and Zarini, F., Solid-Phase Synthesis of Indoles Using the Palladium-Catalysed Coupling of Alkynes with Iodoaniline Derivatives, Tetrahedron Lett., 38 (1997) 2307–2310.

Fan1997 Fancelli, D., Fagnola, M.C., Severino, D. and Bedeschi, A., Solid Phase Synthesis of 2-Substituted Benzofurans via the Palladium-Catalyzed Heteroannulation of Acetylenes, Tetrahedron Lett., 38 (1997) 2311–2314.

Far1979 Farrall, M.J., Durst, T. and Frechet, J.M.J., Polymeric Reagents. IV. Generation of Sulfonium Ylides on Insoluble Resins by Phase Transfer Catalysts, Tetrahedron Lett., (1979) 203–206.

Far1976 Farrall, M.J. and Frechet, J.M.J., Bromination and Lithiation: Two Important Steps in the Functionalization of Polystyrene Resins, J. Org. Chem., 41 (1976) 3877–3882.

Far1978 Farrall, M.J. and Frechet, J.M.J., Reaction of Crosslinked Chloromethyl Polystyrene with 1,4-Butanedithiol: Site-Site Interactions and Their Control, J. Am. Chem. Soc., 100 (1978) 7998.

Fat1996 Fathi, R., Rudolph, M.J., Gentles, R.G., Patel, R., MacMillan, E.W., Reitman, M.S., Pelham, D. and Cook, A.F., Synthesis and Properties of Combinatorial Libraries of Phosphoramidates, J. Org. Chem., 61 (1996) 5600–5609.

Fin1989 Findeis, M.A. and Kaiser, E.T., Nitrobenzophenone Oxime Based Resins for the Solid-Phase Synthesis of Protected Peptide Segments, J. Org. Chem., 54 (1989) 3478–3482.

Fit1997 Fitch, W.L. and Szardenings, A.K., Chemiluminescent Nitrogen Detection for HPLC: An Important New Tool in Organic Analytical Chemistry, Tetrahedron Lett., 38 (1997) 1689–1692.

Fitz1997 Fitzpatrick, L.J. and Rivero, R.A., Solid Phase Synthesis of Substituted Aminosulfonyl Ureas Using a Sulfonylcarbamate Linker, Tetrahedron Lett., 38 (1997) 7479–7482.

Fiv1997 Fivush, A.M. and Willson, T.M., AMEBA: An Acid Sensitive Aldehyde Resin for Solid Phase Synthesis, Tetrahedron Lett., 38 (1997) 7151–7154.

Fle1996 Flegelova, Z. and Patek, M., Bis-Allylic Templates for Pd(0)-Catalyzed Solid Phase Synthesis of Tertiary Amines, J. Org. Chem., 61 (1996) 6735–6738.

Flo1996 Floyd, C.D., Lewis, C.N., Patel, S.R. and Whittaker, M., A Method for the Synthesis of Hydroxamic acids on Solid Phase, Tetrahedron Lett., 37 (1996) 8045–8048.

Form1995 Forman, F.W. and Sucholeiki, I., Solid-Phase Synthesis of Biaryls via the Stille Reaction, J. Org. Chem., 60 (1995) 523–528.

Fral1997 Fraley, M.E. and Rubino, R.S., Two Methods for the Preparation of 2-Cyclohexenones from Resin-Bound 1,3-Cyclohexanedione, Tetrahedron Lett., 38 (1997) 3365–3368.

Frac1995 Franot, C., Stone, G.B., Engeli, P., Spondlin, C. and Waldvogel, E., A Polymer-Bound Oxazaborolidine Catalyst: Enantioselective Borane Reductions of Ketones, Tetrahedron: Asymmetry, 6 (1995) 2755–2766.

Frec1971 Frechet, J.M. and Schuerch, C., Solid-Phase Synthesis of Oligosaccharides. I. Preparation of the Solid Support. Poly[p-(1-propen-3-ol-1-yl)styrene], J. Am. Chem. Soc., 93 (1971) 492–496.

Frec1972a Frechet, J.M. and Schuerch, C., Solid-Phase Synthesis of Oligosaccharides. II. Steric Control by C-6 Substituents in Glucoside Syntheses, J. Am. Chem. Soc., 94 (1972) 604–609.

Frec1972b Frechet, J.M. and Schuerch, C., Solid-Phase Synthesis of Oligosaccharides. III. Preparation of Some Derivatives of Di- and Trisaccharides via a Simple Alcoholysis Reaction, Carbohydr. Res., 22 (1972) 399–412.

Frec1981 Frechet, J.M.J., Synthesis and Applications of Organic Polymers as Supports and Protecting Groups, Tetrahedron, 37 (1981) 663–683.

Frec1979a Frechet, J.M.J., de Smet, M. and Farrall, M.J., Chemical Modification of Crosslinked Resins by Phase Transfer Catalysis: Preparation of Polymer-Bound Dinitriles and Diamines, Tetrahedron Lett., (1979) 137–138.

Frec1979b Frechet, J.M.J., de Smet, M.D. and Farrall, M.J., Application of Phase-Transfer Catalysis to the Chemical Modification of Cross-Linked Polystyrene Resins, J. Org. Chem., 44 (1979) 1774–1779.

Frec1979c Frechet, J.M.J., de Smet, M.D. and Farrall, M.J., Functionalization of Crosslinked Polystyrene Resins: 2. Preparation of Nucleophilic Resins Containing Hydroxyl or Thiol Functionalities, Polymer, 20 (1979) 675–680.

Frec1980 Frechet, J.M.J., Farrall, M.J., Benézra, C. and Cheminat, A., Removal of Allergens from Natural Oils by Selective Binding to Polymer Supports III, Polymer Preprints, 21 (1980) 101–103.

Frec1975a Frechet, J.M.J. and Haque, K.E., Polymeric Reagents. Preparation of Resins Containing Polyvinylperbenzoic Acid Units, Macromolecules, 8 (1975) 130–134.

Frec1975b Frechet, J.M.J. and Haque, K.E., Use of Polymers as Protecting Groups in Organic Synthesis. II. Protection of Primary Alcohol Functional Groups, Tetrahedron Lett., (1975) 3055–3056.

Frec1984 Frechet, J.M.J., Kelly, J. and Sherrington, D.C., Some Novel Polymer-Supported Optically Active Phase Transfer Catalysts: 1. Synthesis, Polymer, 25 (1984) 1491–1498.

Frec1975c Frechet, J.M.J. and Pellé, G., Use of Polymers as Protecting Groups in Organic Synthesis. Preparation of Partially Substituted Derivatives of D-Glucose, J. Chem. Soc., Chem. Commun., (1975) 225–226.

Frec1978 Frechet, J.M.J. and Seymour, E., Use of Polymers as Protecting Groups in Organic Synthesis VII. Preparation of Monobenzoates of Acyclic Triols, Isr. J. Chem., 17 (1978) 253–256.

Fren1994 Frenette, R. and Friesen, R.W., Biaryl Synthesis via Suzuki Coupling on a Solid Support, Tetrahedron Lett., 35 (1994) 9177–9180.

Fret1996 Fretz, H., Solid Phase Synthesis of Peptides Containing Novel L-Phenylalanine Derivatives Substituted with Vicinal Tricarbonyl Moieties, Tetrahedron Lett., 37 (1996) 8479–8482.

Fru1996 Fruechtel, J.S. and Jung, G., Organic Chemistry on Solid Supports, Angew. Chem., Int. Ed. Engl., 35 (1996) 17–42.

Furt1997 Furth, P.S., Reitman, M.S. and Cook, A.F., A Novel Formamidine Linker for use in Solid-Phase Synthesis, Tetrahedron Lett., 38 (1997) 5403–5406.

Fur1997 Furth, P.S., Reitman, M.S., Gentles, R. and Cook, A.M., Solid-Phase Synthesis of Novel Amino-Ether Derivatives, Tetrahedron Lett., 38 (1997) 6643–6646.

Fyl1976 Fyles, T.M. and Leznoff, C.C., The Use of Polymer Supports in Organic Synthesis. V. The Preparation of Monoacetates of Symmetrical Diols, Can. J. Chem., 54 (1976) 935–942.

Fyl1977 Fyles, T.M., Leznoff, C.C. and Weatherston, J., The Use of Polymer Supports in Organic Synthesis. XII. The Total Stereoselective Synthesis of Cis Insect Sex Attractants on Solid Phases, Can. J. Chem., 55 (1977) 4135–4143.

Fyl1978 Fyles, T.M., Leznoff, C.C. and Weatherston, J., The Use of Polymer Supports in Organic Synthesis. Part XV. Bifunctionalized Resins. Applications to the Synthesis of Insect Sex Attractants, Can. J. Chem., 56 (1978) 1031–1041.

Gari1997 Garigipati, R.S., Reagents For Combinatorial Organic Synthesis: Preparation and Uses of Rink-Chloride, Tetrahedron Lett., 38 (1997) 6807–6810.

Gas1995 Gasparrini, F., Misiti, D., Villani, C., Borchardt, A., Burger, M.T. and Still, W.C., Enantioselective Recognition by a New Chiral Stationary Phase at Receptorial Level, J. Org. Chem., 60 (1995) 4314–4315.

Gav1979 Gaviña, F., Gil, P. and Palazón, B., Pericyclic Reactions of Diimine, Tetrahedron Lett., (1979) 1333–1336.

Gay1997 Gayo, L.M. and Suto, M.J., Traceless Linker: Oxidative Activation and Displacement of a Sulfur-Based Linker, Tetrahedron Lett., 38 (1997) 211–214.

Gen1995 Gennari, C., Nestler, H.P., Salom, B. and Still, W.C., Solid-Phase Synthesis of Vinylogous Sulfonyl Peptides, Angew. Chem., Int. Ed. Engl., 34 (1995) 1763–1765.

Geo1996 George, M.H., Hailes, H.C. and Widdowson, D.A., The Synthesis of Precisely Structured Polyurethanes. Part 2. Chain Building Methodology, J. Chem. Soc., Perkin Trans. 1, (1996) 1395–1401.

Geri1990 Gerigk, U., Gerlach, M., Neumann, W.P., Vieler, R. and Weintritt, V., Polymer-Supported Organotin Hydrides as Immobilized Reagents for Free Radical Synthesis, Synthesis, (1990) 448–452.

Gerl1991 Gerlach, M., Jördens, F., Kuhn, H., Neumann, W.P. and Peterseim, M., A Polymer-Supported Organotin Hydride and Its Multipurpose Application in Radical Organic Synthesis, J. Org. Chem., 56 (1991) 5971–5972.

Gil1996 Gilbertson, S.R. and Wang, X., The Combinatorial Synthesis of Chiral Phosphine Ligands, Tetrahedron Lett., 37 (1996) 6475–6478.

Gir1984 Giralt, E., Rizo, J. and Pedroso, E., Application of Gel-Phase ^{13}C NMR to Monitor Solid Phase Peptide Synthesis, Tetrahedron, 40 (1984) 4141–4152.

Gir1997 Girdwood, J.A. and Shute, R.E., Solid-Phase Synthesis of Tri-functionalised, α-substituted Carbamoylmethylhomocysteine Compounds, their Release from the Resin and Subsequent Intramolecular Cyclisation to Give Novel 1,3,3-Trisubstituted Succinimides, J. Chem. Soc., Chem. Commun., (1997) 2307–2308.

Gof1995a Goff, D.A. and Zuckermann, R.N., Solid-Phase Synthesis of Defined 1,4-Benzodiazepine-2,5-dione Mixtures, J. Org. Chem., 60 (1995) 5744–5745.

Gof1995b Goff, D.A. and Zuckermann, R.N., Solid-Phase Synthesis of Highly Substituted Peptoid 1(2H)-Isoquinolinones, J. Org. Chem., 60 (1995) 5748–5749.

Gof1996 Goff, D.A. and Zuckermann, R.N., The Synthesis of 2-Oxopiperazines by Intramolecular Michael Addition on Solid Support, Tetrahedron Lett., 37 (1996) 6247–6250.

Gol1978 Goldwasser, J.M. and Leznoff, C.C., The Use of Polymer Supports in Organic Synthesis. Part XVI. The Solid Phase Synthesis of Monoester Monoamides and Monoester Monoalcohols from Symmetrical Diacid Chlorides, Can. J. Chem., 56 (1978) 1562–1568.

Gop1997 Gopalsamy, A. and Pallai, P.V., Combinatorial Synthesis of Heterocycles: Solid Phase Synthesis of 2-Arylquinoline-4-carboxylic Acid Derivatives, Tetrahedron Lett., 38 (1997) 907–910.

Gor1997a Gordeev, M.F., Gordon, E.M. and Patel, D.V., Solid Phase Synthesis of β-Sultams, J. Org. Chem., 62 (1997) 8177–8181.

Gor1997b Gordeev, M.F., Hui, H.C., Gordon, E.M. and Patel, D.V., A General and Efficient Solid Phase Synthesis of Quinazoline-2,4-diones, Tetrahedron Lett., 38 (1997) 1729–1732.

Gor1996a Gordeev, M.F., Patel, D.V. and Gordon, E.M., Approaches to Combinatorial Synthesis of Heterocycles: A Solid-Phase Synthesis of 1,4-Dihydropyridines, J. Org. Chem., 61 (1996) 924–928.

Gor1996b Gordeev, M.F., Patel, D.V., Wu, J. and Gordon, E.M., Approaches to Combinatorial Synthesis of Heterocycles: Solid Phase Synthesis of Pyridines and Pyrido[2,3-d]pyrimidines, Tetrahedron Lett., 37 (1996) 4643–4646.

Goro1995 Gordon, D.W. and Steele, J., Reductive Alkylation on a Solid Phase: Synthesis of a Piperazinedione Combinatorial Library, Bioorg. Med. Chem. Lett., 5 (1995) 47–50.

Goro1994 Gordon, E.M., Barrett, R.W., Dower, W.J., Fodor, S.P.A. and Gallop, M.A., Applications of Combinatorial Technologies to Drug Discovery. 2. Combinatorial Organic Synthesis, Library Screening Strategies, and Future Directions, J. Med. Chem., 37 (1994) 1385–1401.

Gos1996 Gosselin, F., Di Renzo, M., Ellis, T.H. and Lubell, W.D., Photoacoustic FTIR Spectroscopy, a Nondestructive Method for Sensitive Analysis of Solid-Phase Organic Chemistry, J. Org. Chem., 61 (1996) 7980–7981.

Gou1996 Gouilleux, L., Fehrentz, J.-A., Winternitz, F. and Martinez, J., Solid Phase Synthesis of Chiral 3-Substituted Quinazoline-2,4-diones, Tetrahedron Lett., 37 (1996) 7031–7034.

Gow1997 Gowravaram, M.R. and Gallop, M.A., 'Traceless' Solid Phase Synthesis of Furans via 1,3-Dipolar Cycloaddition Reactions of Isomunchnones, Tetrahedron Lett., 38 (1997) 6973–6976.

Gray1997 Gray, N.S., Kwon, S. and Schultz, P.G., Combinatorial Synthesis of 2,9-Substituted Purines, Tetrahedron Lett., 38 (1997) 1161–1164.

Gre1995 Green, J., Solid Phase Synthesis of Lavendustin A and Analogs, J. Org. Chem., 60 (1995) 4287–4290.

Gru1977 Grubbs, R., Lau, C.P., Cukier, R. and Brubaker, C., Jr., Polymer Attached Metallocenes. Evidence for Site Isolation, J. Am. Chem. Soc., 99 (1977) 4517–4518.

Gru1973 Grubbs, R.H., Gibbons, C., Kroll, L.C., Bonds, W.D., Jr. and Brubaker, C.H., Jr., Activation of Homogeneous Catalysts by Polymer Attachment, J. Am. Chem. Soc., 95 (1973) 2373–2375.

Gru1976 Grubbs, R.H. and Su, S.H., The Preparation of Polymeric Organophosphorus Ligands for Catalyst Attachment, J. Organometallic Chem., 122 (1976) 151–157.

Gud1996 Gude, M., Piarulli, U., Potenza, D., Salom, B. and Gennari, C., A New Method for the Solution and Solid Phase Synthesis of Chiral β-Sulfonopeptides under Mild Conditions, Tetrahedron Lett., 37 (1996) 8589–8592.

Gui1996 Guiles, J.W., Johnson, S.G. and Murray, W.V., Solid-Phase Suzuki Coupling for C-C Bond Formation, J. Org. Chem., 61 (1996) 5169–5171.

Gut1973 Guthrie, R.D., Jenkins, A.D. and Roberts, G.A.F., Synthesis of Oligosaccharides on Polymer Supports. II. Synthesis of β-D-Gentiobiose Derivatives on Soluble Support Copolymers of Styrene and 6-O-(p-Vinylbenzoyl) or 6-O-(p-Vinylphenylsulfonyl) Derivatives of D-Glucopyranose, J. Chem. Soc., Perkin Trans. 1, (1973) 2414–2417.

Halc1994 Halcomb, R.L., Huang, H. and Wong, C.-H., Solution- and Solid-Phase Synthesis of Inhibitors of H. pylori Attachment and E-Selectin-Mediated Leukocyte Adhesion, J. Am. Chem. Soc., 116 (1994) 11315–11322.

Hall1982 Hall, T.W., Greenberg, S., McArthur, C.R., Khouw, B. and Leznoff, C.C., The Solid Phase Synthesis of Unsymmetrical Phthalocyanines, Nouv. J. Chem., 6 (1982) 653–658.

Hal1997 Halm, C., Evarts, J. and Kurth, M.J., A New Attachment/Cleavage Strategy: Polymer-Bound Allylic Sulfones in a Solid-Phase Route to Trisubstituted Olefins, Tetrahedron Lett., 38 (1997) 7709–7712.

Hamp1996 Hamper, B.C., Dukesherer, D.R. and South, M.S., Solid-Phase Synthesis of Proline Analogs via a Three Component 1,3-Dipolar Cycloaddition, Tetrahedron Lett., 37 (1996) 3671–3674.

Han1997 Han, H. and Janda, K.D., Multipolymer-Supported substrate and Ligand Approach to the Sharpless Asymmetric Dihydroxylation, Angew. Chem., Int. Ed. Engl., 36 (1997) 1731–1733.

Hane1974 Hanessian, S., Ogawa, T., Guindon, Y., Kamennof, J.L. and Roy, R., Utilization of Functionalized Polymers in Synthesis: Immobilization of Carbohydrates as Acetals, and Some Chemical Transformations, Carbohyd. Res., 38 (1974) C15-C18.

Hane1996 Hanessian, S. and Yang, R.-Y., Solution and Solid Phase Synthesis of 5-Alkoxyhydantoin Libraries with a Three-Fold Functional Diversity, Tetrahedron Lett., 37 (1996) 5835–5838.

Harr1974 Harrison, C.R. and Hodge, P., Preparation of a Polymer-supported Per-acid and Its Use to Oxidize Olefins to Epoxides, J. Chem. Soc., Chem. Commun., (1974) 1009–1010.

Harr1976 Harrison, C.R. and Hodge, P., Preparation of Polymer-supported Peroxy-acids and their Use to Oxidise Olefins to Epoxides, J. Chem. Soc., Perkin Trans. 1, (1976) 605–609.

Hau1995 Hauske, J.R. and Dorff, P., A Solid Phase CBZ Chloride Equivalent. A New Matrix Specific Linker, Tetrahedron Lett., 36 (1995) 1589–1592.

Hei1997 Heinonen, P. and Lonnberg, H., A Novel Support for Derivatisation and Subsequent N-Alkylation of Secondary Amines: Preparation of N-Alkylated 5- and 6-Alkoxy-1,2,3,4-tetrahydroisoquinolines via Mitsunobu Reaction, Tetrahedron Lett., 38 (1997) 8569–8572.

Hei1972 Heitz, W. and Michels, R., Polymeric Wittig Reagents, Angew. Chem., Int. Ed. Engl., 11 (1972) 298–299.

Herm1996 Hermkens, P.H.H., Ottenheijm, H.C.J. and Rees, D., Solid-Phase Organic Reactions: A Review of the Recent Literature, Tetrahedron, 52 (1996) 4527–4554.

Hern1997 Hernández, A.S. and Hodges, J.C., Solid-Supported tert-Alkoxycarbonylation Reagents for Anchoring of Amines During Solid Phase Organic Synthesis, J. Org. Chem., 62 (1997) 3153–3157.

Hira1983 Hirao, A., Itsuno, S., Hattori, I., Yamaguchi, K., Nakahama, S. and Yamazaki, N., New Method for the Preparation of Benzhydrylamine Resins, J. Chem. Soc., Chem. Commun., (1983) 25–26.

Hird1997 Hird, N.W., Irie, K. and Nagai, K., Solid Phase Synthesis of 2-Aminobutadienes Using a Piperazine Linker, Tetrahedron Lett., 38 (1997) 7111–7114.

Hiro1995 Hiroshige, M., Hauske, J.R. and Zhou, P., Palladium-Mediated Macrocyclization on Solid Support and Its Applications to Combinatorial Synthesis, J. Am. Chem. Soc., 117 (1995) 11590–11591.

Ho1997 Ho, C.Y. and Kukla, M.J., Carbamate Linkers as Latent N-Methylamines in Solid Phase Synthesis, Tetrahedron Lett., 38 (1997) 2792–2802.

Hoc1990 Hocart, S.J., Murphy, W.A. and Coy, D.H., Analogues of Growth Hormone-Releasing Factor (1–29) Amide Containing the Reduced Peptide Bond Isostere in the N-Terminal Region, J. Med. Chem., 33 (1990) 1954–1958.

Hod1985 Hodge, P., Khoshdel, E., Waterhouse, J. and Frechet, J.M.J., Michael Additions Catalyzed by Cinchona Alkaloids Bound via Their Vinyl Groups to Preformed Crosslinked Polymers, J. Chem. Soc., Perkin Trans. 1, (1985) 2327–2331.

Hod1980 Hodge, P. and Sherrington, D.C., Polymer-supported Reactions in Organic Synthesis, Wiley, Chichester, 1980.

Hod1983 Hodge, P. and Waterhouse, J., Preparation of a Polymer-Supported Diol and Its Use in Isolating Aldehydes and Ketones from Mixtures and as a Protecting Group for Aldehydes and Ketones, J. Chem. Soc., Perkin Trans. 1, (1983) 2319–2323.

Hoe1997 Hoekstra, W.J., Greco, M.N. and Yabut, S.C., Solid-Phase Synthesis via N-Terminal Attachment to the 2-Chlorotrityl Resin, Tetrahedron Lett., 38 (1997) 2629–2632.

Hoe1996 Hoekstra, W.J., Maryanoff, B.E., Andrade-Gordon, P., Cohen, J.H., Costanzo, M.J., Damiano, B.P., Haertlein, B.J., Harris, B.D., Kauffman, J.A., Keane, P.M., McComsey, D.F., Villani, F.J. and Yabut, S.C., Solid-Phase Parallel Synthesis Applied to Lead Optimization: Discovery of Potent Analogs of GPIIb/IIIa Antagonist RWJ–50042, Bioorg. Med. Chem. Lett., 6 (1996) 2371–2376.

Hol1995 Holmes, C.P., Chinn, J.P., Look, G.C., Gordon, E.M. and Gallop, M.A., Strategies for Combinatorial Organic Synthesis: Solution and Polymer-Supported Synthesis of 4-Thiazolidinones and 4-Metathiazanones Derived from Amino Acids, J. Org. Chem., 60 (1995) 7328–7333.

Hol1997 Holmes, D., Smith, E. and Nowick, J., Solid-Phase Synthesis of Artificial β-Sheets, J. Am. Chem. Soc., 119 (1997) 7665–7669.

Holy1979 Holy, N.L., Versatile Polymer-Bound Hydrogenation Catalysts. Rhodium(I)-Catalyzed Hydrogenation, J. Org. Chem., 44 (1979) 239–243.

Hor1978 Horiki, K., Igano, K. and Inouye, K., Synthesis of the Merrifield Resin Esters of N-Protected Amino Acids with the Aid of Hydrogen Bonding, Chem. Lett., (1978) 165–168.

Hu1996 Hu, Q.-S., Zheng, X.-F. and Pu, L., The First Optically Active and Sterically Regular Poly(1,1′-bi-2-naphthol)s: Precursors to a New Generation of Polymeric Catalysts, J. Org. Chem., 61 (1996) 5200–5201.

Huan1982 Huang, X., Chan, C.-C. and Zhou, Q.-S., A New Polymeric Reagent – 8-Acyloxyquinoline Polymer for Acylation of Amines, Synth. Commun., 12 (1982) 709–714.

Hug1996 Hughes, I., Application of Polymer-Bound Phosphonium Salts as Traceless Supports for Solid Phase Synthesis, Tetrahedron Lett., 37 (1996) 7595–7598.

Hun1996 Hunt, J.A. and Roush, W.R., Solid-Phase Synthesis of 6-Deoxy Oligosaccharides, J. Am. Chem. Soc., 118 (1996) 9998–9999.

Hut1994 Hutchins, S.M. and Chapman, K.T., A General Method for the Solid Phase Synthesis of Ureas, Tetrahedron Lett., 35 (1994) 4055–4058.

Hut1995 Hutchins, S.M. and Chapman, K.T., A Strategy for Urea Linked Diamine Libraries, Tetrahedron Lett., 36 (1995) 2583–2586.

Hut1996a Hutchins, S.M. and Chapman, K.T., Fischer Indole Synthesis on a Solid Support, Tetrahedron Lett., 37 (1996) 4869–4872.

Hut1996b Hutchins, S.M. and Chapman, K.T., Solid Phase Synthesis of Tetrahydroisoquinolines and Tetrahydroimidazopyridines, Tetrahedron Lett., 37 (1996) 4865–4868.

Isi1982 Isied, S.S., Kuehn, C.G., Lyon, J.M. and Merrifield, R.B., Specific Peptide Sequences for Metal Ion Coordination. 1. Solid-Phase Synthesis of cyclo-$(Gly\text{-}His)_3$, J. Am. Chem. Soc., 104 (1982) 2632–2634.

Ito1975 Ito, H., Takamatsu, N. and Ichikizaki, I., A Promising Modification of the Carbodiimide Method Using an Insoluble Carbodiimide, Chem. Lett., (1975) 577–578.

Its1984 Itsuno, S., Ito, K., Hirao, A. and Nakahama, S., Asymmetric Synthesis Using Chirally Modified Borohydrides. Part

2. Enantioselective Reduction of Ketones with Polymeric (S)-Prolinol-Borane Reagent, J. Chem. Soc., Perkin Trans. 1, (1984) 2887–2893.

Its1987 Itsuno, S., Sakurai, Y., Ito, K., Hirao, A. and Nakahama, S., Asymmetric Reduction of Acetophenone O-Methyloxime with the Reagent Prepared from Borane and Polymer-supported (S)-(–)-2-Amino-3-(4-hydroxyphenyl)-1,1-diphenylpropan-1-ol, Polymer, 28 (1987) 1005–1008.

Iye1996 Iyer, R.P., Devlin, T., Habus, I., Ho, N.-H., Yu, D. and Agrawal, S., N-Pent-4-enoyl Nucleosides: Application in the Synthesis of Support-bound and Free Oligonucleotide Analogs by the H-phosphonate Approach, Tetrahedron Lett., 37 (1996) 1539–1542.

Jac1979 Jacobson, S.E., Mares, F. and Zambri, P.M., Biphase and Triphase Catalysis. Arsonated Polystyrenes as Catalysts in the Baeyer-Villiger Oxidation of Ketones by aqueous Hydrogen Peroxide, J. Am. Chem. Soc., 101 (1979) 6938–6946.

Jam1997 James, I.W., A Compendium of Solid-Phase Chemistry Publications. In Moos, W.H., Pavia, M.R., Ellington, A.D. and Kay, B.K. (Eds.), Annual Reports in Combinatorial Chemistry and Molecular Diversity, vol. 1, ESCOM, Leiden, The Netherlands, 1997, 326–344.

Jang1997 Jang, S.-B., Polymer-Bound Palladium-Catalyzed Cross-Coupling of Organoboron Compounds with Organic Halides and Organic Triflates, Tetrahedron Lett., 30 (1997) 1793–1796.

Jano1982 Janout, V. and Regen, S.L., Polystyryl-Mercury Trifluoroacetate. A Convenient and Mild Reagent for Thiacetal and Thioketal Hydrolysis, J. Org. Chem., 47 (1982) 2212–2213.

Jay1976 Jayalekshmy, P. and Mazur, S., Pseudodilution, the Solid-Phase Immobilization of Benzyne, J. Am. Chem. Soc., 98 (1976) 6710–6711.

John1995 Johnson, C.R. and Zhang, B., Solid Phase Synthesis of Alkenes Using the Horner-Wadsworth-Emmons Reaction and Monitoring by Gel Phase ^{31}P NMR, Tetrahedron Lett., 36 (1995) 9253–9256.

Jud1996 Judice, J.K., Namenuk, A.K. and Burnier, J.P., Rapid Assembly of 'Subtiligase' Substrates to Elucidate Optimal Ligation Junctures, Bioorg. Med. Chem. Lett., 6 (1996) 1961–1966.

Jun1997 Junggebauer, J. and Neumann, W.P., An Improved Synthesis of a Polymer-supported Distannane and Its Application to Radical Formation, Tetrahedron, 53 (1997) 1301–1310.

Kali1974 Kalir, R., Fridkin, M. and Patchornik, A., (4-Hydroxy-3-nitro)benzylated Polystyrene. An Improved Polymeric Nitrophenol Derivative for Peptide Synthesis, Eur. J. Biochem., 42 (1974) 151–156.

Kalj1995 Kaljuste, K. and Unden, A., Solid-Phase Synthesis of 1,2,3,4-Tetrahydro-β-carbolines. Implications for Combinatorial Chemistry, Tetrahedron Lett., 36 (1995) 9211–9214.

Kalj1993 Kaljuste, K. and Undén, A., New Method for the Synthesis of N-Methyl Amino Acids Containing Peptides by Reductive Methylation of Amino Groups on the Solid Phase, Int. J. Pept. Protein Res., 42 (1993) 118–124.

Kalj1996 Kaljuste, K. and Undén, A., Solid-Phase Synthesis of Peptide Aminoalkylamides Using an Allyl Linker, Tetrahedron Lett., 37 (1996) 3031–3034.

Kam1996 Kamahori, K., Ito, K. and Itsuno, S., Asymmetric Diels-Alder Reaction of Methacrolein with Cyclopentadiene Using Polymer-Supported Catalysts: Design of Highly Enantioselective Polymeric Catalysts, J. Org. Chem., 61 (1996) 8321–8324.

Kam1995 Kamahori, K., Tada, S., Ito, K. and Itsuno, S., Synthesis of Polymer-Supported Chiral N-Sulfonylamino Acids and Their Use in Asymmetric Diels-Alder Reaction of Cyclopentadiene with Methacrolein, Tetrahedron: Asymmetry, 6 (1995) 2547–2555.

Kan1994 Kania, R.S., Zuckermann, R.N. and Marlowe, C.K., Free C-Terminal Resin Bound Peptides: Reversal of Peptide Orientation via a Cyclization/Cleavage Protocol, J. Am. Chem. Soc., 116 (1994) 8835–8836.

Kan1997 Kantorowski, E.J. and Kurth, M.J., Solid-Phase Synthesis of Isoxazoles and Isoxazolines: En Route to a New Class of Ionophores, J. Org. Chem., 62 (1997) 6797–6803.

Kate1994 Kates, S.A., de le Torre, B.G., Eritja, R. and Albericio, F., Solid-Phase N-Glycopeptide Synthesis Using Allyl Side-Chain Protected Fmoc-Amino Acids, Tetrahedron Lett., 35 (1994) 1033–1034.

Kato1976 Kato, M., Michels, R. and Heitz, W., Synthesis of Selenium Containing Polymers as Polymer Reagent, Polymer Lett., 14 (1976) 413–415.

Kat1997 Katritzky, A.R., Toader, D., Watson, K. and Kiely, J.S., New Synthesis of SASRIN^{TM} Resin, Tetrahedron Lett., 38 (1997) 7849–7850.

Katr1997 Katritzky, A.R., Xie, L., Zhang, G., Griffith, M., Watson, K. and Kiely, J.S., Synthesis of Primary Amines via Nucleophilic Addition of Organometallic Reagents to Aldimines on Solid Support, Tetrahedron Lett., 38 (1997) 7011–7014.

Kaw1972 Kawana, M. and Emoto, S., Asymmetric Synthesis on an Insoluble Polymer Support, Tetrahedron Lett., (1972) 4855–4858.

Kaw1974 Kawana, M. and Emoto, S., Asymmetric Synthesis with Sugar Derivatives. V. The Synthesis of α-Hydroxy Acids on Insoluble Polymer Supports, Bull. Chem. Soc. Jpn, 47 (1974) 160–165.

Keat1996 Keating, T.A. and Armstrong, R.W., Postcondensation Modifications of Ugi Four-Component Condensation Products: 1-Isocyanocyclohexene as a Convertible Isocyanide. Mechanism of Conversion, Synthesis of Diverse Structures, and Demonstration of Resin Capture, J. Am. Chem. Soc., 118 (1996) 2574–2583.

Kha1996 Khan, N.M., Arumugam, V. and Balasubramanian, S., Solid Phase Reductive Alkylation of Secondary Amines, Tetrahedron Lett., 37 (1996) 4819–4822.

Kic1995 Kick, E.K. and Ellman, J.A., Expedient Method for the Solid-Phase Synthesis of Aspartic Acid Protease Inhibitors Directed toward the Generation of Libraries, J. Med. Chem., 38 (1995) 1427–1430.

Kim1996 Kim, J.-M., Bi, Y., Paikoff, S.J. and Schultz, P.G., The Solid Phase Synthesis of Oligoureas, Tetrahedron Lett., 37 (1996) 5305–5308.

Kim1997 Kim, S.W., Ahn, S.Y., Koh, J.S., Lee, J.H., Ro, S. and Cho, H.Y., Solid Phase Synthesis of Hydantoin Library Using a Novel Cyclisation and Traceless Cleavage Step, Tetrahedron Lett., 38 (1997) 4603–4606.

Kin1979 King, R.B. and Hanes, R.M., Polymer-Anchored Palladium(II) Complexes as Catalysts for the Conversion of Quadricycline to Norbornane, J. Org. Chem., 44 (1979) 1092–1094.

Kis1997 Kiselyov, A.S. and Armstrong, R.W., Solid Support Synthesis of Tetrahydroquinolines via the Grieco Three Component Condensation, Tetrahedron Lett., 38 (1997) 6163–6166.

Kli1996 Klinguer, C., Melnyk, O., Loing, E. and Gras-Masse, H., Synthesis of Hydrazinopeptides Using Solid Phase N-Amination. Application To Chemical Ligation, Tetrahedron Lett., 37 (1996) 7259–7262.

Kob1996a Kobayashi, S., Hachiya, I. and Yasuda, M., Aldol Reactions on Solid Phase. $Sc(OTf)_3$-Catalyzed Aldol Reactions of Polymer-Supported Silyl Enol Ethers with Aldehydes Providing Convenient Methods for the Preparation of 1,3-Diol, β-Hydroxy Carboxylic Acid, and β-Hydroxy Aldehyde Libraries, Tetrahedron Lett., 37 (1996) 5569–5572.

Kob1996b Kobayashi, S., Moriwaki, M., Akiyama, R., Suzuki, S. and Hachiya, I., Parallel Synthesis Using Mannich-type Three-Component Reactions and 'Field Synthesis' for the Construction of an Amino Alcohol Library, Tetrahedron Lett., 37 (1996) 7783–7786.

Kob1996c Kobayashi, S. and Nagayama, S., A New Methodology for Combinatorial Synthesis. Preparation of Diverse Quinoline Derivatives Using a Novel Polymer-Supported Scandium Catalyst, J. Am. Chem. Soc., 118 (1996) 8977–8978.

Koh1996 Koh, J.S. and Ellman, J.A., Palladium-Mediated Three-Component Coupling Strategy for the Solid-Phase Synthesis of Tropane Derivatives, J. Org. Chem., 61 (1996) 4494–4495.

Kol1996 Kolodziej, S.A. and Hamper, B.C., Solid-Phase Synthesis of 5,6-Dihydropyrimidine-2,4-diones, Tetrahedron Lett., 37 (1996) 5277–5280.

Kow1996 Kowalski, J. and Lipton, M.A., Solid Phase Synthesis of a Diketopiperazine Catalyst Containing the Unnatural Amino Acid (S)-Norarginine, Tetrahedron Lett., 37 (1996) 5839–5840.

Kra1971 Kraus, M.A. and Patchornik, A., Reactive Species Mutually Isolated on Insoluble Polymeric Carriers. II. The Alkylation of Esters, Isr. J. Chem., 9 (1971) 269–271.

Krc1995 Krchnak, V., Flegelova, Z., Weichsel, A.S. and Lebl, M., Polymer-Supported Mitsunobu Ether Formation and Its Use in Combinatorial Chemistry, Tetrahedron Lett., 36 (1995) 6193–6196.

Krc1997a Krchnak, V. and Weichsel, A.S., Polymer-Supported Synthesis of Diverse Perhydro-1,4,-Diazepine-2,5-diones, Tetrahedron Lett., 38 (1997) 7299–7302.

Krc1997b Krchnak, V., Weichsel, A.S., Lebl, M. and Felder, S., Automated Solid-Phase Organic Synthesis in Micro-Plate Wells. Synthesis of N-(Alkoxy-acyl)amine Alcohols, Bioorg. Med. Chem. Lett., 7 (1997) 1013–1016.

Kro1997 Kroll, F.E.K., Morphy, R., Rees, D. and Gani, D., Resin-Immobilised Benzyl and Aryl Vinyl Sulfones: New Versatile Traceless Linkers for Solid-Phase Organic Synthesis, Tetrahedron Lett., 38 (1997) 8573–8576.

Kur1994 Kurth, M.J., Ahlberg Randall, L.A., Chen, C., Melander, C., Miller, R.B., McAlister, K., Reitz, G., Kang, R., Nakatsu, T. and Green, C., Library-Based Lead Compound Discovery: Antioxidants by an Analogous Synthesis/Deconvolutive Assay Strategy, J. Org. Chem., 59 (1994) 5862–5864.

Kur1996 Kurth, M.J., Randall, L.A.A. and Takenouchi, K., Solid-Phase Combinatorial Synthesis of Polyisoxazolines: A Two-Reaction Iterative Protocol, J. Org. Chem., 61 (1996) 8755–8761.

Lam1997 Lam, K.S., Lebl, M. and Krchnak, V., The 'One-Bead-One-Compound' Combinatorial Library Method, Chem. Rev., 97 (1997) 411–448.

Lar1996 Larhed, M., Lindeberg, G. and Hallberg, A., Rapid Microwave-Assisted Suzuki Coupling on Solid-Phase, Tetrahedron Lett., 37 (1996) 8219–8222.

Las1987 Laszlo, P., Preparative Chemistry Using Supported Reagents, Academic Press, San Diego, CA, 1987.

LeB1996 Le Bec, C. and Wickstrom, E., Stereospecific Grignard-Activated Solid Phase Synthesis of DNA Methylphosphonate Dimers, J. Org. Chem., 61 (1996) 510–513.

LeH1997 Le Hete, C., David, M., Carreaux, F., Carboni, B. and Sauleau, A., Synthesis of Functionalized γ- and δ-Lactones via Polymer-Bound Epoxides, Tetrahedron Lett., 38 (1997) 5153–5156.

Lee1996 Lee, J., Griffin, J.H. and Nicas, T.I., Solid-Phase Synthesis of Bacitracin A, J. Org. Chem., 61 (1996) 3983–3986.

Ley1995 Ley, S.V., Mynett, D.M. and Koot, W.-J., Solid Phase Synthesis of Bicyclo[2.2.2]octane Derivatives via Tandem Michael Addition Reactions and Subsequent Reductive Amination, Synlett, (1995) 1017–1020.

Lez1977a Leznoff, C.C. and Dixit, D.M., The Use of Polymer Supports in Organic Synthesis. XI. The Preparation of Monoethers of Symmetrical Dihydroxy Aromatic Compounds, Can. J. Chem., 55 (1977) 3351–3355.

Lez1976 Leznoff, C.C. and Fyles, T.M., Synthesis of Insect Sex Attractants on Solid Phases, J. Chem. Soc., Chem. Commun., (1976) 251–252.

Lez1977b Leznoff, C.C., Fyles, T.M. and Weatherston, J., Use of Polymer Supports in Organic Synthesis. VIII. Solid Phase Synthesis of Insect Sex Attractants, Can. J. Chem., 55 (1977) 1143–1153.

Lez1977c Leznoff, C.C. and Goldwasser, J.M., The Use of Insoluble Polymer Supports as Monoblocking Groups of Symmetrical Diacid Chlorides, Tetrahedron Lett., (1977) 1875–1878.

Lez1982 Leznoff, C.C. and Hall, T.W., The Synthesis of a Soluble, Unsymmetrical Phthalocyanine on a Polymer Support, Tetrahedron Lett., 23 (1982) 3023–3026.

Lez1993 Leznoff, C.C., McArthur, C.R. and Qin, Y., Phthalocyanine-Modified Silica Gels and Their Application in the Purification of Unsymmetrical Phthalocyanines, Can. J. Chem., 71 (1993) 1319–1326.

Lez1978 Leznoff, C.C. and Svirskaya, P.I., Solid-Phase Synthesis of Unsymmetrical Tetraarylporphyrins, Angew. Chem. Int. Ed. Engl., 17 (1978) 947.

Lez1977d Leznoff, C.C. and Sywanyk, W., Use of Insoluble Polymer Supports in Organic Synthesis. 9. Synthesis of Unsymmetrical Carotenoids on Solid Phases, J. Org. Chem., 42 (1977d) 3203–3205.

Lez1972 Leznoff, C.C. and Wong, J.Y., Use of Polymer Supports in Organic Synthesis. Synthesis of Monotrityl Ethers of Symmetrical Diols, Can. J. Chem., 50 (1972) 2892–2893.

Lez1973 Leznoff, C.C. and Wong, J.Y., Use of Polymer Supports in Organic Synthesis. III. Selective Chemical Reactions on One Aldehyde Group of Symmetrical Dialdehydes, Can. J. Chem., 51 (1973) 3756–3764.

Lez1980 Leznoff, C.C. and Yedidia, V., The Solid Phase Synthesis of Tertiary Hydroxyesters from Symmetrical Diacid Chlorides using Organomanganese Reagents, Can. J. Chem., 58 (1980) 287–290.

Li1997 Li, W. and Yan, B., A Direct Comparison of the Mixing Efficiency in Solid-Phase Organic Synthesis by Single Bead IR and Fluorescence Spectrometry, Tetrahedron Lett., 38 (1997) 6485–6488.

Liu1995 Liu, G. and Ellman, J.A., A General Solid-Phase Synthesis Strategy for the Preparation of 2-Pyrrolidinemethanol Ligands, J. Org. Chem., 60 (1995) 7712–7713.

Loh1996 Lohray, B.B., Nandanan, E. and Bhushan, V., Catalytic Asymmetric Dihydroxylation of Alkenes Using Silica Gel Supported Cinchona Alkaloid, Tetrahedron: Asymmetry, 7 (1996) 2805–2808.

Loo1994 Look, G.C., Holmes, C.P., Chinn, J.P. and Gallop, M.A., Methods for Combinatorial Organic Synthesis: The Use of Fast ^{13}C NMR Analysis for Gel Phase Reaction Monitoring, J. Org. Chem., 59 (1994) 7588–7590.

Loo1995 Look, G.C., Murphy, M.M., Campbell, D.A. and Gallop, M.A., Trimethylorthoformate: A Mild and Effective Dehydrating Reagent for Solution and Solid Phase Imine Formation, Tetrahedron Lett., 36 (1995) 2937–2940.

Loo1996 Look, G.C., Schullek, J.R., Holmes, C.P., Chinn, J.P., Gordon, E.M. and Gallop, M.A., The Identification of Cyclooxygenase-1 Inhibitors from 4-Thiazolidinone Combinatorial Libraries, Bioorg. Med. Chem. Lett., 6 (1996) 707–712.

Lor1996 Lorsbach, B.A., Miller, R.B. and Kurth, M.J., Reissert-Based 'Traceless' Solid-Phase Synthesis: Isoquinoline and Isoxazoline-Containing Heterocycles, J. Org. Chem., 61 (1996) 8716–8717.

MacD1996 MacDonald, A.A., DeWitt, S.H., Hogan, E.M. and Ramage, R., A Solid Phase Approach to Quinolones Using the DIVERSOMER Technology, Tetrahedron Lett., 37 (1996) 4815–4818.

Mae1992 Maeji, N.J., Tribbick, G., Bray, A.M. and Geysen, H.M., Simultaneous Multiple Synthesis of Peptide-Carrier Conjugates, J. Immunol. Methods, 146 (1992) 83–90.

Marl1993 Marlowe, C.K., Peptide Cyclization on TFA Labile Resin Using the Trimethylsilyl (TMSE) Ester as an Orthogonal Protecting Group, Bioorg. Med. Chem. Lett., 3 (1993) 437–440.

Marq1996 Marquais, S. and Arlt, M., Aryl-Aryl Cross Coupling on a Solid Support Using Zinc Organic Reagents and Palladium Catalysis, Tetrahedron Lett., 37 (1996) 5491–5494.

Mars1970 Marshall, D.L. and Liener, I.E., A Modified Support for Solid-Phase Peptide Synthesis Which Permits the Synthesis of Protected Peptide Fragments, J. Org. Chem., 35 (1970) 867–878.

Mari1997 Marti, R.E., Bleicher, K.H. and Bair, K.W., Solid Phase Synthesis of β-Peptides via Arndt-Eistert Homologation of Fmoc-Protected Amino Acid Diazoketones, Tetrahedron Lett., 38 (1997) 6145–6148.

Marx1997 Marx, M.A., Grillot, A.-L., Louer, C.T., Beaver, K.A. and Bartlett, P.A., Synthetic Design for Combinatiorial Chemistry. Solution and Polymer-Supported Synthesis of Polycyclic Lactams by Intramolecular Cyclisation of Azomethine Ylides, J. Am. Chem. Soc., 119 (1997) 6153–6167.

Marz1996 Marzinzik, A.L. and Felder, E.R., Solid Support Synthesis of Highly Functionalized Pyrazoles and Isoxazoles; Scaffolds for Molecular Diversity, Tetrahedron Lett., 37 (1996) 1003–1006.

Mas1978 Masuda, T. and Stille, J.K., Transition Metal Catalyzed Asymmetric Organic Syntheses via Polymer-Attached Optically Active Phosphine Ligands. Synthesis of R Amino Acids by Hydrogenation with a Polymer Catalyst Containing Optically Active Alcohol Sites, J. Am. Chem. Soc., 100 (1978) 268–272.

Mata1997 Mata, E.G., β-Lactams on Solid Support: Mild and Efficient Removal of Penicillin Derivatives from Merrifield Resin using Aluminium Chloride, Tetrahedron Lett., 38 (1997) 6336–6338.

Mat1997 Matthews, J. and Rivero, R.A., Base-Promoted Solid-Phase Synthesis of Substituted Hydantoins and Thiohydantoins, J. Org. Chem., 62 (1997) 6090–6092.

May1996a Mayer, J.P., Bankaitis-Davis, D., Zhang, J., Beaton, G., Bjergarde, K., Andersen, C.M., Goodman, B.A. and Herrera, C.J., Application of the Pictet-Spengler Reaction in Combinatorial Chemistry, Tetrahedron Lett., 37 (1996) 5633–5636.

May1995 Mayer, J.P., Heil, J.R., Zhang, J. and Munson, M.C., An Alternative Solid-Phase Approach to C1-Oxytocin, Tetrahedron Lett., 36 (1995) 7387–7390.

May1997 Mayer, J.P., Lewis, G.S., Curtis, M.J. and Zhang, J., Solid Phase Synthesis of Quinazolines, Tetrahedron Lett., 38 (1997) 8445–8448.

May1996b Mayer, J.P., Zhang, J., Bjergarde, K., Lenz, D.M. and Gaudino, J.J., Solid-Phase Synthesis of 1,4-Benzodiazepine-2,5-diones, Tetrahedron Lett., 37 (1996) 8081–8084.

Maz1979 Mazur, S. and Jayalekshmy, P., Chemistry of Polymer-Bound o-Benzyne. Frequency of Encounter between Substituents on Cross-Linked Polystyrenes, J. Am. Chem. Soc., 101 (1979) 677–683.

McA1982 McArthur, C.R., Worster, P.M., Jiang, J.L. and Leznoff, C.C., Polymer Supported Enantioselective Reactions. II. α-Methylation of Cyclohexanone, Can. J. Chem., 60 (1982) 1836–1841.

McK1972 McKinley, S.V. and Rakshys, J.W. Jr., Wittig Resins. Preparation and Application of Insoluble Polymeric Phosphoranes, J. Chem. Soc., Chem. Commun., (1972) 134–135.

Mel1994 Meldal, M., Auzanneau, F.-I., Hindsgaul, O. and Palcic, M.M., A PEGA Resin for Use in the Solid-Phase Chemical-Enzymic Synthesis of Glycopeptides, J. Chem. Soc., Chem. Commun., (1994) 1849–1850.

Mel1997a Meldal, M., Juliano, M.A. and Jansson, A.M., Azido Acids in a Novel Method of Solid-Phase Peptide Synthesis, Tetrahedron Lett., 38 (1997) 2531–2534.

Mel1997b Mellor, S.L. and Chan, W.C., 4-[2,4-Dimethoxyphenyl(N-fluoren-9-ylmethoxycarbonyl-N-alkylaminooxy)-methyl]phenoxymethlyl polystyrene: A Multiple Solid-Phase Approach to N-Alkylhydroxamic Acids, J. Chem. Soc., Chem. Commun., (1997) 2005–2006.

Mell1997 Mellor, S.L., McGuire, C. and Chan, W.C., N-Fmoc-Aminooxy-2-Chlorotrityl Polystyrene Resin: A Facile Solid-Phase Methodology for the Synthesis of Hydroxamic Acids, Tetrahedron Lett., 38 (1997) 3311–3314.

Mer1963 Merrifield, R.B., Solid Phase Peptide Synthesis. I. The Synthesis of a Tetrapeptide, J. Am. Chem. Soc., 85 (1963) 2149–2154.

Meu1995 Meutermans, W.D.F. and Alewood, P.F., The Solid Phase Synthesis of Dihydro- and Tetrahydroisoquinolines, Tetrahedron Lett., 36 (1995) 7709–7712.

Mey1995 Meyer, J.-P., Davis, P., Lee, K.B., Porreca, F., Yamamura, H.I. and Hruby, V.J., Synthesis Using a Fmoc-Based Strategy and Biological Activities of Some Reduced Peptide Bond Pseudopeptide Analogues of Dynorphin A, J. Med. Chem., 38 (1995) 3462–3468.

Meye1995 Meyers, H.V., Dilley, G.J., Durgin, T.L., Powers, T.S., Winssinger, N.A., Zhu, H. and Pavia, M.R., Multiple Simultaneous Synthesis of Phenolic Libraries, Mol. Div., 1 (1995) 13–20.

Mil1997 Miller, S.C. and Scanlan, T.S., Site-Selective N-Methylation of Peptides on Solid Support, J. Am. Chem. Soc., 119 (1997) 2301–2302.

Mitc1978 Mitchell, A.R., Kent, S.B.H., Engelhard, M. and Merrifield, R.B., A New Synthetic Route to tert-Butyloxycarbonylaminoacyl-4-(oxymethyl)phenylac etamidomethylresin, an Improved Support for Solid–Phase Peptide Synthesis, J. Org. Chem., 43 (1978) 2845–2852.

Mja1996 Mjalli, A.M.M., Sarshar, S. and Baiga, T.J., Solid Phase Synthesis of Pyrroles Derived From a Four Component Condensation, Tetrahedron Lett., 37 (1996) 2943–2946.

Moh1996 Mohan, R., Chou, Y.-L. and Morrissey, M.M., Pictet-Spengler Reaction of Solid Support: Synthesis of 1,2,3,4-Tetrahydro-β-carboline Libraries, Tetrahedron Lett., 37 (1996) 3963–3966.

Moo1992 Moon, H.S., Schore, N.E. and Kurth, M.J., A Polymer-Supported Chiral Auxiliary Applied to the Iodolactonization Reaction: Preparation of γ-Butyrolactones, J. Org. Chem., 57 (1992) 6088–6089.

Moo1994 Moon, H.-S., Schore, N.E. and Kurth, M.J., A Polymer-Supported C_2-Symmetric Chiral Auxiliary: Preparation of Non-Racemic 3,5-Disubstituted-γ-butyrolactones, Tetrahedron Lett., 35 (1994) 8915–8918.

Morp1996 Morphy, J.R., Rankovic, Z. and Rees, D.C., A Novel Linker Strategy for Solid-Phase Synthesis, Tetrahedron Lett., 37 (1996) 3209–3212.

Mue1995 Mueller, C., Kitas, E. and Wessel, H.P., Novel Oligosaccharide Mimetics by Solid-Phase Synthesis, J. Chem. Soc.,

Chem. Commun., (1995) 2425–2426.

Mur1995 Murphy, M.M., Schullek, J.R., Gordon, E.M. and Gallop, M.A., Combinatorial Organic Synthesis of Highly Functionalized Pyrrolidines: Identification of a Potent Angiotensin Converting Enzyme Inhibitor from a Mercaptoacyl Proline Library, J. Am. Chem. Soc., 117 (1995) 7029–7030.

Nan1997 Nandanan, E., Sudalai, A. and Ravindranathan, T., New Polymer Supported Cinchona Alkaloids for Heterogeneous Catalytic Asymmetric Dihydroxylation of Olefins, Tetrahedron Lett., 38 (1997) 2577–2580.

Nef1997a Nefzi, A., Ostresh, J.M. and Houghten, R.A., The Current Status of Heterocyclic Combinatorial Libraries, Chem. Rev., 97 (1997) 449–472.

Nef1997b Nefzi, A., Ostresh, J.M. and Houghten, R.A., Solid Phase Synthesis of 1,3,4,7-Tetrasubstituted Perhydro-1,4-diazepine-2,5-diones, Tetrahedron Lett., 38 (1997) 4943–4946.

Nef1997c Nefzi, A., Ostresh, J.M., Meyer, J.-P. and Houghten, R.A., Solid Phase Synthesis of Heterocyclic Compounds from Linear Peptide: Cyclic Ureas and Thioureas, Tetrahedron Lett., 38 (1997) 931–934.

Nel1996 Nelson, J.C., Young, J.K. and Moore, J.S., Solid-Phase Synthesis of Phenylacetylene Oligomers Utilizing a Novel 3-Propyl-3-(benzyl-supported) Triazene Linkage, J. Org. Chem., 61 (1996) 8160–8168.

Nes1994 Nestler, H.P., Bartlett, P.A. and Still, W.C., A General Method for Molecular Tagging of Encoded Combinatorial Chemistry Libraries, J. Org. Chem., 59 (1994) 4723–4724.

Ngu1997a Ngu, K. and Patel, D.V., A New and Efficient Solid Phase Synthesis of Hydroxamic Acids, J. Org. Chem., 62 (1997) 7088–7089.

Ngu1997b Ngu, K. and Patel, D.V., Preparation of Acid-Labile Resins with Halide Linkers and Their Utility in Solid Phase Organic Synthesis, Tetrahedron Lett., 38 (1997) 973–976.

Nguy1995 Nguyen, S.T. and Grubbs, R.H., The Synthesis and activities of Polystyrene-Supported Olefin Metathesis Catalysts based on $Cl_2(PR_3)_2Ru{=}CH{-}CH{=}CPh_2$, J. Organomet. Chem., 497 (1995) 195–200.

Ni1996 Ni, Z.-J., Maclean, D., Holmes, C.P., Murphy, M.M., Ruhland, B., Jacobs, J.W., Gordon, E.M. and Gallop, M.A., Versatile Approach To Encoding Combinatorial Organic Syntheses Using Chemically Robust Secondary Amine Tags, J. Med. Chem., 39 (1996) 1601–1608.

Nic1997a Nicalaou, K.C., Winssinger, N., Pastor, J., Ninkovic, S., Sarabia, F., He, Y., Vourloumis, D., Yang, Z., Li, T., Giannakakou, P. and Hamel, E., Synthesis of Epthilones A and B in Solid and Solution Phase, Nature, 387 (1997) 268–272.

Nic1997b Nicolaou, K.C., Vourloumis, D., Li, T., Pastor, J., Winssinger, N., He, Y., Ninkovic, S., Sarabia, F., Vallberg, H., Roschangar, F., King, N.P., Ray, M., Finlay, V., Giannakakou, P., Verdier-Pinard, P. and Hamel, E., Designed Epothilones: Combinatorial Synthesis, Tubulin Assembly Properties, and Cytotoxic Action Against Taxol-Resistant Tumor Cells, Angew. Chem., Int. Ed. Engl., 36 (1997) 2097–2103.

Nic1997c Nicolaou, K.C., Winssinger, N., Pastor, J. and DeRoose, F., A General and Highly Efficient Solid Phase Synthesis of Oligosaccharides. Total Synthesis of a Heptasaccharide Phytoalexin Elicitor (HPE), J. Am. Chem. Soc., 119 (1997) 449–450.

Nie1996 Nielsen, J. and Lyngsoe, L.O., Combinatorial Solid-Phase Synthesis of Balanol Analogs, Tetrahedron Lett., 37 (1996) 8439–8442.

Nieu1976 Nieuwstad, T.J., Kieboom, A.P.G., Breijer, A.J., van der Linden, J. and van Bekkum, H., Reversible Binding of Sulfur Dioxide by Polymer-Bound 1,3-Diene Acceptor Systems, Rec. Trav. Chim. Pays-Bas, 95 (1976) 225–231.

Nor1996 Norman, T.C., Gray, N.S., Koh, J.T. and Schultz, P.G., A Structure-Based Library Approach to Kinase Inhibitors, J. Am. Chem. Soc., 118 (1996) 7430–7431.

Nug1997a Nugiel, D.A., Cornelius, L.A.M. and Corbett, J.W., Facile Preparation of 2,6-Disubstituted Purines Using Solid-Phase Chemistry, J. Org. Chem., 62 (1997) 201–203.

Nug1997b Nugiel, D.A., Wacker, D.A. and Nemeth, G.A., Facile Preparation of Chloromethylaryl Solid Supports Using Methanesulfonyl Chloride and Hunig's Base, Tetrahedron Lett., 38 (1997) 5789–5790.

O'D1996 O'Donnell, M.J., Zhou, C. and Scott, W.L., Solid-Phase Unnatural Peptide Synthesis (UPS), J. Am. Chem. Soc., 118 (1996) 6070–6071.

O'D1997 O'Donnell, M.J., Lugar, C.W., Pottorf, R.S. and Zhou, C., Solid-Phase Synthesis of Unnatural Amino Acids using Unactivated Alkyl Halides, Tetrahedron Lett., 38 (1997) 7163–7166.

Oht1995 Ohtani, N., Inoue, Y., Inagaki, Y., Fukuda, K. and Nishiyama, T., Reactions of Amino Acid Decyl Esters with Nucleophiles Catalyzed by Polymer-Supported Amine-Metal Complexes, Bull. Chem. Soc. Jpn, 68 (1995) 1669–1675.

Oji1994 Ojima, I., Tsai, C.-Y. and Zhang, Z., Catalytic Asymmetric Synthesis of Peptides on Polymer Support, Tetrahedron Lett., 35 (1994) 5785–5788.

Ost1994 Ostresh, J.M., Husar, G.M., Blondelle, S., Dorner, B., Weber, P.A. and Houghten, R.A., 'Libraries from Libraries': Chemical Transformation of Combinatorial Libraries to Extend the Range and Repertoire of Chemical Diversity, Proc. Natl. Acad. Sci. USA, 91 (1994) 11138–11142.

Pai1996 Paikoff, S.J., Wilson, T.E., Cho, C.Y. and Schultz, P.G., The Solid Phase Synthesis of N-Alkylcarbamate Oligomers, Tetrahedron Lett., 37 (1996) 5653–5656.

Pan1996 Panek, J.S. and Zhu, B., Synthesis of Aromatic 1,2-Diazines by Inverse Electron Demand Diels-Alder Reaction of Polymer-Supported 1,2,4,5-Tetrazines, Tetrahedron Lett., 37 (1996) 8151–8154.

Park1997 Park, B.-D., Lee, H.-I., Ryoo, S.-J. and Lee, Y.-S., Convenient Method for Preparing Polystyrene having a β-Hydroxy Group: Its Application to the Synthesis of Polyethylene Glycol-Grafted Polystyrene Resin, Tetrahedron Lett., 38 (1997) 591–594.

Parl1997 Parlow, J.J., Mishke, D.A. and Woodard, S.S., Utility of Complementary Molecular Reactivity and Molecular Recognition (CMR/R) Technology and Polymer-Supported Reagents in the Solution-Phase Synthesis of Heterocyclic Carboxamides, J. Org. Chem., 62 (1997) 5908–5919.

Patc1970 Patchornik, A. and Kraus, M.A., Reactive Species Mutually Isolated on Insoluble Polymeric Carriers. I. The Directed Monoacylation of Esters, J. Am. Chem. Soc., 92 (1970) 7587–7589.

Patc1975 Patchornik, A. and Kraus, M.A., The Use of Polymeric Reagents in Organic Synthesis, Pure Appl. Chem., 43 (1975) 503–526.

Patk1994 Patek, M., Drake, B. and Lebl, M., All-Cis Cyclopentane Scaffolding for Combinatorial Solid Phase Synthesis of Small Non-Peptide Compounds, Tetrahedron Lett., 35 (1994) 9169–9172.

Patk1995 Patek, M., Drake, B. and Lebl, M., Solid-Phase Synthesis of 'Small' Organic Molecules Based on Thiazolidine Scaffold, Tetrahedron Lett., 36 (1995) 2227–2230.

Patl1996 Patel, D.V. and Gordon, E.M., Applications of Small-Molecule Combinatorial Chemistry to Drug Discovery, Drug Discovery Today, 1 (1996) 134–144.

Pea1997 Pearson, W.H. and Clark, R.B., Solid Phase Synthesis of Pyrrolidines Via 2-Azaallyl Anion Cycloadditions with Alkenes, Tetrahedron Lett., 38 (1997) 7669–7672.

Pei1997 Pei, Y., Houghten, R.A. and Kiely, J., Synthesis of 4-Amino-3,4-dihydro-2(1H)-quinolines via β-Lactam Intermediates on the Solid Phase, Tetrahedron Lett., 38 (1997) 3349–3352.

Pei1994 Pei, Y. and Moos, W.H., Post-Modification of Peptoid Side Chains: [3+2] Cycloaddition of Nitrile Oxides with Alkenes and Alkynes on the Solid-Phase, Tetrahedron Lett., 35 (1994) 5825–5828.

Pet1995 Petrucci, M.G.L. and Kakkar, A.K., A Convenient Molecular Self-Assembly Route to Thin Films Containing Terminal Donor Ligands and Anchored Organotransition-Metal Complexes for Heterogenized Homogeneous Catalysis, J. Chem. Soc., Chem. Commun., (1995) 1577–1578.

Phi1996 Phillips, G.B. and Wei, G.P., Solid Phase Synthesis of Benzimidazoles, Tetrahedron Lett., 37 (1996) 4887–4890.

Pie1975 Pietta, P. and Brenna, O., p-(Aminomethyl)phenoxymethyl Polymer for Solid Phase Synthesis of Protected Peptide Amides, J. Org. Chem., 40 (1975) 2995–2996.

Pie1970 Pietta, P.G. and Marshall, G.M., Amide Protection and Amide Supports in Solid-Phase Peptide Synthesis, J. Chem. Soc., Chem. Commun., (1970) 650–651.

Piet1997 Piettre, S.R. and Baltzer, S., A New Approach to the Solid-Phase Suzuki Coupling Reaction, Tetrahedron Lett., 38 (1997) 1197–1200.

Pis1997 Piscopio, A.D., Miller, J.F. and Koch, K., Solid Phase Heterocyclic Synthesis via Ring Closing Metathesis: Traceless Linking and Cyclative Cleavage Through a Carbon-Carbon Double Bond, Tetrahedron Lett., 38 (1997) 7143–7146.

Pit1972 Pittman, C.U., Jr, Grube, P.L., Ayers, O.E., MacManus, S.P., Rausch, M.D. and Moser, G.A., Organometallic Polymers. XVI. Copolymerization of Styrenetricarbonylchromium and the Reaction of Polystyrene with Chromium Hexacarbonyl, J. Polymer Sci., 10 (1972) 379–386.

Pit1974 Pittman, C.U., Jr. and Hanes, R.M., Aldehyde and Ketone Synthesis using Polymer-Bound Rhodium Catalysts. III, Ann. N. Y. Acad. Sci., 239 (1974) 76–87.

Pit1975a Pittman, C.U., Jr. and Smith, L.R., Selective, High-Yield, Linear Dimerization of 1,3-Butadiene Catalyzed by $(PPh_3)_2NiBr_2$ and $NaBH_4$ and Its Polymer-Bound Ni(0) Analog, J. Am. Chem. Soc., 97 (1975) 341–344.

Pit1975b Pittman, C.U., Jr. and Smith, L.R., Sequential Multistep Reactions Catalyzed by Polymer-Anchored Homogeneous Catalysts, J. Am. Chem. Soc., 97 (1975) 1749–1754.

Pit1975c Pittman, C.U., Jr., Smith, L.R. and Hanes, R.M., Catalytic Reactions Using Polymer-Bound vs. Homogeneous Complexes of Nickel, Rhodium, and Ruthenium, J. Am. Chem. Soc., 97 (1975) 1742–1748.

Plu1995a Plunkett, M.J. and Ellman, J.A., A Silicon-Based Linker for Traceless Solid-Phase Synthesis, J. Org. Chem., 60 (1995) 6006–6007.

Plu1995b Plunkett, M.J. and Ellman, J.A., Solid-Phase Synthesis of Structurally Diverse 1,4-Benzodiazepine Derivatives Using the Stille Coupling Reaction, J. Am. Chem. Soc., 117 (1995) 3306–3307.

Plu1997 Plunkett, M.J. and Ellman, J.A., Germanium and Silicon Linking Strategies for Traceless Solid-Phase Synthesis, J. Org. Chem., 62 (1997) 2885–2893.

Pop1997 Pop, I.E., Déprez, B.P. and Tartar, A.L., Versatile Acylation of N-Nucleophiles Using a New Polymer-Supported 1-Hydroxybenzotriazole Derivative, J. Org. Chem., 62 (1997) 2594–2603.

Pop1996 Pop, I.E., Dhalluin, C.F., Deprez, B.P., Melnyk, P.C., Lippens, G.M. and Tartar, A.L., Monitoring of a Three-Step Solid Phase Synthesis Involving a Heck Reaction Using Magic Angle Spinning NMR Spectroscopy, Tetrahedron, 52 (1996) 12209–12222.

Pul1993 Pulley, S.R. and Hegedus, L.S., Solid-Phase, Solution and Segment Condensation Peptide Syntheses Incorporating Chromium Carbene Complex-Derived Nonproteinogenic ('Unnatural') Amino Acid Fragments, J. Am. Chem. Soc., 115 (1993) 9037–9047.

Pur1996 Pursch, M., Schlotterbeck, G., Tseng, L.-H., Albert, K. and Rapp, W., Monitoring the Reaction Progress in Combinatorial Chemistry: ^{1}H MAS NMR Investigations on Single Macro Beads in the Suspended State, Angew. Chem., Int. Ed. Engl., 35 (1996) 2867–2869.

Rad1996 Rademann, J. and Schmidt, R.R., A New Method for the Solid Phase Synthesis of Oligosaccharides, Tetrahedron Lett., 37 (1996) 3989–3990.

Raf1978 Rafalko, J.J., Lieto, J., Gates, B.C. and Schrader, G.L., Jr., Polymer-bound Tetranuclear Iridium Carbonyl Catalyst, J. Chem. Soc., Chem. Commun., (1978) 540–541.

Raj1997a Raju, B. and Kogan, P.T., Use of Halomethyl Resins to Immobilise Amines: An Efficient Method for Synthesis of Sulphonamides and Amides on a Solid Support, Tetrahedron Lett., 28 (1997) 4965–4968.

Raj1997b Raju, B. and Kogan, T.P., Solid Phase Synthesis of Sulfonamides Using a Carbamate Linker, Tetrahedron Lett., 38 (1997) 3373–3376.

Ran1995 Rano, T.A. and Chapman, K.T., Solid Phase Synthesis of Aryl Ethers via the Mitsunobu Reaction, Tetrahedron Lett., 36 (1995) 3789–3792.

Rao1997 Rao, Y.V.S., De Vos, D.E. and Jacobs, P.A., Practical Heterogenisation of an Active Manganese Triazacyclononane Epoxidation Catalyst via Surface Glycidylation, J. Chem. Soc., Chem. Commun., (1997) 355–356.

Rav1994 Ravindranathan, T., Chavan, S.P. and Awachat, M.M., Polymer Supported Nitrobenzaldehyde: Efficient, Highly Selective Catalytic Deprotection of Oxathioacetals, Tetrahedron Lett., 35 (1994) 8835–8838.

Reb1974 Rebek, J. and Gaviña, F., The Three-Phase Test for Reaction Intermediates. Cyclobutadiene, J. Am. Chem. Soc., 96 (1974) 7112–7114.

Reb1975a Rebek, J., Jr. and Gaviña, F., The Three Phase Test for Reaction Intermediates. Evidence for Monomeric Metaphosphate, J. Am. Chem. Soc., 97 (1975) 3221–3222.

Reb1975b Rebek, J., Jr. and Gaviña, F., Three-Phase Test. Detection of Free Cyclobutadiene, J. Am. Chem. Soc., 97 (1975) 3453–3456.

Reg1977 Regen, S.L. and Lee, D.P., Influence of Solvent on Site Isolation in Polystyrene Resins, Macromolecules, 10 (1977) 1418–1419.

Regg1996 Reggelin, M. and Brenig, V., Towards Polyketide Libraries: Iterative, Asymmetric Aldol Reactions on a Solid Support, Tetrahedron Lett., 37 (1996) 6851–6852.

Rell1974 Relles, H.M. and Schluenz, R.W., Chemical Transformations with Regenerable, Polymer-Supported Trisubstituted Phosphine Dichlorides. The Efficacious Incorporation of Phosphorus Reagents on Polymer Supports, J. Am. Chem. Soc., 96 (1974) 6469–6475.

Ric1997 Richter, L.S. and Desai, M.C., A TFA-Cleavable Linkage for Solid-Phase Synthesis of Hydroxamic Acids, Tetrahedron Lett., 38 (1997) 321–322.

Ric1994a Richter, L.S. and Gadek, T.R., A Surprising Observation About Mitsunobu Reactions in Solid Phase Synthesis, Tetrahedron Lett., 35 (1994) 4705–4706.

Ric1994b Richter, L.S., Tom, J.Y.K. and Burnier, J.P., Peptide-Cyclizations on Solid Support: A Fast and Efficient Route to Small Cyclopeptides, Tetrahedron Lett., 35 (1994) 5547–5550.

Ric1995 Richter, L.S. and Zuckermann, R.N., Synthesis of Peptide Nucleic Acids (PNA) by Submonomer Solid Phase Synthesis, Bioorg. Med. Chem. Lett., 5 (1995) 1159–1162.

Rob1995 Roberge, J.Y., Beebe, X. and Danishefsky, S.J., A Strategy for a Convergent Synthesis of N-Linked Glycopeptides on a Solid Support, Science, 269 (1995) 202–204.

Robl1994 Robles, J., Pedroso, E. and Grandas, A., Stepwise Solid-Phase Synthesis of the Nucleopeptide Phac-Phe-Val-Ser($p^{3'}$ACT)-Gly-OH, J. Org. Chem., 59 (1994) 2482–2486.

Roc1996 Rockwell, A., Melden, M., Copeland, R.A., Hardman, K., Decicco, C.P. and DeGrado, W.F., Complementarity of Combinatorial Chemistry and Structure-Based Ligand Design: Application to the Discovery of Novel Inhibitors of Matrix Metalloproteinases, J. Am. Chem. Soc., 118 (1996) 10337–10338.

Rod1997a Rodebaugh, R., Fraser-Reid, B. and Geysen, H.M., A New o-Nitrobenzyl Photocleavable Linker for Solid Phase Synthesis, Tetrahedron Lett., 38 (1997) 7653–7656.

Rod1997b Rodebaugh, R., Joshi, S., Fraser-Reid, B. and Gesen, H.M., Polymer-Supported Oligosaccharides via n-Pentenyl Glycosides: Methodology for a Carbohydrate Library, J. Org. Chem., 62 (1997) 5660–5661.

Roe1996 Roelfing, K., Thiel, M. and Kuenzer, H., 1,2,6-Trisubstituted Tetrahydroisoquinoline Derivatives by Solid-Phase Synthesis, Synlett, (1996) 1036–1038.

Rol1975 Rollman, L.D., Porous, Polymer-Bonded Metalloporphyrins, J. Am. Chem. Soc., 97 (1975) 2132–2136.

Rot1996 Rotella, D.P., Solid Phase Synthesis of Olefin and Hydroxyethylene Peptidomimetics, J. Am. Chem. Soc., 118 (1996) 12246–12247.

Rou1974 Roush, W.R., Feitler, D. and Rebek, J., Polymer-Bound Tosyl Azide, Tetrahedron Lett., (1974) 1391–1392.

Rous1997 Roussel, P., Bradley, M., Matthews, I. and Kane, P., A Linker for Amides in Solid Phase Synthesis, Tetrahedron Lett., 38 (1997) 4861–4864.

Rout1997a Routledge, A., Abell, C. and Balasubramanian, S., The Use of a Dithiane Protected Benzoin Photolabile Safety Catch Linker for Solid-Phase Synthesis, Tetrahedron Lett., 38 (1997) 1227–1230.

Rout1997b Routledge, A., Abell, C. and Balasubramanian, S., An Investigation into Solid-Phase Radical Chemistry – Synthesis of Furan Rings, Synlett, (1997) 61–62.

Rou1997 Routledge, A., Stock, H.T., Flitsch, S.L. and Turner, N.J., New Fluoride-Labile Linkers for Solid Phase Organic Synthesis, Tetrahedron Lett., 38 (1997) 8287–8290.

Rov1991 Rovero, P., Quartara, L. and Fabbri, G., Synthesis of Cyclic Peptides on Solid Support, Tetrahedron Lett., 32 (1991) 2639–2642.

Ruh1996 Ruhland, B., Bhandari, A., Gordon, E.M. and Gallop, M.A., Solid-Supported Combinatorial Synthesis of Structurally Diverse β-Lactams, J. Am. Chem. Soc., 118 (1996) 253–254.

Ruh1997 Ruhland, B., Bombrun, A. and Gallop, M.A., Solid-Phase Synthesis of 4-Arylazetidin-2-ones via Suzuki and Heck Cross-Coupling Reactions, J. Org. Chem., 62 (1997) 7820–7826.

Sah1993 Saha, A.K., Sardaro, M., Waychunas, C., Delecki, D., Kruse, L.I., Kutny, R., Cavanaugh, P., Yawman, A. and Upson, D.A., Diisopropylsilyl-linked Oligonucleotide Analogs: Solid-Phase Synthesis and Physicochemical Properties, J. Org. Chem., 58 (1993) 7827–7831.

Sara1997 Sarantakis, D. and Bicksler, J.J., Solid Phase Synthesis of Sec-Amides and Removal from the Polymeric Support Under Mild Conditions, Tetrahedron Lett., 38 (1997) 7325–7328.

Sar1996 Sarshar, S., Siev, D. and Mjalli, A.M.M., Imidazole Libraries on Solid Support, Tetrahedron Lett., 37 (1996) 835–838.

Sas1987 Sasaki, Y., Murphy, W.A., Heiman, M.L., Lance, V.A. and Coy, D.H., Solid-Phase Synthesis and Biological Properties of $\psi[CH_2 2NH]$ Pseudopeptide Analogues of a Highly Potent Somatostatin Octapeptide, J. Med. Chem., 30 (1987) 1162–1166.

Schl1997 Schleyer, A., Meldal, M., Manat, R., Paulsen, H. and Bock, K., Direct Solid Phase Glycosylations of Peptide Templates on a Novel PEG-Based Resin, Angew. Chem., Int. Ed. Engl., 36 (1997) 1976–1978.

Scho1990 Schore, N.E. and Najdi, S.D., Pauson-Khand Cycloadditions of Polymer-linked Substrates, J. Am. Chem. Soc., 112 (1990) 441–442.

Sch1996 Schuster, M., Pernerstorfer, J. and Blechert, S., Ruthenium-Catalyzed Metathesis of Polymer-Bound Olefins, Angew. Chem., Int. Ed. Engl., 35 (1996) 1979–1980.

Sch1994 Schuster, M., Wang, P., Paulson, J.C. and Wong, C.-H., Solid-Phase Chemical-Enzymatic Synthesis of Glycopeptides and Oligosaccharides, J. Am. Chem. Soc., 116 (1994) 1135–1136.

Sci1996 Scialdone, M.A., Diisocyanates as Scaffolds for Combinatorial Libraries. The Solid-Phase Synthesis of Bis[ureas] from Polymer-Supported Diisocyanates, Tetrahedron Lett., 37 (1996) 8141–8144.

See1997 Seeberger, P.H., Beebe, X., Sukenick, G.D., Pochapsky, S. and Danishefsky, S.J., Monitoring the Progress of Solid-Phase Oligosaccharide Synthesis by High-Resolution Magic Angle Spinning NMR: Observations of Enhanced Selectivity for β-Glycoside Formation from α-1,2-Anhydrosugar Donors in Solid-Phase Couplings, Angew. Chem., Int. Ed. Engl., 36 (1997) 491–493.

Sei1997 Seitz, O. and Wong, C., Chemoenzymatic Solution- and Solid-Phase Synthesis of O-Glycopeptides of the Mucin Domain of MAdCAM-1. A General Route to O-LacNAc, OSialyl-LacNAc and O-Sialyl-Lewis-X-Peptides, J. Am. Chem. Soc., 119 (1997) 8766–8776.

Sem1995 Semenov, A.N. and Gordeev, K.Y., A Novel Oxidation-Labile Linker For Solid-Phase Peptide Synthesis, Int. J. Pept. Protein Res., 45 (1995) 303–304.

Sen1996 Seneci, P., Sizemore, C., Islam, K. and Kocis, P., Combinatorial Chemistry and Natural Products. Teicoplanin Aglycon as a Molecular Scaffold for Solid Phase Synthesis of Combinatorial Libraries, Tetrahedron Lett., 37 (1996) 6319–6322.

Set1994 Setty-Fichman, M., Blum, J., Sasson, Y. and Eisen, M., Polystyrene-Supported $RhCl_3$-Quaternary Ammonium Ion Pair as a Long-Lived, Efficient and Recyclable Catalyst, Tetrahedron Lett., 35 (1994) 781–784.

Sey1976 Seymour, E. and Frechet, J.M.J., Use of Polymers as Protecting Groups in Organic Synthesis. IV. Applications of a Polystyrylboronic Acid Resin to the Selective Functionalization of Some Glycosides, Tetrahedron Lett., (1976) 1149–1152.

Sham1973 Shambhu, M.B. and Digenis, G.A., Insoluble Resins in Organic Syntheses I. Preparation and Reactions of Polymeric Anhydrides, Tetrahedron Lett., (1973) 1627–1629.

Shao1995 Shao, J. and Tam, J.P., Unprotected Peptides as Building Blocks for the Synthesis of Peptide Dendrimers with Oxime, Hydrazone, and Thiazolidine Linkages, J. Am. Chem. Soc., 117 (1995) 3893–3899.

Shap1994a Shapiro, G. and Buechler, D., Mild and Rapid Azide-Mediated, Palladium Catalyzed Cleavage of Allylester Based Protecting Groups, Tetrahedron Lett., 35 (1994) 5421–5424.

Shap1994b Shapiro, G., Swoboda, R. and Stauss, U., Fmoc Solid Phase Synthesis of Serine Phosphopeptides via Selective Protection of Serine and On Resin Phosphorylation, Tetrahedron Lett., 35 (1994) 869–872.

Shap1996 Shapiro, M.J., Kumaravel, G., Petter, R.C. and Beveridge, R., ^{19}F NMR Monitoring of a S_NAr Reaction on Solid Support, Tetrahedron Lett., 37 (1996) 4671–4674.

Shar1997 Sharma, S. and Pasha, S., Rapid Anchoring of Cesium Salts of Boc-Protected Amino Acids On Chloromethyl Polystyrene Resin Using Dibenzo-18-Crown-6 as Catalyst, Bioorg. Med. Chem. Lett., 7 (1997) 2077–2080.

Shar1996 Sharma, S.K., Wu, A.D. and Chandramouli, N., Maleimide-Assisted On-Resin Macrocyclization, Tetrahedron Lett., 37 (1996) 5665–5668.

Shea1978 Shea, K.J. and Thompson, E.A., Template Syntheses of Macromolecules. Selective Functionalization of an Organic Polymer, J. Org. Chem., 43 (1978) 4253–4254.

Shi1996 Shimizu, H., Ito, Y., Kanie, O. and Ogawa, T., Solid Phase Synthesis of Polylactosamine Oligosaccharide, Bioorg. Med. Chem. Lett., 6 (1996) 2841–2846.

Shin1997 Shin, J.A., Specific DNA Binding Peptide-Derivatised Solid Support, Bioorg. Med. Chem. Lett., 7 (1997) 2367–2372.

Sho1996 Short, K.M., Ching, B.W. and Mjalli, A.M.M., The Synthesis of Hydantoin 4-Imides on Solid Support, Tetrahedron Lett., 37 (1996) 7489–7492.

Sieb1987 Sieber, P., A New Acid-Labile Anchor Group for the Solid-Phase Synthesis of C-Terminal Peptide Amides by the Fmoc Method, Tetrahedron Lett., 28 (1987) 2107–2110.

Simo1992 Simon, R.J., Kania, R.S., Zuckermann, R.N., Huebner, V.D., Jewell, D.A., Banville, S., Ng, S., Wang, L., Rosenberg, S., Marlowe, C.K., Spellmeyer, D.C., Tan, R., Frankel, A.D., Santi, D.V., Cohen, F.E. and Bartlett, P.A., Peptoids: A Modular Approach to Drug Discovery, Proc. Natl. Acad. Sci. USA, 89 (1992) 9367–9371.

Simo1994 Simon, R.J., Martin, E.J., Miller, S.M., Zuckermann, R.N., Blaney, J.M. and Moos, W.H., Using Peptoid Libraries [Oligo N-Substituted Glycines] for Drug Discovery, Tech. Prot. Chem., 5 (1994) 533–539.

Smi1996a Smith, A.L., Thomson, C.G. and Leeson, P.D., An Efficient Solid Phase Synthetic Route to 1,3-Disubstituted (1H,3H)-Quinazoline-2,4-diones Suitable for Combinatorial Synthesis, Bioorg. Med. Chem. Lett., 6 (1996) 1483–1486.

Smi1996b Smith, J., Liras, J.L., Schneider, S.E. and Anslyn, E.V., Solid and Solution Phase Organic Syntheses of Oligomeric Thioureas, J. Org. Chem., 61 (1996) 8811–8818.

Sof1996 Sofia, M.J., Generation of oligosaccharide and glycoconjugate libraries for drug discovery, Drug Discovery Today, 1 (1996) 27–34.

Step1997 Stephensen, H. and Zaragoza, F., Resin-Bound Isothiocyanates and Their Synthetic Equivalents as Intermediates for the Solid Phase Synthesis of Substituted Thiophenes, J. Org. Chem., 62 (1997) 6096–6097.

Ste1996 Stevens, S.Y., Bunin, B.A., Plunkett, M.J., Swanson, P.C., Ellman, J.A. and Glick, G.D., Non Nucleic Acid Inhibitors of Protein-DNA Interactions Identified through Combinatorial Chemistry, J. Am. Chem. Soc., 118 (1996) 10650–10651.

Stra1997 Stranix, B.R., Liu, H.Q. and Darling, G.D., Functional Polymers from (Vinyl)polystyrene. Recyclable Polymer-Supported Organosilicon Protecting Groups for Solid-Phase Synthesis, J. Org. Chem., 62 (1997) 6183–6186.

Str1974 Strathdee, G. and Given, R., The Synthesis and Application of Anchored Catalysts in Hydrogen Isotope Exchange Systems. D_2-Protic Solvent Exchange, Can. J. Chem., 52 (1974) 3000–3007.

Suc1994 Sucholeiki, I., Solid-Phase Photochemical C-S Bond Cleavage of Thioethers – A New Approach To The Solid-Phase Production of Non-Peptide Molecules, Tetrahedron Lett., 35 (1994) 7307–7310.

Sut1996 Sutherlin, D.P., Stark, T.M., Hughes, R. and Armstrong, R.W., Generation of C-Glycoside Peptide Ligands for Cell Surface Carbohydrate Receptors Using a Four-Component Condensation on Solid Support, J. Org. Chem., 61 (1996) 8350–8354.

Svi1984 Svirskaya, P.I. and Leznoff, C.C., Syntheses of Unconjugated (Z,Z)-Diolefinic Insect Pheromones on Insoluble Polymer Supports, J. Chem. Ecol., 10 (1984) 321–333.

Svi1987 Svirskaya, P.I., Leznoff, C.C. and Steinman, M., Trifluoroacetylation of Amines and Amino Acids by Polymer-Bound Trifluoroacetylation Reagents, J. Org. Chem., 52 (1987) 1362–1364.

Svi1979 Svirskaya, P.I., Leznoff, C.C., Weatherston, J. and Laing, J.E., Syntheses of Trans Alken-1-ols as Candidates for Insect Sex Attractants, J. Chem. Eng. Data, 24 (1979) 152–155.

Swa1997 Swayze, E.E., The Solid-Phase Synthesis of Trisubstituted 1,4-Diazabicyclo[4.3.0]nonan-2-one. Scaffolds: On Bead Monitoring of Heterocycle Forming Reactions Using ^{15}N NMR, Tetrahedron Lett., 38 (1997) 8643–8646.

Syl1997 Sylvain, C., Wagner, A. and Mioskowski, C., Ozone: A Versatile Reagent for Solid Phase Synthesis, Tetrahedron Lett., 38 (1997) 1043–1044.

Sza1996 Szardenings, A.K., Burkoth, T.S., Look, G.C. and Campbell, D.A., A Reductive Alkylation Procedure Applicable to Both Solution- and Solid-Phase Synthesis of Secondary Amines, J. Org. Chem., 61 (1996) 6720–6722.

Tak1967 Takagi, T., Epoxidation with Peracid Type Polymer, J. Polymer Sci., Part B, Polymer Lett., 5 (1967) 1031–1035.

Tam1985 Tam, J.P., A Gradative Deprotection Strategy for the Solid-Phase Synthesis of Peptide Amides Using p-(Acyloxy)benzhydrylamine Resin and the S_N2 Deprotection Method, J. Org. Chem., 50 (1985) 5291–5298.

Tan1995 Tan, W., Iyer, R.P., Jiang, Z., Yu, D. and Agrawal, S., An Efficient Synthesis of Radioisotopically Labelled Oligodeoxyribonucleotides Through Direct Solid-Phase 5′-Phosphitylation, Tetrahedron Lett., 36 (1995) 5323–5326.

Tay1982 Taylor, R.T., Cassell, R.A. and Flood, L.A., Electrophilic Additions to Mercurated Polystyrene, Ind. Eng. Chem. Prod. Res. Dev., 21 (1982) 462–464.

Tay1981 Taylor, R.T., Crawshaw, D.B., Paperman, J.B., Flood, L.A. and Cassell, R.A., Preparation and Halogenation of Silylated Polystyrene, Macromolecules, 14 (1981) 1134–1135.

Tay1983 Taylor, R.T. and Flood, L.A., Polystyrene-Bound Phenylselenic Acid: Catalytic Oxidations of Olefins, Ketones, and Aromatic Systems, J. Org. Chem., 48 (1983) 5160–5164.

Tem1997 Tempest, P.A. and Armstrong, R.W., Cyclobutenedione Derivatives on Solid Support: Toward Multiple Core Structure Libraries, J. Am. Chem. Soc., 119 (1997) 7607–7608.

Tem1996 Tempest, P.A., Brown, S.D. and Armstrong, R.W., Solid-Phase, Parallel Syntheses by Ugi Multicomponent Condensation, Angew. Chem., Int. Ed. Engl., 35 (1996) 640–642.

Ter1995 Terrett, N.K., Bojanic, D., Brown, D., Bungay, P.J., Gardner, M., Gordon, D.W., Mayers, C. and Steele, J., The Combinatorial Synthesis of a 30,752-Compound Library: Discovery of SAR Around the Endothelin Antagonist, FR-139317, Bioorg. Med. Chem. Lett., 5 (1995) 917–922.

Thi1997 Thieret, N., Alsina, J., Giralt, E., Guibe, F. and Albericio, F., Use of Alloc-amino Acids in Solid-Phase Peptide Synthesis. Tandem Deprotection-Coupling Reactions Using Neutral Conditions, Tetrahedron Lett., 38 (1997) 7275–7278.

Tho1994 Thompson, L.A. and Ellman, J.A., Straightforward and General Method for Coupling Alcohols to Solid Supports, Tetrahedron Lett., 35 (1994) 9333–9336.

Tie1996a Tietze, L.F., Hippe, T. and Steinmetz, A., Solid-Phase Three-Component Domino Reactions. Combinatorial Approach to Substituted 3,4-Dihydro-2H-pyrans, Synlett, (1996) 1043–1044.

Tie1996b Tietze, L.F. and Steinmetz, A., A General and Expedient Method for the Solid-Phase Synthesis of Structurally Diverse 1-Phenyl-pyrazolone Derivatives, Synlett, 1996 (1996) 667–668.

Tie1996c Tietze, L.F. and Steinmetz, A., Stereoselective Solid-Phase Synthesis of Cyclopentane and Cyclohexane Derivatives by Two-Component Domino Reactions: Generation of Combinatorial Libraries, Angew. Chem. Int. Ed. Engl., 35 (1996) 651–652.

Tie1997 Tietze, L.F., Steinmetz, A. and Balkenhohl, F., Solid-Phase Synthesis of Polymer-Bound β-Ketoesters and Their Application in the Synthesis of Structurally Diverse Pyrazolones, Bioorg. Med. Chem. Lett., 7 (1997) 1303–1306.

Tor1996 Tortolani, D.R. and Biller, S.A., A Solid-Phase Synthesis of Miconazole Analogs via an Iodoetherification Reaction, Tetrahedron Lett., 37 (1996) 5687–5690.

Tun1978 Tundo, P., Easy and Economical Synthesis of Widely Porous Resins: Very Efficient Supports for Immobilized Phase-Transfer Catalysts, Synthesis, (1978) 315–316.

Uoz1997 Uozumi, Y., Danjo, H. and Hayashi, T., New Amphiphilic Palladium-Phosphine Complexes Bound to Solid Supports: Preparation and Use for Catalytic Allylic Substitution in Aqueous Media, Tetrahedron Lett., 38 (1997) 3557–3560.

Val1996a Valerio, R.M., Bray, A.M. and Patsiouras, H., Multipin Solid Phase Synthesis of Ethers Using Modified Mitsunobu Chemistry, Tetrahedron Lett., 37 (1996) 3019–3022.

Val1996b Valerio, R.M., Bray, A.M. and Stewart, K.M., Multipin Solid-Phase Synthesis of Acyl 2,3-Diaminopropionic Acid Oligomers, Int. J. Pept. Protein Res., 47 (1996) 414–418.

van1997 van den Ancker, T.R., Hanson, G.R., Lee, F.-C. and Raston, C.L., Polymer Supported Alkali Metal Naphthalenides: Application in the Generation of Lithium and Sodium Reagents, J. Chem. Soc., Chem. Commun., (1997) 125–126.

Vane1975 Vanest, J.-M., Gorsane, M., Libert, V., Pecher, J. and Martin, R.H., Helicenes: The Photochemical Ring-Closure of Polymer-Supported 1,2-Diarylethylenes: Hexahelicene- and Benzo[c]phenanthrene-2-carbaldehyde, Chimia, 29 (1975) 343–344.

Veen1987 Veeneman, G.H., Notermans, S., Liskamp, R.M.J., van der Marel, G.A. and van Boom, J.H., Solid-Phase Synthesis of a Naturally Occurring β-(1$\rightarrow$5)-Linked D-Galactofuranosyl Heptamer Containing the Artificial Linkage Arm L-Homoserine, Tetrahedron Lett., 28 (1987) 6695–6698.

Ver1993 Verduyn, R., van der Klein, P.A.M., Douwes, M., van der Marel, G.A. and van Boom, J.H., Polymer-Supported Solution Synthesis of a Heptaglucoside Having Phytoalexin Elicitor Activity, Recl. Trav. Chim. Pays-Bas., 112 (1993) 464–466.

Vet1995 Vetter, D., Tumelty, D., Singh, S.K. and Gallop, M.A., A Versatile Solid-Phase Synthesis of N-Linked Glycopeptides, Angew. Chem., Int. Ed. Engl., 34 (1995) 60–63.

Vir1994 Virgilio, A.A. and Ellman, J.A., Simultaneous Solid-Phase Synthesis of β-Turn Mimetics Incorporating Side-Chain Functionality, J. Am. Chem. Soc., 116 (1994) 11580–11581.

Vir1996 Virgilio, A.A., Schurer, S.C. and Ellman, J.A., Expedient Solid-Phase Synthesis of Putative β-Turn Mimetics Incorporating the i+1, i+2, and i+3 Sidechains, Tetrahedron Lett., 37 (1996) 6961–6964.

Vla1997 Vlattas, I., Dellureficio, J., Dunn, R., Sytwu, I.I. and Stanton, J., The Use of Thioesters in Solid Phase Organic Chemistry, Tetrahedron Lett., 38 (1997) 7321–7324.

Vo1997 Vo, N.H., Eyermann, C.J. and Hodge, C.N., Transformations of Resin-Bound Pyridinium Ylides: I. A Stereoselective Synthesis of 2,2,3-Trisubstituted Cyclopropanecarboxylates, Tetrahedron Lett., 38 (1997) 7951–7954.

Wall1997 Wallace, O.B., Solid Phase Synthesis of Ketones from Esters, Tetrahedron Lett., 38 (1997) 4939–4942.

Wan1997a Wang, F. and Hauske, J.R., Solid Phase Synthesis of Benzoxazoles via Mitsunobu Reaction, Tetrahedron Lett., 38 (1997) 6529–6532.

Wan1997b Wang, F. and Hauske, J.R., Solid-Phase Synthesis of 3,4-Dihydroquinazoline, Tetrahedron Lett., 38 (1997) 8651–8654.

Wan1997c Wang, G.T., Chen, Y., Wang, S., Sciotti, R. and Sowin, T., Solid Phase Synthesis of Ureas of Secondary Amines via Carbamoyl Chloride, Tetrahedron Lett., 38 (1997) 1895–1898.

Wan1994 Wang, P.-W. and Fox, M.A., A Polymer-Bound Bidentate-Phosphine-Palladium Complex as a Catalyst in the Heck Arylation, J. Org. Chem., 59 (1994) 5358–5364.

War1996 Ward, Y.D. and Farina, V., Solid Phase Synthesis of Aryl Amines via Palladium Catalyzed Amination of Resin-Bound Aromatic Bromides, Tetrahedron Lett., 37 (1996) 6993–6996.

Wei1975 Weinshenker, N.M., Crosby, G.A. and Wong, J.Y., Polymeric Reagents. IV. Synthesis and Utilization of an Insoluble Polymeric Organotin Dihydride Reagent, J. Org. Chem., 40 (1975) 1966–1971.

Wei1972 Weinshenker, N.M. and Shen, C.M., Polymeric Reagents I. Synthesis of an Insoluble Polymeric Carbodiimide, Tetrahedron Lett., (1972) 3281–3284.

Wei1988 Weinshenker, N.M., Shen, C.M. and Wong, J.Y., Polymeric Carbodiimide. Preparation, Organic Synthesis, Coll. Vol. VI, (1988) 951–954.

Whit1997 Whitehouse, D.L., Nelson, K.H., Jr, Savinov, S.N. and Austin, D.J., A Chemoselective Rhodium(II) Mediated Solid Phase 1,3-Dipolar Cycloaddition and its Application to a Thermally Self-Cleaving Furan Scaffold, Tetrahedron Lett., 38 (1997) 7139–7142.

Wij1996 Wijkmans, J.C.H.M., Meeuwenoord, N.J., Bloemhoff, W., van der Marel, G.A. and van Boom, J.H., Solid Phase Synthesis of Alkylphosphonopeptides, Tetrahedron, 52 (1996) 2103–2112.

Wil1995 Williard, R., Jammalamadaka, V., Zava, D., Benz, C.C., Hunt, C.A., Kushner, P.J. and Scanlan, T.S., Screening and Characterization of Estrogenic Activity from a Hydroxystilbene Library, Chem. Biol., 2 (1995) 45–51.

Will1996 Willoughby, C.A. and Chapman, K.T., Solid Phase Synthesis of Aryl Amines, Tetrahedron Lett., 37 (1996) 7181–7184.

Wip1995 Wipf, P. and Cunningham, A., A Solid Phase Protocol of the Biginelli Dihydropyrimidine Synthesis Suitable for Combinatorial Chemistry, Tetrahedron Lett., 36 (1995) 7819–7822.

Wip1997 Wipf, P. and Henninger, T.C., Solid-Phase Synthesis of Peptide Mimetics with (E)-Alkene Amide Bond Replacements Derived from Alkenylaziridines, J. Org. Chem., 62 (1997) 1586–1587.

Won1973 Wong, J.Y. and Leznoff, C.C., The Use of Polymer Supports in Organic Synthesis. II. Syntheses of Monoethers of Symmetrical Diols, Can. J. Chem., 51 (1973) 2452–2456.

Won1974 Wong, J.Y., Manning, C. and Leznoff, C.C., Application of Insoluble Polymeric Carriers to Organic Syntheses. 4. Photocyclization of Acylpolyenes. 8. Solid Phase Synthesis and Photochemistry of 4,4′-Stilbenedicarboxaldehyde, Angew. Chem., Int. Ed. Engl., 13 (1974) 666–667.

Woo1997 Woolard, F.X., Paetsch, J. and Ellman, J.A., A Silicon Linker for Direct Loading of Aromatic Compounds to Supports. Traceless Synthesis of Pyridine-Based Tricyclics, J. Org. Chem., 62 (1997) 6102–6103.

Wor1979 Worster, P.M., McArthur, C.R. and Leznoff, C.C., Asymmetric Synthesis of 2-Alkylcyclohexanones on Solid Carriers, Angew. Chem., Int. Ed. Engl., 18 (1979) 221–222.

Xia1997 Xiao, X., Ngu, K., Chao, C. and Patel, V., Selective Solid Phase Synthesis of Ureas and Hydantoins from Common Phenyl Carbamate Intermediates, J. Org. Chem., 62 (1997) 6968–6973.

Xiao1997 Xiao, X.-Y., Parandoosh, Z. and Nova, M.P., Design and Synthesis of a Taxoid Library Using Radiofrequency Encoded Combinatorial Chemistry, J. Org. Chem., 62 (1997) 6029–6033.

Xu1983 Xu, Z.H., McArthur, C.R. and Leznoff, C.C., The Use of Polymer Supports in Organic Synthesis. XXVIII. The Monoblocking of Symmetrical Diketones on Insoluble Polymer Supports, Can. J. Chem., 61 (1983) 1405–1409.

Yan1996a Yan, B., Fell, J.B. and Kumaravel, G., Progression of Organic Reactions on Resin Supports Monitored by Single Bead FTIR Microspectroscopy, J. Org. Chem., 61 (1996) 7467–7472.

Yan1996b Yan, B. and Gstach, H., An Indazole Synthesis on Solid Support Monitored by Single Bead FTIR Microspectroscopy, Tetrahedron Lett., 37 (1996) 8325–8328.

Yan1996c Yan, B. and Kumaravel, G., Probing Solid-phase Reactions by Monitoring the IR Bands on a Single 'Flattened' Resin Bead, Tetrahedron, 52 (1996) 843–848.

Yan1995 Yan, B., Kumaravel, G., Anjaria, H., Wu, A., Petter, R.C., Jewell, C.F., Jr. and Wareing, J.R., Infrared Spectrum of a Single Resin Bead for Real-Time Monitoring of Solid-Phase Reactions, J. Org. Chem., 60 (1995) 5736–5738.

Yan1994 Yan, L., Taylor, C.M., Goodnow, R., Jr. and Kahne, D., Glycosylation on the Merrifield Resin Using Anomeric Sulfoxides, J. Am. Chem. Soc., 116 (1994) 6953–6954.

Yang1997 Yang, L. and Chiu, K., Solid Phase Synthesis of Fmoc N-Methyl Amino Acids: Application of the Fukuyama Amine Synthesis, Tetrahedron Lett., 38 (1997) 7307–7310.

Yed1980 Yedidia, V. and Leznoff, C.C., The Use of Polymer Supports in Organic Synthesis. Part XXII. Regioselectivity in Cycloaddition Reactions on Solid Phases, Can. J. Chem., 58 (1980) 1144–1150.

Yip1971 Yip, K.F. and Tsou, K.C., A New Polymer-Support Method for the Synthesis of Ribooligonucleotide, J. Am. Chem. Soc., 93 (1971) 3272–3276.

Yoo1997 Yoo, S.-E., Seo, J.-S., Yi, K.-Y. and Gong, Y.-D., Solid Phase Synthesis of Biphenyltetrazole Derivatives, Tetrahedron Lett., 38 (1997) 1203–1206.

You1994 Young, J.K., Nelson, J.C. and Moore, J.S., Synthesis of Sequence Specific Phenylacetylene Oligomers on an Insoluble Solid Support, J. Am. Chem. Soc., 116 (1994) 10841–10842.

Yman1997 Youngman, M.A. and Dax, S.L., Mannich Reactions of a Resin-Bound Terminal Alkyne, Tetrahedron Lett., 38 (1997) 6347–6350.

Yu1994 Yu, K.-L., Deshpande, M.S. and Vyas, D.M., Heck Reactions in Solid Phase Synthesis, Tetrahedron Lett., 35 (1994) 8919–8922.

Yun1996 Yun, W. and Mohan, R., Heck Reaction on Solid Support: Synthesis of Indole Analogs, Tetrahedron Lett., 37 (1996) 7189–7192.

Zam1994 Zambias, R.A., Boulton, D.A. and Griffin, P.R., Microchemical Structural Determination of a Peptoid Covalently Bound to a Polymeric Bead by Matrix-Assisted Laser Desorption Ionization Time-of-Flight Mass Spectrometry, Tetrahedron Lett., 35 (1994) 4283–4286.

Zar1996a Zaragoza, F., Solid-Phase Synthesis of Substituted 3-Aminothiophenes and 2-Methylene-2,3-dihydrothiazoles, Tetrahedron Lett., 37 (1996) 6213–6216.

Zar1996b Zaragoza, F. and Peterson, S.V., Solid-Phase Synthesis of Substituted 1,2,3-Triazoles, Tetrahedron, 52 (1996) 10823–10826.

Zeh1973 Zehavi, U. and Patchornik, A., Oligosaccharide Synthesis on a Light-Sensitive Solid Support. I. The Polymer and Synthesis of Isomaltose (6-O-α-D-Glucopyranosyl-D-glucose), J. Am. Chem. Soc., 95 (1973) 5673–5677.

Zha1996 Zhang, C., Moran, E.J., Woiwode, T.F., Short, K.M. and Mjalli, A.M.M., Synthesis of Tetrasubstituted Imidazoles via α-(N-Acyl-N-alkylamino)-β-ketoamides on Wang Resin, Tetrahedron Lett., 37 (1996) 751–754.

Zha1997a Zhang, H.-C., Brumfield, K.K. and Maryanoff, B.E., Synthesis of Trisubstituted Indoles on the Solid Phase via Palladium-Mediated Heteroannulation of Internal Alkynes, Tetrahedron Lett., 38 (1997) 2439–2442.

Zha1997b Zhang, H.-C. and Maryanoff, B.E., Construction of Indole and Benzofuran Systems on the Solid Phase via Palladium-Mediated Cyclizations, J. Org. Chem., 62 (1997) 1804–1809.

Zik1995 Zikos, C.C. and Ferderigos, N.G., Preparation of High Capacity Aminomethyl-Polystyrene Resin, Tetrahedron Lett., 36 (1995) 3741–3744.

Zuc1994a Zuckermann, R.N. and Goff, D.A., Synthesis of (N-Substituted)-Glycine Polymers of Defined Length and Sequence, Polymer Preprints, 35 (1994) 975–976.

Zuc1992 Zuckermann, R.N., Kerr, J.M., Kent, S.B.H. and Moos, W.H., Efficient Method for the Preparation of Peptoids [Oligo(N-Substituted Glycines)] by Submonomer Solid-Phase Synthesis, J. Am. Chem. Soc., 114 (1992) 10646–10647.

Zuc1994b Zuckermann, R.N., Martin, E.J., Spellmeyer, D.C., Stauber, G.B., Shoemaker, K.R., Kerr, J.M., Figliozzi, G.M., Goff, D.A., Siani, M.A., Simon, R.J., Banville, S.C., Brown, E.G., Wang, L., Richter, L.S. and Moos, W.H., Discovery of Nanomolar Ligands for 7-Transmembrane G-Protein-Coupled Receptors from a Diverse N-(Substituted)glycine Peptoid Library, J. Med. Chem., 37 (1994) 2678–2685.

Zup1975 Zupan, M. and Pollak, A., Preparation of a Polymer-supported Aryliodine(III) Difluoride and its Use to Fluorinate Olefins to Difluorides, J. Chem. Soc., Chem. Commun., (1975) 715–716.

Author Index

Subject Index

MIX
Papier aus verantwortungsvollen Quellen
Paper from responsible sources
FSC® C105338

If you have any concerns about our products,
you can contact us on
ProductSafety@springernature.com

In case Publisher is established outside the EU,
the EU authorized representative is:
Springer Nature Customer Service Center GmbH
Europaplatz 3, 69115 Heidelberg, Germany

Printed by Libri Plureos GmbH
in Hamburg, Germany